S0-AVG-573

"Faster, Better, Cheaper" in the History of Manufacturing

From the Stone Age to Lean Manufacturing and Beyond

"Faster, Better, Cheaper" in the History of Manufacturing

From the Stone Age to Lean Manufacturing and Beyond

Christoph Roser

CRC Press
Taylor & Francis Group
Boca Raton London New York

CRC Press is an imprint of the
Taylor & Francis Group, an **informa** business
A PRODUCTIVITY PRESS BOOK

·Hand axe by Michel-Georges Bernard and licensed under the Creative Commons Attribution-Share Alike 3.0 Unported license.

·Smith from "Hausbuch der Mendelschen Zwölfbrüderstiftung", around 1425.

·Girl working in an American textile mill 1910. Photo by Lewis Hine.

·Automotive spot welding factory. Image by KUKA Systems GmbH and licensed under the Creative Commons Attribution-Share Alike 3.0 Unported license.

·Ford Magneto Assembly Line 1913.

·Steam engine designed by James Watt. Image from Meyers Konversations-Lexikon 1885-1890.

CRC Press
Taylor & Francis Group
6000 Broken Sound Parkway NW, Suite 300
Boca Raton, FL 33487-2742

Printed on acid-free paper
Version Date: 20160524

International Standard Book Number-13: 978-1-4987-5630-3 (Hardback)

Library of Congress Cataloging-in-Publication Data

Names: Roser, Christoph, author.
Title: "Faster, better, cheaper" in the history of manufacturing : from the Stone Age to lean manufacturing and beyond / Christoph Roser.
Description: 1 Edition. | Boca Raton : CRC Press, 2016. | Includes bibliographical references and index.
Identifiers: LCCN 2016003463 | ISBN 9781498756303
Subjects: LCSH: Manufacturing processes--History.
Classification: LCC TS183 .R68 2016 | DDC 670--dc23
LC record available at https://lccn.loc.gov/2016003463

Visit the Taylor & Francis Web site at
http://www.taylorandfrancis.com

and the CRC Press Web site at
http://www.crcpress.com

Printed and bound in the United States of America by Publishers Graphics, LLC on sustainably sourced paper.

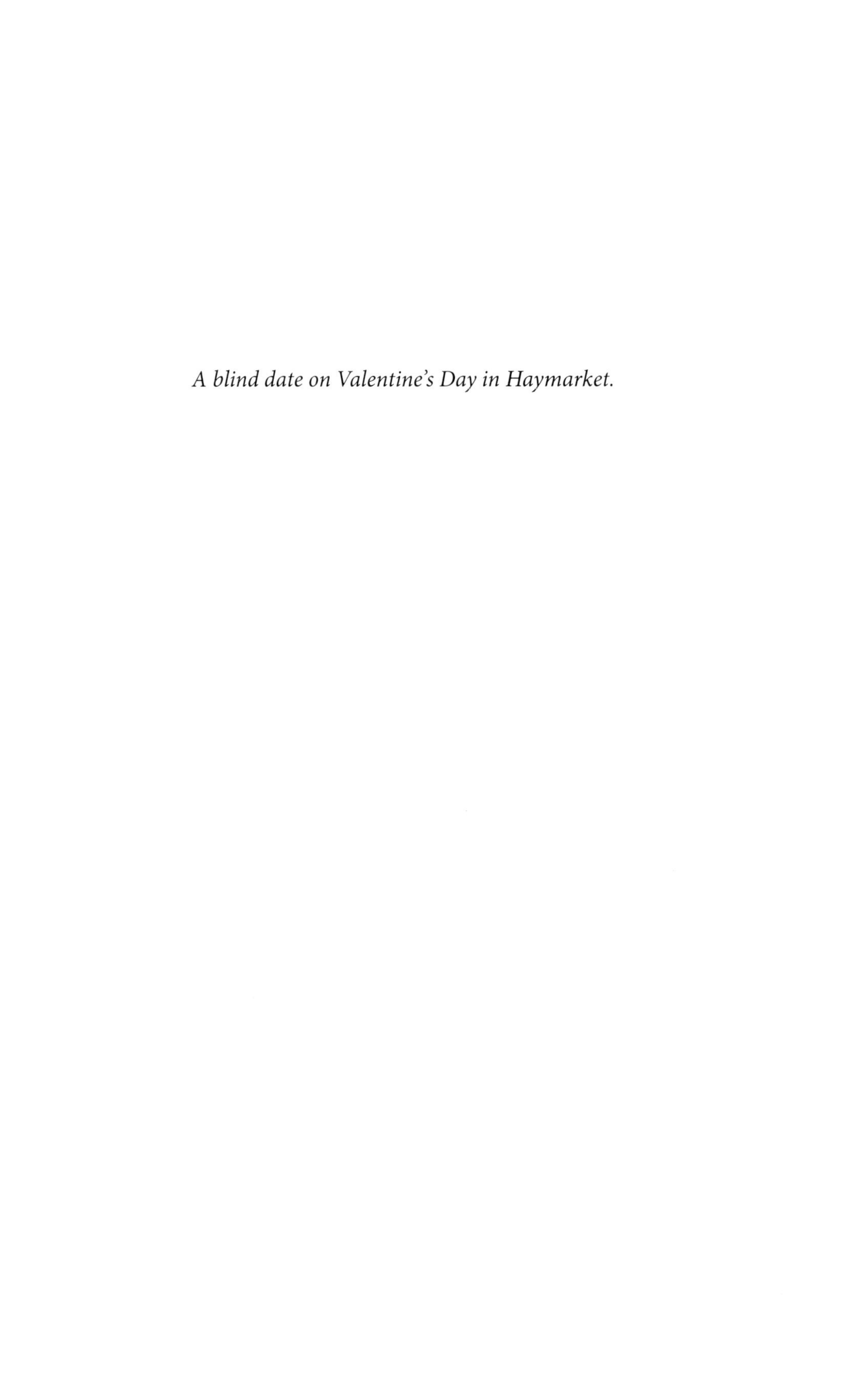

A blind date on Valentine's Day in Haymarket.

Contents

SECTION III Modern Times—Mass Production for the Masses

List of Figures

Preface

Manufacturing: Originated in Middle French, from Medieval Latin manufactura, which originated from Latin manu for *hand* and factus for *make*. First Known Use: 1567. (1) The act or process of producing something; (2) the process of making wares by hand or by machinery especially when carried on systematically with division of labor; (3) a productive industry using mechanical power and machinery.

"manufacturing" Merriam-Webster.com, 2016

MY GRANDPARENTS' WORKSHOP

As a child, I was frequently playing in my grandfather's workshop. My grandfather, Hermann Kurz, was a silversmith in Schwäbisch Gmünd, southern Germany, and his workshop was a six-year old's dream. In every corner, there was something new to discover. Many large machines were powered by a belt from an overhead shaft that was driven by an electric motor, which I never got to work. Tools and parts were in every corner. A bag of close to 50 lion paw-shaped silver feet for teapots was collecting dust in one corner, and a box of long out-of-date nonmetric screws in another corner, mixed with a generous amount of parts for candleholders everywhere. However, what amazed me the most was his large collection of hammers. More than 400 different hammers were hanging in special cabinets. These hammers had all kinds of heads you could imagine for different purposes that were a mystery to me, and sizes ranging from so big that I could barely lift it to tiny hammers probably more suited to a clockmaker. A modern hardware store would probably have fewer hammers than my grandfather's workshop. Overall, my grandfather was a skilled silversmith, and his workshop was a child's dream.

My grandmother, Emilie Kurz, on the other hand, chose the most unusual profession for a woman who was born during World War I. She was a fully trained *Meisterin* (master craftswoman) for galvanization, a craft that involved acids, poisons, and (back then very modern) electricity, combined with shiny, polished gold and silverware. These dangerous components were probably the main reason why I was never allowed to play in her workshop. She was the first and, for a long time, the only female master craftswoman in galvanization far and wide and one of the very few women of her time with any kind of technical master craftsman degree.

I visited her workshop only once and was struck by the radical difference compared to my grandfather's workshop. Her workshop was very organized and clean. There were no random things lying around. The few things that were in the workshop had the feel that they were placed there on purpose, rather than the chaotic placement of my grandfather's workshop. Overall, it was very boring for a six-year old.

Just as their workshops differed, so did their working styles. My grandfather enjoyed tinkering with metal and making stuff, while my grandmother's goal was to make money. It was my grandmother who ran the family business and made sure my grandfather's tinkering also turned a profit. Especially during the hard times during World War II and the devastation afterward, my family was starving frequently. Back then, my grandmother provided most of the income. She spoke some English—another thing unusual in Germany for her time—and therefore was able to do business with the American soldiers, when about 15,000 were stationed in the area of Schwäbisch Gmünd after World War II. American GIs were pretty much the only people with some purchasing power after the German reichsmark lost its value. For the soldiers interested in buying silverware, my grandmother was the only silversmith in the area who spoke English. Hence, she soon attracted lots of business. My family accepted nearly any form of payment. The U.S. dollar was, of course, much preferred over the near-worthless reichsmark, but she also bartered her goods for Pall Mall cigarettes, Hershey's chocolate, Maxwell coffee, or anything else a GI could get his hands on.

While the Americans had much more purchasing power than Germans after the war, they were still common soldiers. Like in most wars, the men crawling through the dirt and dying in combat were usually not from the wealthy upper class. Hence, the real money was elsewhere. My grandmother found out that the British upper class also liked silverware. Yet, silver goods in Britain were taxed with a punitive luxury tax. Therefore, the British gentry loaded up on tax-free silverware during their holiday visits to the English-speaking Mediterranean island of Malta. Again, my grandmother, with her English skills, was able to get into business with jewelers in Malta's capital, Valetta. British silverware was usually made from

Sterling silver, containing 92.5% silver. Common German silverware, on the other hand, contained only 83.5% silver. Hence, German goods contained about 10% less silver and were therefore about 10% cheaper. Even though my grandparents' goods were clearly marked as 835 silver, British vacationers bought them just the same. Soon my family was preparing weekly shipments of silverware to Malta, which quickly turned into the most significant source of income and helped us to put food on the table after the war.

Since my childhood, I have learned much about manufacturing systems and Lean manufacturing. Hence, I now understand that while my grandfather was a skilled silversmith, my grandmother was a much better manufacturer. She wasted no effort or material, did not indulge in a large number of tools (who in the world needs 400 hammers?), had her workshop clearly structured, and was intensely focused on turning her work into money. In my view, manufacturing without the intent to make money is just a hobby, not a profession.

This book is not about hobbies; it is about making money through manufacturing. Industry is on a never ending quest to produce faster, better, and cheaper. The book shows how different developments over time raised efficiency and allowed production of more with less effort and materials. This brought us a large part of the wealth and prosperity we enjoy today. Whenever possible, the stories of real inventors and industrialists are told, including not only their successes but also their problems and failures. The effect of good or bad management on manufacturing is a recurring theme in many chapters, as is the fight for intellectual property. This is not only a story of successes but also about failures; not only about technology but also about social aspects. *It is not a book about machines but about people!*

Acknowledgments

It took me almost a decade to write this book about the history of manufacturing. However, this book would not have been possible without the help of many others. Throughout my career, I was lucky to have many different mentors who went above and beyond their duty to help me. For this, I thank my mentors Professor Dr. Ernst Kühner, Professor Dr. David Kazmer, Professor Dr. Masaru Nakano, and Dr. Gernot Strube.

In the process of writing this book, I also received much helpful feedback from reviewers. Hence, I would like to thank Christoph Albrecht, Werner Bergholz, Stefan Bleiweis, Karl-Ludwig Blocher, Machiko Hoshino-Roser, David Kazmer, Roland Kurz, Hartmut Lorentzen, Masaru Nakano, Anke Roser, Jochen Roser, Cheong Tsang, and many more for their valuable input. I also thank Manisha Rath for her dedication in organizing my notes.

As it turns out, writing a book is one thing; getting it published is another. Hence, I also thank Mark Warren for making the connection to CRC Press and pointing out some juicy bits of history. I would like to thank my agent, Maryann Karinch, and the acquiring editor at Productivity Press, Michael Sinocchi, for sorting out the details of the publishing contract.

Finally, I would like to thank my parents and my beloved wife for their support.

1

Failure and Success in Manufacturing—The General Motors–Toyota NUMMI Joint Venture

I'm part of the team; I don't have a team. Let people maintain their own personality.

Rick Madrid
Team leader at NUMMI

The General Motors (GM) Fremont plant, around 1980, was one of the worst automotive plants ever to operate in the United States (Womack 1990, p. 81). Even the United Auto Workers (UAW) union representatives—who rarely see problems with its members—considered Fremont to have the worst workforce ever (Siegel 2010). A host of problems plagued it, mostly stemming from very low morale of the employees.

The quality of GM was already mediocre at best, but the quality of the vehicles produced in Fremont was the worst within GM (Shook 2010). Workers intentionally sabotaged many vehicles. For example, empty cola bottle caps were left in the doors with the sole purpose to annoy the customer through its clanking sound. Or worse, half-eaten tuna sandwiches were welded in (Childress 2013, p. 219). Cars were intentionally scratched. Screws on safety-critical parts were deliberately left loose. The employees wanted to hurt the company by hurting the customer.

Even if the workers tried to produce quality products, the system worked against them. The most important rule in the Fremont plant was to never, ever stop the assembly line. Even in the case of accidents, the line did not stop. Hence, all problems were simply pushed downstream, resulting in miserable quality at the end of the line. Sometimes, engines were installed backward; other cars had brakes or steering wheels missing. Some cars would be half of one model and another half of a completely different model. Many cars arrived at the end of the line incomplete or defective, unable to drive, and they had to be towed (Glass 2010).

All kinds of illegal activities were also readily available within the plant. Sex, drugs, and gambling were widespread. A large part of the workforce was drunk, if they showed up at all. On an average day, one out of five people simply did not show up, with Mondays being considerably worse. Yet, the assembly line must not stop. Some days, when there were too few people to start the line, management simply hired people from the bar across the street on the spot so they could start the line. Of course, this worsened quality even more (Siegel 2010).

People hated working there, yet due to a lack of alternatives, they had to stay with their secure jobs. The only thing they did with enthusiasm was quarrel and fight. Both workers and management were constantly occupied with disputes, mostly through formal grievances, even for minor disagreements. Strikes, including wildcat, strikes were common. Management did not help either by belittling and micromanaging its workers.

Finally, GM had enough and put this sorry excuse of a plant out of its misery. The plant closed in March 1982, and 5000 people lost their jobs with only three weeks' notice (Turner 1990).

Two years later, the plant reopened as a Toyota–GM joint venture named New United Motor Manufacturing, Inc., better known as NUMMI. Production restarted in the same buildings as the old Fremont plant (adding only a stamping plant). The manufacturing technology was also not much different (Adler, Goldoftas, and Levine 1998). Incredibly, they rehired mostly former employees of the old Fremont plant. With virtually the same machines, materials, and manpower, it looked like NUMMI was destined to repeat the previous miserable manufacturing system.

To everybody's surprise, the plant turned out to be the best automotive plant in the United States. Quality by far exceeded that of any other GM plant. In fact, quality exceeded that of any other U.S. automotive plant and was very close to the then-legendary quality of the Toyota plants in Japan.* Productivity also soared, and the new NUMMI plant produced almost

* Quality is usually measured in defects per 100 vehicles. In 1992, the average U.S. automotive maker had 136 defects per 100 vehicles. European cars were even worse with 158. Japan was far ahead with only 105 defects per 100 vehicles (not only Toyota but all Japanese carmakers). NUMMI, with only 83 defects per 100 vehicles, reached quality levels similar to those of Toyota (Adler 1995).

twice as many vehicles with the same labor force as before.* The cost per vehicle was reduced by $750 (Keller 1989, p. 130). Even employee morale, previously the biggest handicap to success, reached amazingly high levels.† Absenteeism, previously at 20%, reached very low levels, below 3%. Annual employee turnover was also less than 6% (Adler 1995). Many workers, for the first time in their lives, enjoyed coming to work.

What happened? What turned the pathetic old Fremont plant into the stellar outperforming NUMMI plant, and that with the same workforce, to boot?

Toyota changed the culture in the plant. The manufacturing system used the same hardware, but the *software* of running a manufacturing system was very different. Toyota was able to implement its highly successful Toyota Production System at NUMMI. There were too many changes to list them all here, but probably the biggest change was that Toyota worked with the workers, not against them. Toyota treated its workers with respect, valued their input, and not only allowed but even encouraged them to make decisions related to their workplace. While Fremont was based on conflict, NUMMI was based on teamwork, and everybody was treated with respect and fairness.

Another major difference was the focus on quality. At Fremont, the quality control department checked quality after production. At NUMMI, every worker was responsible for quality. At Fremont, the most important rule was to never, ever stop the assembly line. At NUMMI, workers not only were allowed to stop the line, they even *had* to stop the line if there was a quality problem. As a result, Fremont needed 12% of its space for rework, whereas NUMMI used only 7% (Keller 1989, p. 132).

The hiring of new employees also received much more significance. On average, NUMMI personnel spent 35 hours with every potential new employee during the hiring process. This was not only to find good employees but also to instill a sense of importance into the potential employees (Keller 1989, p. 132). In comparison, Fremont sometimes simply picked up new temporary employees in the bar across the street.

---

* The old Fremont plant needed 29.1 salaried hours per produced vehicle, and the comparable GM Framingham plant needed 30.8. NUMMI required only 19.6 hours per vehicle, comparable to the best plants of the world, for example, the Toyota Takaoka plant in Japan, with 18.0 hours per vehicle. Note that the NUMMI workforce on average was 10 years older than the Takaoka workforce (Adler 1995). Inventory levels were also reduced from two weeks at Fremont to two days at NUMMI, while Takaoka had only two hours' worth of inventory (Womack 1990, p. 81ff).

† Worker satisfaction in 1991 was very high at 90%. Similarly, satisfaction with job security reached 89% in 1991 (Adler 1995).

There were many other changes. For example, the workers' assignments were more flexible, as the number of job categories was reduced from 183 to 4. There was a strong focus on reducing waste and streamlining the manufacturing process. Floor space was used much more efficiently (Keller 1989, p. 129). Many things were done differently, creating a new superior manufacturing system that made NUMMI the best automotive plant in America.

One of the goals of GM in this joint venture was to learn the secret of Toyota. They wanted to understand the magic with which the Japanese produced superior cars at lower cost. Only, there was no secret. It was respect, common sense, and hard work to improve lots of little details.

GM wanted to apply the learnings from NUMMI to its other plants. Except, they did not get it. They did not get it at all. The former Fremont plant had a culture of oppression, threats, and distrust. NUMMI was based on respect, trust, and teamwork. However, GM management used the only management style they knew and wanted to force and bully their employees at other plants into trusting them. Incredibly, they set up a competition between two plants, Van Nuys near Los Angeles and Norwood in Cincinnati. The loser would be shut down.

Unsurprisingly, this *trust me or get fired* approach did not work. From the beginning, most employees were highly suspicious about this new NUMMI thing. Union representatives of these plants called it *the most dangerous scheme ever conjured up by GM to rob workers of their union.* They were suspicious that all of this was just a tool for layoffs. Actually, they were right, as GM indeed wanted to reduce the headcount by 25% (Keller 1989, p. 137f). Norwood rejected the NUMMI approach altogether, and nothing was implemented. Van Nuys reluctantly played along, but their heart was never in it.

Van Nuys mechanically implemented many of the successful approaches at NUMMI, but without the trust of the workers, these did not work. For example, the ability to stop the line soon led to the line being stopped so often that both productivity and quality declined (Keller 1989, p. 139). Both management and employees resisted the change, and any ideas connected with NUMMI often met stiff resistance in other plants simply because it was from NUMMI.

This lack of understanding and disinterest also extended to the upper management. A Toyota manager said that GM understood the changes *as far as the hardware and the plant layout are concerned. But I'm afraid that GM upper management doesn't understand the basic concept* (Inkpen

2008). Toyota executives visiting NUMMI were very interested in the details on the shop floor. GM executives, on the other hand, did a *five minute fly-by*, looking for a magic bullet that they can delegate others to implement in other plants. A GM boardroom meeting on *the secret* of NUMMI drew on 25 studies and lots of data but resulted only in a huge discussion without any conclusion (Keller 1989, p. 142).

Toyota, on the other hand, learned successfully how to work with U.S. unions (Clarke 2002), preparing them for further ventures into U.S.-based Toyota plants. They also kept their NUMMI graduates in larger groups of 30 to 60 people, whereas GM diluted their impact by having them fight the GM culture in other plants alone (Finkelstein 2004). Toyota also proved that the Toyota Production System is not dependent on the Japanese culture but can also be successfully applied in other countries. Nowadays, Toyota has plants in more than 20 countries (Reingold 1999, p. 64) and produces in excess of 50% of all Toyota vehicles outside of Japan (Marsh 2012, p. 56).

At GM, Van Nuys eventually won the deplorable competition, and GM closed Norwood in 1987. However, Van Nuys did not have much time to enjoy its victory, as GM also closed it only five years later in 1992.

Overall, GM learned little. Especially, top management resisted the change. While GM did improve some aspects eventually (Sato 2008, p. 255), it was not enough to turn things around. In 2009, GM had the biggest bankruptcy in U.S. history, costing taxpayers $50 billion (Siegel 2010). Toyota, on the other hand, became the largest carmaker in the world and is highly profitable.

BIBLIOGRAPHY

Adler, P.S., 1995. Democratic Taylorism: The Toyota Production System at NUMMI, in: *Lean Work: Empowerment and Exploitation in the Global Auto Industry*. Wayne State University Press, Detroit, MI.

Adler, P.S., Goldoftas, B., Levine, D.I., 1998. Stability and Change at NUMMI, in: *Between Imitation and Innovation: Transfer and Hybridization of Production Models in the International Automobile Industry*. Oxford University Press, Oxford, UK.

Childress, J.R., 2013. *Leverage: The CEO's Guide to Corporate Culture*. Principia Associates, London.

Clarke, C., 2002. Forms and Functions of Standardisation in Production Systems of the Automotive Industry: The Case of Mercedes-Benz (Doctoral). Freie Universität Berlin, Berlin.

Finkelstein, S., 2004. *Why Smart Executives Fail: And What You Can Learn from Their Mistakes*. Portfolio, New York.

Glass, I., 2010. "NUMMI." *This American Life*. Radio broadcast from Public Radio International.

Inkpen, A.C., 2008. Knowledge Transfer and International Joint Ventures: The Case of NUMMI and General Motors. *Strategic Management Journal* 29, 447–453. doi:10.1002/smj.663.

Keller, M., 1989. *Rude Awakening: The Rise Fall and Struggle for Recovery of General Motors*. William Morrow & Co, New York.

Marsh, P., 2012. *The New Industrial Revolution*. Yale University Press, New Haven, CT.

Reingold, E.M., 1999. Toyota—A Corporate History. Penguin, London.

Sato, M., 2008. *The Toyota Leaders: An Executive Guide*. Vertical, New York.

Shook, J., 2010. How to Change a Culture: Lessons from NUMMI. *MIT Sloan Management Rev*. 51(2), 63–67.

Siegel, R., 2010. "The End of the Line for GM–Toyota Joint Venture." *All Things Considered*. Radio broadcast from National Public Radio.

Turner, L., 1990. NUMMI—Japanische Produktionskonzepte in den USA. Articles and Chapters. In M. Muster & U. Richter (Eds.), *Mit Vollgas in den Stau: Automobilproduktion, Unternehmensstrategien und die Perspektiven eines ökologischen Verkehrssystems*, 78–87, VSA-Verlag, Hamburg, Germany.

Womack, J.P., 1990. *The Machine That Changed the World: Based on the Massachusetts Institute of Technology 5-Million-Dollar 5-Year Study on the Future of the Automobile*. Later Printing. Scribner, Rawson Associates, New York.

Section I

The Age of the Artisan

2

The Stone Age

Man is a Tool-using Animal. Weak in himself, and of small stature, he stands on a basis, at most for the flattest-soled, of some half-square foot, insecurely enough; has to straddle out his legs, lest the very wind supplant him. Feeblest of bipeds! Three quintals are a crushing load for him; the steer of the meadow tosses him aloft, like a waste rag. Nevertheless he can use Tools, can devise Tools: with these the granite mountain melts into light dust before him; seas are his smooth highway, winds and fire his unwearying steeds. Nowhere do you find him without Tools; without Tools he is nothing, with Tools he is all.

Thomas Carlyle (1795–1881)
Philosopher and writer, in Sartor Resartus

The vast majority of the objects we interact with in our daily lives are products of manufacturing. This ranges from highly complex products like our cars or our computers to the clothes we wear, the chair we sit on, the roof over our head, and even most of the food we eat*—all these are products of manufacturing. Yet all the manufacturing steps needed to create this vast range of products can be based on a combination of fundamental manufacturing techniques (Beitz and Küttner 1995). These techniques are often summarized into six groups†:

- **Cutting:** to cut off a part from an object. Common examples are drilling, sawing, filing, milling, turning, and chiseling.
- **Changing material properties:** to change the properties of a material. Common examples are hardening, annealing, and magnetizing.

* If you doubt that, read the list of ingredients of your frozen pizza, or almost any supermarket food for that matter.

† According to the German norm DIN8580.

- **Joining:** to merge or assemble one or more objects into a new object. Common examples are welding, gluing, screwing, and bolting.
- **Coating:** to cover the surface of an object with a firmly attached previously formless mass. Common examples are painting, galvanizing, and powder coating.
- **Casting and molding:** to give shape to a previously formless mass, which may be a liquid, a gas, or a paste. Common examples are casting, injection molding, and sintering.
- **Forming:** to change the shape of an object without adding or removing parts. Common examples are forging, bending, and deep drawing.

All manufacturing operations depend on one or more of these six basic techniques. Moreover, while all of these techniques now exist in a multitude of variations, all can be traced back to very simple examples first used during the Stone Age.

2.1 THE FIRST MANUFACTURING TECHNIQUE—CUTTING

The history of manufacturing starts with the history of mankind. Somewhere in Africa 2.6 million years ago, the earliest representative of the genus *Homo* carefully chose a pair of suitable rocks, placed one of them on the ground, and then repeatedly struck it with the other rock. Through this, he created, piece by piece, a cutting edge (Plummer 2004). This step distinguished man from animal. The creature believed to be responsible for this act was the first species of the genus *Homo* and was named after its work, *Homo habilis*, the handyman (Figure 2.1).* Hence, the first species of the genus *Homo* was named after its ability to fashion tools. While the tool, to the untrained eye, would look like nothing but a broken rock, with this broken rock, humanity entered the Stone Age and introduced the world to its first manufacturing technology: cutting.

* *Habilis* is loosely derived from Latin, intended to mean skillful and apt. Since then, another earlier toolmaker may have been found, *Australopithecus garhi*, but this is disputed.

FIGURE 2.1
Artist image of a *Homo habilis.*

These tools are the *Oldowan* stone tools (Figure 2.2)* from the lower Paleolithic era and have been found at numerous sites in Africa, Asia, and Europe. It is not known if these artisans specialized in the production of these tools full time or if they also participated in the normal hunting and foraging work with the other members of the tribe. What is known is that the tools were used to cut meat and bones of animals, that the manufacturers were mostly right-handed, and that they opened the stage for the history of manufacturing.†

Technology evolved around 1.7 million years ago, and the *Oldowan* technology was replaced by the more advanced *Acheulean* technology. These stone tools were produced up to 100,000 years ago and are easy to identify from their well-defined shape. Production of these tools was also more refined. These tools were found in Africa, Europe, and central Asia. One of the oldest manufacturing sites is Olorgesailie, Kenya. Raw

* Named after the first finding in 1930 in the Oldowan Gorge in Tanzania.

† There are theories that even earlier hominids manufactured tools from wood, leaves, and animal parts based on the manufacture and use of tools by modern-day apes and also on morphological comparisons (Panger et al. 2002). These theories date the first usage of tools five to eight million years back, although due to the fragile nature of plant and animal products, no traces of these have yet been found.

FIGURE 2.2
1.8-million-year-old Oldowan hand axe from Olduvai Gorge, Tanzania. (Photo by author, dated April 2010.)

material was carried from two different mountains about 20 kilometers apart to a manufacturing site between these two mountains. The production of an *Acheulean* tool easily took hours. It also appears that the manufacturing site was organized. For example, based on stone fragments there seems to have been one area for making new axes and another for resharpening worn-out axes (Bryson 2004, p. 291). Hence, this site may be one of the first examples of specialization, where multiple individuals focused on different tasks. In any case, the site was in operation for 1 million years, making it easily the longest-running manufacturing site in the world.

Overall, the Paleolithic period of the Stone Age represents the first known manufacture of tools by man, and from 2.6 million to 10,000 years ago, stone tools were quite literally cutting-edge technology. Over the millennia, the technique of making stone products was refined, becoming ever more sophisticated. Other available materials such as wood and bones were also cut. Nevertheless, for the next two million years, manufacturing was limited to shaping objects by cutting or breaking off parts from a workpiece.

2.2 THE SECOND MANUFACTURING TECHNIQUE—CHANGING MATERIAL PROPERTIES

The next step in manufacturing technology was heat treatment of materials. Unfortunately, this is difficult to detect, and the data are even sketchier than with the other manufacturing techniques. The earliest example may be heat treatment of rocks, which may have happened as early as 164,000 years ago and is certain to have happened regularly 72,000 years ago in South Africa. One problem in stone tool manufacturing was the availability of a suitable raw material that could be processed into stone tools. At this location at the coastline of South Africa, a certain type of stone (*silcrete*) was available but unfortunately ill-suited for processing into stone tools. However, if that stone was put into a campfire, it not only changed color but also became much more suitable for making stone tools by chipping away flakes (Brown et al. 2009). Hence, this technique was soon used to prepare stones for further processing.

Another possible early example of heat treatment could also be a spear tip found in Lehringen in Germany, still stuck in the ribcage of a straight-tusked elephant that died 120,000 years ago (Tyldesley and Bahn 1983). Spears have been found dating back 400,000 years (Bower 1997). At a first glance, this spear tip looks remarkably similar to many other spear tips of the preceding 300,000 years, made of yew* and with a very sharp pointy end. However, closer inspection reveals that the tip was slowly charred over a fire, thereby removing moisture from the wood. This process is called fire hardening, which makes the wood harder and tougher. However, as heat treatment is difficult to detect on a millennia-old piece of wood, this result is also disputed (Weiner 2003).

Hence, sometime between 164,000 and 72,000 years ago, the next step in the history of manufacturing happened: changing the physical properties of the product by certain treatments, in this case charring it in or over a fire.

* Yew is a type of tree whose wood is very hard and hence prized for making weapons, especially longbows. However, the demand for wooden longbows has decreased significantly since the Middle Ages. Additionally, yew is rather poisonous. Hence, nowadays, the tree is an endangered species.

2.3 THE THIRD MANUFACTURING TECHNIQUE—JOINING

The next breakthrough in manufacturing happened during the Middle Paleolithic, just as *Homo sapiens* and *Homo neanderthalensis* started to compete over who would rule the earth. A quick look in the mirror will confirm for you that our species, *Homo sapiens*, is still around, whereas the Neanderthal is extinct. Scientists still discuss whether we killed off all the Neanderthals or mated with them until no more full-blooded Neanderthals were left. In any case, it definitely helped us that we were the first to develop composite weapons. A particular set of communities near Howiesons Poort in South Africa was technologically way ahead of all contemporaries by about 20,000 years (Vishnyatsky 1994). These communities started to use heat treatment on ochre and plant gum to attach stone heads to arrow shafts 72,000 years ago (Viegas 2008), thereby introducing the technique of joining to manufacturing. Naturally, this technique was soon used to kill off our fellow hominids.

One such known murder case is that of a 50-year-old arthritic but still strong Neanderthal from Shanidar Cave in northern Iraq. This fellow Neanderthal encountered members of our species, *Homo sapiens*, about 50,000 to 70,000 years ago. Unfortunately for the Neanderthal, the contact was made using a composite spear thrown pointy end forward with full force, hitting him in the torso and puncturing his rib. The injured Neanderthal was able to crawl back into his cave but soon died from his injuries (Churchill et al. 2009).

Despite their larger brains, it took Neanderthals about 30,000 years longer to catch up (Ramirez Rozzi and Bermudez de Castro 2004). The oldest Neanderthal composite spear dates back 40,000 years to Königsaue in Germany, even including the impression of a Neanderthal finger in birch resin (Pettitt 2000). A few millenia later, *Homo sapiens* had taken another step forward and invented the spear thrower 27,000 years ago, thereby greatly increasing the strength of the throw and, subsequently, the distance. While the Neanderthals had much greater strength, they were anatomically not suited to throw spears over any significant distances. Hence, they used spears only as a stabbing weapon (Rhodes and Churchill 2009). Therefore, the spear thrower allowed *Homo sapiens* to pick off Neanderthals from a safe distance out of reach of their strong thrusts or short throws. For that reason, Neanderthals are extinct, while we are still around, even if you sometimes feel like you just saw a Neanderthal in the subway. The last traces of the Neanderthals date back 24,000 years ago, without them ever having used spear throwers (Tzedakis et al. 2007). After that, they were history.

2.4 THE FOURTH MANUFACTURING TECHNIQUE—COATING

While the last Neanderthals were still around, *Homo sapiens* introduced another manufacturing technique. Around 30,000 to 32,000 years ago, they entered deep into a cave that is now known as Chauvet Cave in southern France (Figure 2.3), carrying along ochre, charcoal, and clay (Valladas et al. 2001). Furthermore, they used a torch for lighting, banging the torch against the wall frequently along the nearly 500-meter-long cave to increase the brightness. After reaching a suitable spot, they scraped the wall clean of cave muck, dipped their fingers into the pigments, and started drawing on the cave wall, thereby introducing the new technology of coating into the world of manufacturing. While this may not look like much, it is the first known use of the manufacturing technology of coating. A coat (in this case, a pigment) was applied to a substrate (the cave rock) to improve surface properties (in this case, its color). Besides finger painting, art was also created by spraying pigment from the mouth and by blending and shading charcoal just as artists do today.

The number and quality of the paintings in Chauvet are extraordinary. Hundreds of individual images depict primarily animals that are potentially dangerous to humans, like woolly rhinoceroses, cave lions, mammoths, cave bears, and a panther. Additionally, the images show horses, wisents, reindeer, aurochs, and an owl, plus a number of unidentified shapes and dots. Finally, there are also positive and negative hand imprints, the latter created by spraying paint over a hand on the wall. Often, the natural shape of the cave was used to achieve a three-dimensional feel to the image, and the walls were even engraved to further enhance this effect.

Of course, humans would not be humans if the cave did not include what researchers modestly call *Venus*: detailed depictions of female genitalia including pubic hair and the vulvae slit—leading some researchers to speculate that the ancient artists were male (Sentenac, n.d.). Many researchers attribute such images to a fertility cult, although they do

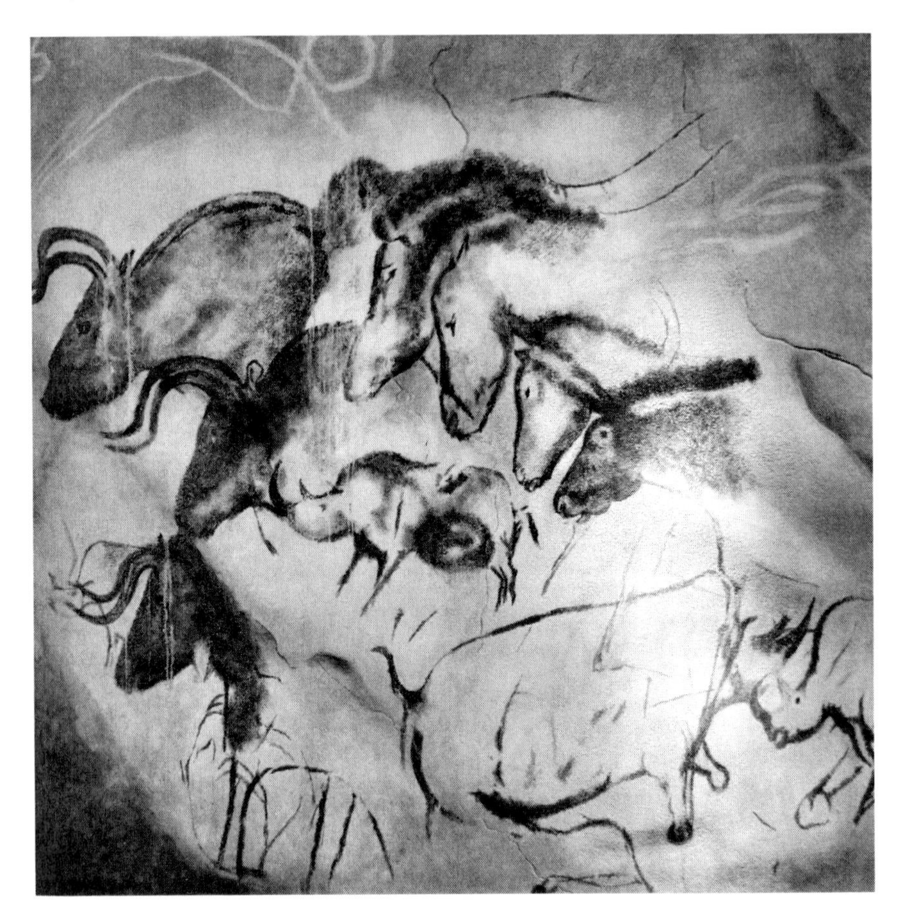

FIGURE 2.3
Chauvet Cave paintings of horses. (Image by Thomas T. and licensed under the Creative Commons Attribution-ShareAlike 2.0 Generic license.)

not have any evidence that they were not drawn for the same reason that very similar images are still frequently found in public toilets.

The cave is also remarkable for the oldest dated human footprints, where over a distance of 70 meters, the barefooted tracks of an adult and an approximately 12-year-old child can be followed. The tracks can be dated because the younger person carried the torch, hitting it against the wall regularly to increase brightness. Carbon-dating this ash determined the footprints to be around 26,000 years old.* These visitors were also among

* Also, the child had a so-called Greek foot, where the second toe is longer than the first one. Nowadays, one-sixth of the population have Greek feet.

the last visitors to the cave, as the entrance was blocked by a rockfall shortly thereafter. This sealed the cave and created a time capsule of ancient art and technology that was not seen again until speleologists rediscovered the cave in 1994.

2.5 THE FIFTH MANUFACTURING TECHNIQUE—CASTING AND MOLDING

The next fundamental manufacturing technique of casting and molding was developed only a few thousand years later. Around 29,000 to 25,000 years ago, a previously formless mass was given a permanent shape near modern Brno in the Czech Republic. In this case, the formless mass was a mixture of ground animal bones and loess, and the shape it was given was the form of a nude female, the Venus of Dolni Vestonice (Figure 2.4). After shaping, the figure was put into a fire, thereby creating the first known ceramics in the world. The figure also most likely cracked in half during the firing process (Vandiver et al. 1989).

Curiously enough, as with the drawings in Chauvet (see Section 2.4), another youngster was involved in the making of this figure, as the fingerprint of a 7- to 15-year-old child was found in the back of the figure. However, researchers believe that the child only handled the figure before burning rather than creating the figure (Králík, Novotný, and Oliva 2002). The figure was found with some other animal figures and burnt lumps of loess of similar age, near two ovens. Similar to other nude female figures and drawings, researchers dubbed this the Venus of Dolni Vestonice.

Recent research also adds more substance to the idea that this figure was more than just ancient erotica. Most male researchers still believe that their male ancestors were the primary providers of meat on the Stone Age table by bravely staring into the eye of a wild mammoth armed only with spears and sticks. However, other researchers found traces of nets and traps, indicating a much safer way to provide meat that is also much more effective than facing a wild elephant head on. Using nets and snares, the whole Stone Age family could hunt. Furthermore, collecting grain, vegetables, and fruits was also a major contributor to the diet. Hence, the Venus of Dolni Vestonice may also be a cult object created by women and intentionally broken in half during or shortly after firing as part of a religious ceremony (Pringle 1998).

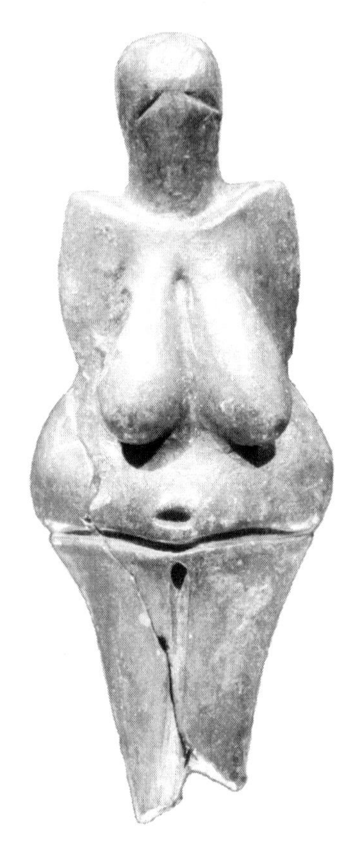

FIGURE 2.4
Venus of Dolni Vestonice dated to be 29,000 to 25,000 years old. (Image by Petr Novák and licensed under the Creative Commons Attribution-ShareAlike 2.5 Generic license.)

2.6 THE SIXTH MANUFACTURING TECHNIQUE—FORMING

The last of the six fundamental manufacturing techniques is forming—changing the shape of a previously shaped object. This obviously requires the object to be made out of a shapeable material, most commonly a metal. Brittle material, such as stone or ceramics, cannot be shaped, as they break before achieving any noticeable change of shape. Metals, on the other hand, can easily be shaped. The Bronze Age—associated with the first significant use of metal, namely, bronze—started around 3300 BCE.

However, even before that, humans developed smelting of ore and casting of metal with lower melting points, namely, tin and lead. These melt in an average campfire and therefore are much easier to extract than copper and bronze alloys, which have a melting point way beyond the temperatures of a campfire. Therefore, the first evidence of smelting is lead beads, dating back to about 6500 BCE in a settlement nowadays called Çatalhöyük in Turkey (Heskel 1983).

However, in nature, metals can also be found in their metallic forms, either pure or as alloys. These metal deposits are called native metals. Since they are subject to corrosion, only corrosion-resistant metals like gold, silver, copper, and platinum can be found in any useable quantities. Hence, humans were able to form native metals before they were able to smelt metal from ore, and the first forming techniques were already used during the Stone Age. The first known use of native copper is a pendant found in the Shanidar Cave in northern Iraq (Solecki and Agelarakis 2004). This almond-shaped pendant was made from native copper and hammered into shape with two holes cut into the ends so that the pendant could be worn around the neck.

This hammering of the pendant is the first known use of the manufacturing technique of forming. The pendant was found in the grave of an adult woman referred to as Burial 5 and dating back to 8700 BCE, almost 2000 years before lead was smelted. Unfortunately, the date of the manufacture of the copper pendant cannot be determined, and it may have been handed down over many generations before being buried with Miss Burial 5. By coincidence, Miss Burial 5 was residing in the same Shanidar Cave as the unfortunate Neanderthal that was killed off by her great-great-great-... grandfather with a composite stone–wood projectile 50,000 years earlier (see Section 2.3).

2.7 MANUFACTURING AT THE END OF THE STONE AGE

Starting with the first intentionally broken rock, humankind developed all six fundamental manufacturing techniques over a span of 2.6 million years. While we do know much about when these ancient products were made and what techniques our ancestors used, we know very little about how they structured their work. It can be assumed from the level of artisanship demonstrated in many of these products that their creator had considerable skill and experience in making them. This leads to a second assumption that not everybody in the tribe was able to make all products but that some members specialized in certain products.

What we do not know is if these early artisans specialized in making only a certain product or if it was a side business besides their main work. For example, was there a master stonecutter who specialized in making stone tools full time, or was this a side business, and the artisan was expected to still join the hunting and gathering? In addition, if the artisan was specialized, did he/she barter his/her products in exchange for food, shelter, and protection? Or was he/she merely contributing to the common resources while at the same time taking other resources from the commons for his/her own use?

Another unanswered question is if there was a division of labor. Were the finely crafted composite spears made by a single artisan? Or were multiple members of the tribe involved in different steps? Did they have a specialized stonecutter and a specialized wood worker delivering suitably shaped stones and spear shafts to another specialized joiner who attached the stones firmly to the wooden shaft? If so, were these components made to order for a specific spear for a specific customer, or were they made for stock from which the joiner assembled his products? The layout of the *Acheulean* manufacturing site (see Section 2.1) indicates that different members focused on different products, leading to specialization, but this is by no means certain. We know very little about how Stone Age humans organized themselves and even less about how they organized their manufacturing.

What we do know is that at the end of the Stone Age, mankind evolved from a nomadic hunter-gatherer society to a settled-down agricultural society that produced rather than collected. In the next chapter, we will look in more detail at these changes during this urban revolution.

BIBLIOGRAPHY

Beitz, W., Küttner, K.-H. (Eds.), 1995. *Dubbel: Taschenbuch für den Maschinenbau*, 18. Auflage. Springer, Berlin, Heidelberg, Germany.

Bower, B., 1997. German Mine Yields Ancient Hunting Spears. *Science News* 151, 134.

Brown, K.S., Marean, C.W., Herries, A.I.R., Jacobs, Z., Tribolo, C., Braun, D., Roberts, D.L., Meyer, M.C., Bernatchez, J., 2009. Fire as an Engineering Tool of Early Modern Humans. *Science* 325, 859–862. doi:10.1126/science.1175028.

Bryson, B., 2004. *A Short History of Nearly Everything*, 1st ed. Broadway Books, New York.

Churchill, S.E., Franciscus, R.G., McKean-Peraza, H.A., Daniel, J.A., Warren, B.R., 2009. Shanidar 3 Neandertal Rib Puncture Wound and Paleolithic Weaponry. *Journal of Human Evolution* 57, 163–178. doi:10.1016/j.jhevol.2009.05.010.

Heskel, D.L., 1983. A Model for the Adoption of Metallurgy in the Ancient Middle East. *Current Anthropology* 24, 362–366. doi:10.2307/2742674.

Králík, M., Novotný, V., Oliva, M., 2002. Fingerprint on the Venus of Dolní Věstonice I. *Anthropologie* 40/2, 107–113.

Panger, M.A., Brooks, A.S., Richmond, B.G., Wood, B., 2002. Older Than the Oldowan? Rethinking the Emergence of Hominin Tool Use. *Evolutionary Anthropology* 11, 235–245.

Pettitt, P., 2000. Odd Man Out: Neanderthals and Modern Humans. *British Archaeology* 51, 1357–4442.

Plummer, T., 2004. Flaked Stones and Old Bones: Biological and Cultural Evolution at the Dawn of Technology. *Yearbook of Physical Anthropology* 47, 118–164.

Pringle, H., 1998. New Women of the Ice Age. *Discover* 19, 62–70.

Ramirez Rozzi, F.V., Bermudez de Castro, J.M., 2004. Surprisingly Rapid Growth in Neanderthals. *Nature* 428, 936–939. doi:10.1038/nature02428.

Rhodes, J.A., Churchill, S.E., 2009. Throwing in the Middle and Upper Paleolithic: Inferences from an Analysis of Humeral Retroversion. *Journal of Human Evolution* 561, 1–10.

Sentenac, A., n.d. Witnesses: André Sentenac, Biologist in France [WWW Document]. The Cave of Chauvet-Pont-d'Arc. Available at http://www.culture.gouv.fr/fr/arcnat/chauvet/en/temoi9.htm (accessed August 16, 2009).

Solecki, R.L., Agelarakis, A.P., 2004. *The Proto-Neolithic Cemetery in Shanidar Cave*. Texas A&M University Press, College Station, TX.

Tyldesley, J.A., Bahn, P.G., 1983. Use of Plants in the European Palaeolithic: A Review of the Evidence. *Quaternary Science Reviews* 2, 53–81.

Tzedakis, P.C., Hughen, K.A., Cacho, I., Harvati, K., 2007. Placing Late Neanderthals in a Climatic Context. *Nature* 449, 206–208. doi:10.1038/nature06117.

Valladas, H., Clottes, J., Geneste, J.-M., Garcia, M.A., Arnold, M., Cachier, H., Tisnerat-Laborde, N., 2001. Palaeolithic Paintings: Evolution of Prehistoric Cave Art. *Nature* 413, 479. doi:10.1038/35097160.

Vandiver, P.B., Soffer, O., Klima, B., Svoboda, J., 1989. The Origins of Ceramic Technology at Dolni Vestonice, Czechoslovakia. *Science* 246, 1002–1008. doi:10.1126/science.246.4933.1002.

Viegas, J., 2008. Early Weapon Evidence Reveals Bloody Past. *Discovery News*.

Vishnyatsky, L.B., 1994. Running Ahead of Time in the Development of Palaeolithic Industries. *Antiquity* 68, 134–140.

Weiner, J., 2003. Kenntnis-Werkzeug-Rohmaterial. Ein Vademekum zur Technologie der steinzeitlichen Holzbearbeitung. *Archäologische Informationen* 26, 407–426.

3

The Urban Revolution—The Emergence of Society

There must be first a farmer, secondly a builder, thirdly a weaver, to which may be added a cobbler. Four or five citizens at least are required to make a city.

Plato (ca. 424–348 BCE)
Philosopher, in The Republic, Book II

At the end of the Middle Stone Age, humankind knew all six fundamental manufacturing techniques but otherwise had not evolved beyond nomadic groups of hunter-gatherers. They stayed in one spot only as long as there was enough food around, and when the food supply started to dwindle, they moved on to another site with more food. This style of living was actually quite efficient and, most of the time, provided enough food for the group. On the downside, it required a large area to sustain a group, and subsequently, the population density was quite low. Furthermore, food needed to be available all year round to support the group.

However, 11,000 years ago, the climate in the world changed. The Ice Age was taking a break, and the Holocene warm period started. Temperatures rose, on average, about 6°C; winters became less harsh; and a lot of ice melted. In fact, so much ice melted that the ocean level rose 110 meters* (IPCC 2007). Among other things, this also led to the flooding of a great depression between modern-day Ukraine and Turkey. This depression was roughly the size of California and is nowadays called the Black Sea (Ballard, Coleman, and Rosenberg 2000). The resulting waterfall was certainly bigger than any

* In comparison, modern global warming is expected to cause a sea level rise of less than 1 meter (Harman 2002).

other waterfall we know nowadays, and its remains are now the Bosporus strait. Altogether, the water levels in this depression rose 150 meters in a very short period, making a nasty surprise for the local inhabitants.

Undoubtedly, something big had happened. Summer was getting warmer and wetter, leading to a major change in the ecosystem. Many large animals were unable to cope with this change. Mammoths cave bears, saber-toothed cats, and ground sloths became extinct. Even horses became locally extinct in America. The warmer climate also caused annual dry seasons, making it difficult for plants that live multiple years to survive. On the other hand, annual plants that die off during the dry season and sprout again in the wet season had a significant advantage.

Humans were deeply influenced by these changes. Their previous hunter-gatherer lifestyle was becoming increasingly difficult. While there was still more than enough food available in the summer, getting something to eat during the winter proved to be more and more of a challenge. Humankind needed to change, and change we did. We adapted over the course of a few thousand years, which is quite a rapid change compared to the previous changes during the Stone Age, and hence, this change is nowadays known as the Neolithic revolution.*

3.1 THE NEOLITHIC REVOLUTION—FROM NOMADIC HUNTERS TO SETTLED FARMERS

One of the main problems caused by warming was the problem of obtaining food. Hunting and gathering works only so long as there is food around to be hunted or gathered. Yet, while there was sufficient food during the wet season, there was not enough during the dry season. Therefore, humans needed to find other ways to put food on the table (Harman 2002). This led to the emergence of agriculture and the domestication of animals around 10,000 BCE in the so-called Fertile Crescent in the Middle East (Gupta 2004) . Figure 3.1 shows one of the earliest tools for grinding grain. Agriculture and domestication also appeared independently later in different parts of Africa and America.

* The term, together with the term *urban revolution*, was coined by Australian archaeologist Gordon Childe (1892–1957), whose groundbreaking works studied the social structure of mankind in early history.

FIGURE 3.1
One of the oldest tools for grinding grain. From Syria, around 9500–9000 BCE. (Photo by author, dated April 2010.)

The first domesticated plants in the world are the so-called Neolithic founder crops: the cereals emmer, einkorn, and barley; the pulses lentil, pea, chickpea, and bitter vetch; and finally, the fiber flax (Brown et al. 2009). The first domesticated animals were sheep, goats, pigs, and cattle* (Diamond 1997). By growing their own food, farmers were able to produce much more food than hunters were. Hence, they were able not only to store food for the dry season but also to have more children and to support a much larger population. Agriculture, on the other hand, also had its drawbacks. The variety of foods available was reduced to domesticated species, and early farmers suffered from malnutrition more frequently than hunters and gatherers. Additionally, herding and farming were much harder labor compared to hunting and gathering.

Since agriculture and domestication of animals were incompatible with the nomadic life of hunting and gathering, people settled. These settlements soon evolved into villages. One of the largest of these villages was Çatalhöyük in modern Anatolia, Turkey. Yet, it was still only a village, and research suggests that there was not yet any higher degree of specialization (Balter 1998). For example, it appears that all the buildings were built slightly differently, presumably by their owners rather than specialized builders. Additionally, most houses contained obsidian flakes, indicating again that these products were crafted in these houses rather than in a centralized workshop by a specialist. Overall, it is believed that there was no division of labor except as caused by age or gender. While there may have been some villagers better at some tasks than others, all still needed to farm in order to make ends meet. This style of manufacturing remained unchanged for the next few thousand years until the end of the Stone Age, around 3300 BCE.

* Only dogs were domesticated earlier, around 15,000 BCE.

3.2 EARLY DIVISION OF LABOR—THE EMERGENCE OF THE ARTISAN DURING THE BRONZE AGE

In the millennia leading up to the Bronze Age, humankind made numerous inventions and discoveries. The plough was invented around 6000 BCE, irrigation around 5000 BCE, sailboats around 4800 BCE (Ben-Menahem 2009), and the wheel around 5000–4000 BCE (Bulliet 2016). All this increased food production and hence allowed a higher population density. Soon, the first villages grew into cities, with some of the first cities originating around 3500 BCE in Mesopotamia (modern Iraq). At that time, Uruk was the largest city in the world, with a population of about 14,000 inhabitants, quickly growing to 50,000 (Modelski 1997).

While the rapid improvements in agriculture and domestication of animals were probably aided by manufacturing the tools necessary for these tasks, they were not a breakthrough for manufacturing itself. However, the increase in food production and population density allowed, for the first time in history, some members of society not to focus on food but to specialize in other topics. New full-time occupations appeared in areas like leadership, warfare, religion, transport, trade, the service industry, and—most importantly within this book—manufacturing.

The two key developments that definitely happened at the latest during the Bronze Age were the division of labor and specialization. Division of labor is the distribution of different tasks to different people in a society. For example, in order to create a dress, some individuals spin the yarn, while others weave the cloth. Others tailor the cloth into a dress, transport the dress, and sell the dress to the final customer. Advantages of division of labor are a faster learning curve, and thereby higher quality, and also higher efficiency combined with less transport or waiting time. Disadvantages are less self-sufficiency, and therefore a higher dependence on others and a higher sensitivity to changes in the economic system.

Specialization is the long-term focus of an individual in society on a certain set of tasks, for example, a tailor who does not only make clothes part time besides other businesses (for instance, farming or weaving) but focuses exclusively on tailoring as his/her primary occupation. *Specialization* is often used synonymously with *division of labor*. However, division of labor does not necessarily require specialization. For example, a day laborer may have a different task every day as part of the division of labor but in no way specializes in the task. Similarly, craftsmen can specialize in simple products that require no division of labor. Benefits of specialization are a higher skill level and, subsequently, higher quality and faster development of new techniques.

Division of labor and specialization go hand in hand with metalworking in the Bronze Age. While the first copper–arsenic bronze artifacts may date back to 4200 BCE, larger-scale manufacturing of higher-quality copper–tin bronze products started in 3300 BCE in the Middle East, giving a name for the whole era, the Bronze Age.

Yet, this technology was also much more demanding than previous technologies. For example, all you needed to make stone tools were suitable stones and some time to fiddle around until you found a good way to shape them. Similarly for woodworking, you could start any time and work on it as much as you had time to spare. However, the hurdles for bronze products were much higher. First, you needed a tool highly advanced for its time: a smelter. Constructing a smelter was time-consuming and complicated, and required significant up-front investment of time. Furthermore, bronze is an alloy of copper and tin. While copper or copper ore was frequently available, tin was available only in a few regions in the Middle East. This required significant planning and trade connections to obtain the necessary raw materials.

In addition, while you could interrupt your work with wood or stone at any time to do other work, once you fired up your smelter, you had to stick with it, or the whole process would have failed. Leaving the smelter to go farming was simply not an option. Finally, while most woodwork or stone products required dexterity, the relation between the raw material and the finished product was always clear. On the other hand, bronze production required significant specialized knowledge. You needed to know how much tin to mix with the bronze, how long and how hot to heat the smelter, and a whole lot of other details that are not obvious before you are able to produce any useful product. Therefore, due to the high up-front investment, the advanced logistics of the raw materials, the time needed for producing a batch, and the skill level required, making bronze part time was simply not an option. If you wanted to be in the bronze business, you had to be specialized full time on bronze making and have other people provide you with the necessities, for example, tin for your bronze or food for yourself. Bronze making is impossible without specialization and division of labor.

Another specialization of Bronze Age artisans was pottery. While pottery was available for many centuries beforehand (see Section 2.5), the invention of the potter's wheel around 3000 BCE greatly increased the output of the potters. In northern Palestine, for example, there are a large number of pottery shards of large storage jars known as pithos from around 2500 BCE. Many of these shards are stamped with what are presumed to be the seals of the potter. Judging from the small number of different seals found on a large number of shards in many different locations in northern Palestine, it is safe to say that this was the output of a few specialized potters who probably traveled across the region to fulfill the local demand (Philip and Baird 2000, p. 304).

3.3 AT THE BOTTOM OF THE SOCIAL HIERARCHY—SLAVE LABOR AND ITS SUPERVISION

Stone Age hunter-gatherers had very few social distinctions. There probably was a leader of the group, possibly also a shaman or healer, but usually, everybody joined the primary activities of the group—hunting and gathering. This changed dramatically with the emergence of societies. Different members of the group specialized in different professions, leading to a division of labor and specialization, as described in Section 3.2. Subsequently, some individuals in society had more power than others and worked actively on increasing their power base. An easy way to increase your own power is to take away power from others. This leads, in its extreme, to one person having all power over another person, including where to live, what to do, with whom to mate, and whether to live or to die altogether. Hence, a person basically was the property of another person, who could do with him/her as he/she wished. In sum, this person was a slave.

Slavery was widespread, and there were very few cultures in ancient times that did not have slaves. Mesopotamian tablets with signs for slave girls date back to 3000 BCE; similar tablets for male slaves* date only slightly later (Adams 2005, p. 96). Figure 3.2 shows for example Assyrian

* Many societies had initially more female slaves, since male slaves were harder to control. Captured male enemies were often killed or sometimes may have been blinded to make them more docile (Adams 2005, p. 97).

FIGURE 3.2
Assyrian prisoners of war, around 700–692 BCE. (Photo by author, dated April 2010.)

prisoners of war, a popular source of slaves. There were different ways for a person to become a slave: at birth, through capture by enemies, as punishment for criminals, to pay off debts, or through sale by parents.

Slavery in various forms was a significant part of most ancient economies, and many societies in ancient Europe were 30% slaves or more. Additionally, parts of the population may have officially been *free*, but indebtedness or dependence made them little more than slaves. Besides the cost of obtaining a slave, there was the cost of feeding and clothing him/her, a considerable expense for society considering the high percentage of slaves. To be worth the expense, slaves were used to work in many areas, such as agriculture, service, construction, and manufacturing. Slave labor was probably mainly used for unpleasant, demanding, or repetitive tasks, or to assist the master. It is difficult to generalize, but one can safely assume that a large percentage of slaves had little influence on the way they worked. They were often driven by fear of reprisal if they did not do what they were told. In order to keep them working, a high level of supervision was necessary—the proverbial slave driver, who instructed the slave, observed the work, and punished the slave if the outcome was not satisfactory. This made slave drivers the first supervisors in manufacturing.

3.4 TRANSITION FROM THE BRONZE AGE TO THE IRON AGE

The Iron Age followed the Bronze Age from the 12th century BCE. The material iron is more widely available than copper, but requires a much higher melting point. Bronze melts at around 950°C, slightly lower than pure copper at 1084°C; this could be done in a larger charcoal campfire. Iron melts at 1535°C, although the reduction of iron ore to iron bloom* was possible already at 1200°C. Nevertheless, a more advanced technology was needed to extract iron from the ore.

Iron can also be used to make steel, a material much harder and stronger for the same weight compared to bronze. However, this requires careful control of the carbon content in the iron, which again required sophisticated technology. Steel making probably dates from the first or second millennium BCE, often involving more luck than skill. Nevertheless, most of the iron used up to medieval times was wrought iron with a very low carbon content and softer than bronze.

Iron, however, had two major advantages over bronze tools—it did not require tin or other exotic metals to make, and it could be sharpened, whereas bronze needed to be recast. Subsequently, many warriors were equipped with inferior but cheaper and easier-to-maintain iron weapons.

From a manufacturing point of view, iron objects did not require a constant supply of rare and expensive raw materials. Once you have mastered the required technology to extract iron, you can make much less expensive products compared to bronze. In turn, due to the lower cost, there were a larger number of customers who could afford these products, and you were able to sell much more iron products compared to bronze. Please note that we are talking about relatively lower cost, and iron was still a very expensive product, although much less so than bronze. Subsequently, bronze production declined, as did the trade associated with bronze making and societies based on bronze tools.†

Of course, development was not the same across the world; some regions chose different paths, giving them a technological advantage or

* Iron bloom is a spongy mass of iron with slag and other impurities. This could be separated mechanically and hammered (or *wrought*) into a lump of wrought iron.

† Some historians see this decline as rather violent and quick and describe this as the collapse of the Bronze Age.

disadvantage. The following chapter looks at advances during antiquity and selected regional developments in more detail.

BIBLIOGRAPHY

Adams, R.M., 2005. *The Evolution of Urban Society: Early Mesopotamia and Prehispanic Mexico*. Aldine Transaction, New Brunswick, NJ.

Ballard, R.D., Coleman, D.F., Rosenberg, G.D., 2000. Further Evidence of Abrupt Holocene Drowning of the Black Sea Shelf. *Marine Geology* 170, 253–262.

Balter, M., 1998. The First Cities: Why Settle Down? The Mystery of Communities. *Science* 282, 1442. doi:10.1126/science.282.5393.1442.

Ben-Menahem, A., 2009. *Historical Encyclopedia of Natural and Mathematical Sciences*, 1st ed. Springer, New York.

Brown, T.A., Jones, M.K., Powell, W., Allaby, R.G., 2009. The Complex Origins of Domesticated Crops in the Fertile Crescent. *Trends in Ecology & Evolution* 24(2), 103–109.

Bulliet, R.W., 2016. *The Wheel: Inventions and Reinventions*. Columbia University Press, New York.

Diamond, J., 1997. *Guns, Germs, and Steel. The Fates of Human Societies*, 1st ed. W.W. Norton and Company, New York.

Gupta, T.P., 2004. Origin of Agriculture and Domestication of Plants and Animals Linked to Early Holocene. *Current Science* 87(1), 54–59.

Harman, C., 2002. *A People's History of the World*, 2nd ed. Bookmarks Publications Ltd, London.

IPCC, 2007. Climate Change 2007: The Physical Science Basis. Contribution of Working Group I to the Fourth Assessment Report of the Intergovernmental Panel on Climate Change, in: Summary for Policymakers.

Modelski, G., 1997. Cities of the ancient world: An Inventory (−3500 to −1200). Working paper.

Philip, G., Baird, D., 2000. *Ceramics and Change in the Early Bronze Age of the Southern Levant*. Sheffield Academic Press, Sheffield, UK.

4

Advances during Antiquity

In a small city the same man must make beds and chairs and ploughs and tables, and often build houses as well; and indeed he will be only too glad if he can find enough employers in all trades to keep him. [...] But in the great cities, owing to the wide demand for each particular thing, a single craft will suffice for a means of livelihood. [...] Or one artisan will get his living merely by stitching shoes, another by cutting them out, a third by shaping the upper leathers, and a fourth will do nothing but fit the parts together. Necessarily the man who spends all his time and trouble on the smallest task will do that task the best.

Xenophon of Athens (ca. 430–354 BCE)
Soldier and writer, in Cyropaedia

The Neolithic revolution fundamentally changed the life of mankind. In terms of magnitude of change, it probably is second only to the industrial revolution. Settling down allowed farming, which eventually provided a much larger supply of food than hunting within the same territory. Increased food availability allowed division of labor and specialization. Through this, small bands of people grew into tribes, chiefdoms, and eventually, states. Knowledge was accumulated. New technology was invented. It was a slow process, but it laid the foundations for the rise of manufacturing during the Middle Ages. First and foremost, it increased man's power far beyond his own muscles.

4.1 BREAKING THE ENERGY CONSTRAINT—ANIMAL, WATER, AND WIND POWER

During the Stone Age, man's ability to produce was mostly limited to his own muscle power. The only additional energy source used in manufacturing was the heat of fire to harden wood, create pottery, and later, also to melt metals. Everything else was created by the power of human muscles. Yet, man's conversion of food into muscle power is very inefficient. It is estimated that less than 25% of the power from food is available for movement; everything else is required just to keep us alive. Furthermore, compared to other animals of similar size, man is weak. An ox can create seven times as much power than a man, and a horse, ten times. Yet, all pales in comparison with the enormous tireless energy of waterpower (Wikander 2008, p. 136ff). While we have significantly more manual dexterity than almost all other animals and mental abilities to learn and invent far beyond all other animals, we lack brute strength.

The harnessing of animal power started with the domestication of animals during the Neolithic revolution. Probably one of the first uses of animals for power was transport, first as a pack animals and later by pulling a sled. Similar to animal power, waterpower was also first used for transportation through floats and boats. Sailboats originated during the third millennium BCE. Soon, animals helped in agriculture, for example, by pulling a plough (see Section 3.2). However, animal, water, and wind power work only in one direction, yet manufacturing requires, in its simplest form, either a rotary motion or a back-and-forth motion. The linear motions of animals, water, and wind in their unaltered form were unsuitable for this purpose.

Therefore, it was necessary to develop technology to convert these linear motions into a rotary or back-and-forth motion before these power sources could be utilized for manufacturing purposes. Due to the lack of a suitable power source, the earliest protomachines were tools powered by hand. These simple machines include levers, winches, pulleys, wedges, wheels, and axles. Probably the earliest power sources in manufacturing besides human muscle were Afghan windmills from the second millennium BCE as shown in Figure 4.1 (Rossi, Russo, and Russo 2009, p. 63; Tapper and McLachlan 2003, p. 92). In these windmills, the sails are on a vertical axis that is directly connected to the grindstone, avoiding the use of any kind of complicated gears. Similar technology could have also been used to power water mills, but such mills were not known until the third century BCE.

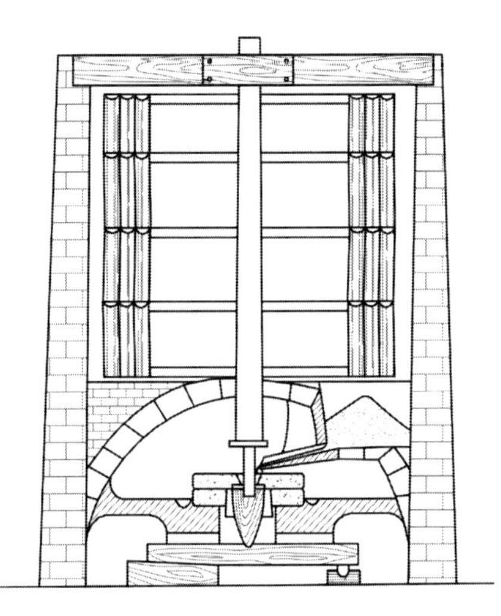

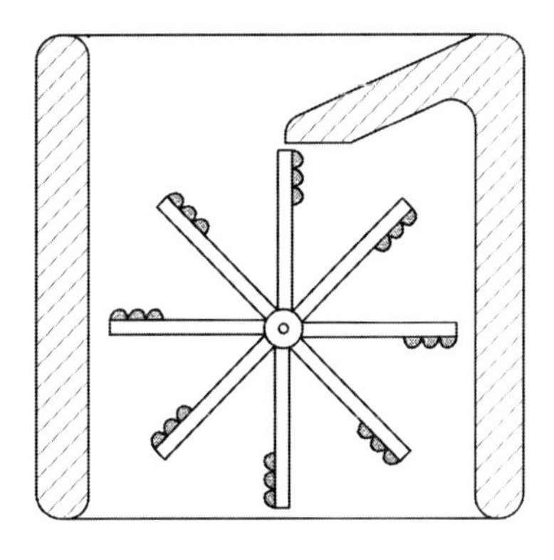

FIGURE 4.1
Diagram of a Persian windmill. (Image by Kaboldy and licensed under the Creative Commons Attribution-ShareAlike 3.0 Unported license.)

During antiquity, more complex machines* were created through the combination of protomachines like levers, winches, pulleys, wedges, wheels, and axles. Egyptians knew how to use drills, potter wheels, and lathes. A crucial step in transforming the direction and the strength of movement was the development of gears around 270 BCE in Greece. This was followed by Archimedes' invention of the worm drive and thereby also the screw around 250 BCE. Apollonius of Perge (262–190 BCE) described right-angle gears by 240 BCE.

This technology to transfer linear motion into rotary motion through wheels, shafts, and gears was then used to lift water for irrigation in agriculture. Animal-powered water lifting was known from around 300 BCE. Water-powered irrigation is mentioned by the Greek writer Philo of Byzantium (280–220 BCE) in his treatise *Mechanike syntaxis* (see Section 4.2), but such waterwheels may also have existed earlier (Wilson 2008b, p. 352ff).

The first known uses of water power for manufacturing were mills to process grain into flour during the third century BCE. The oldest known water mills were found in Chemtou and Testour in modern Tunisia, using a horizontal waterwheel and hence not requiring any gears. Nevertheless, the large size and complexity of these mills strongly indicate a significant

* While this book focuses primarily on manufacturing, Greeks and Romans also used machines for agriculture, construction, warfare, and entertainment.

development that is, so far, undiscovered by archaeologists. Only a few years later, technology evolved further in the form of water mills using a vertical waterwheel and therefore requiring a right-angle gear. This type of vertical mill became widespread and was the prime producer of energy for manufacturing purposes from Roman times up to the beginning of the industrial revolution. By the first century BCE, animal power was also used in manufacturing, and animal-powered mills for grinding flour and olives were common in Pompeii and Ostia, as were animal-powered dough-kneading machines.

So far, manufacturing powered by water or animals was limited to rotary movements, primarily in the form of grain milling. The next significant step in powered manufacturing was the development of noncircular motion. The first devices that utilized an up-and-down motion were simple pestles to grind grain and hammers to crush ore for easier processing from around 70 CE. These devices were powered by a waterwheel driving a shaft with a peg sticking out. This peg hit one end of a seesaw-like shaft, pushing it down and lifting the other end up. After the peg loses contact, the weight on the other end falls down, crushing whatever is underneath. A more complex back-and-forth motion using a crank did not emerge until 300 years later, and the first sawmills for wood date back to 370–390 CE. These were soon followed by stone mills, and Ausonius (ca. 310–ca. 395 CE) reports of stone sawmills on the river Kyll in the Eifel, Germany, in the third century.

Before the use of animal, wind, or water power, all of the aforementioned processes like milling and sawing were also done by humans on a smaller scale. The complex machines were only necessary to convert animal or water power into a form that could be utilized for manufacturing. If a human had powered the process, no complex machinery would have been necessary. For example, the complex machinery of a water mill serves only to drive the millstone, something a human could have done on a smaller scale without the complex mechanisms.

However, there were also complex machines that helped humans to do work they otherwise would have had difficulty with. To increase strength, presses were used from the sixth century BCE to extract liquids from fruits and seeds (Muljevic 2001). These wine and oil presses were based first on levers and, from the first century CE, also on screws as an improvement over manually stacking weights on top of the produce. Other machines helped humans with processes they would have had difficulty doing themselves otherwise, for example, the lathe, which was used from the third century BCE. Another example is the loom. While primitive warp-weighted looms were already in use during the Neolithic, more complex and efficient two-beam vertical looms appeared in the first century CE. The two-beam loom creates tension through a rotatable beam to wind the yarn at the bottom.

Water and animals were the main power sources of the classical world. This not only freed mankind from dull repetitive and strenuous tasks but also greatly increased the efficiency of processes, requiring much less food and feed for the same quantity of material. Other machines also reduced workload and increased efficiency. However, water power was used primarily to process food, and to a lesser extent, for mining-related operations, but there was little development to apply this power source for other uses. Additionally, windmills were common in Europe only from the Middle Ages (Cameron and Neal 2002, p. 42).

4.2 ACCUMULATING KNOWLEDGE—SCIENCE AND ENGINEERING WRITINGS

In ancient times, artisans passed on technical knowledge orally from generation to generation, and therefore, anything that was not remembered by the next generation was lost. Accumulation of knowledge was severely limited by the ability to remember, and progress in science and technology was slow. In order to build a larger body of knowledge, it was necessary to store this knowledge so others could understand and expand this knowledge. Hence, writing was a prerequisite for science.

The first writing emerged during the fourth millennium BCE, and the Sumerian cuneiform script is considered to be the first writing system.* The system was created by to keep track of taxes and other records, making it one of the few instances in history where a government bureaucracy actually invented something useful. This system soon evolved, and the first coherent text dates back to 2600 BCE. The first alphabet where a character represents not a word but merely a sound dates back to 1800 BCE. This allowed storing a person's memory or other data in written form, hence not only allowing the copying of data and making it more readily available but also preserving it for future generations to build on.

* However, there are some disputed indications of Chinese writing that is 2000 years older.

One of the first explicit mentions of manufacturing-related topics comes from the *Code of Hammurabi* from around 1790 BCE. This legal text explained the general laws in ancient Babylon, showing us how knowledge was passed on through apprenticeship and that artisans were available for hire for fixed prices:

> If an artisan take a son for adoption and teach him his handicraft, one may not bring claim for him. If he do not teach him his handicraft, that adopted son may return to his father's house.
> [...]
> If any one hire a skilled artisan, he shall pay as wages [...] of the potter five gerahs, of a tailor five gerahs, [...] of a ropemaker four gerahs [...] per day...
>
> ***Code of Hammurabi*, paragraphs 188, 189, and 274**

However, most of the surviving ancient texts deal with religious issues, record keeping, and poetry. There are few mentioning manufacturing details, although there are a number of documents of legal quarrels related to manufacturing. For example, workers complained about working conditions and demanded better tools, food, and more slaves to help (Cuomo 2008, p. 20). The first scientific texts do not look at earthly details but, rather, gaze at the stars and explore the realm of mathematics. For example, problems related to the theorem of Pythagoras are found on clay tablets from 1800 BCE. Astronomy was widely studied in Mesopotamia and Persia. Egypt was also well versed in medicine, mathematics, and astronomy; the Chinese were outstanding astronomers and physicians; and Indians had, besides mathematics and astronomy, a particular interest in linguistics.

The Greek philosopher Thales of Miletus (around 624–546 BCE) was one of the first to explore philosophical topics, for example, the origin of the world, while most other parts of the world looked to religion for answers to these questions. Soon, Greek philosophers were trying to understand the behavior of nature based on deductive reasoning. Eventually scientists and philosophers created a large body of knowledge that led to a great number of inventions.

The first known book focusing on technical aspects is *Mechanike syntaxis* by Philo of Byzantinum (ca. 280–220 BCE), containing sections on mathematics, mechanics, architecture, and warfare technology, although not all of these have survived. The oldest completely surviving work on technology is *De architectura* by Marcus Vitruvius Pollio (ca. 80–70 BCE to after 15 BCE), focusing on architecture but also containing one large section on manufacturing technology, detailing the transformation from rotary to linear motion and describing a water mill. The following is its description of the right-angle gear for a water mill:

> Water mills are turned on the same principle, and are in all respects similar, except that at one end of the axis they are provided with a drum-wheel, toothed and framed fast to the said axis; this being placed vertically on the edge turns round with the wheel. Corresponding with the drum-wheel a larger horizontal toothed wheel is placed, working on an axis whose upper head is in the form of a dovetail, and is inserted into the mill-stone. Thus the teeth of the drum-wheel which is made fast to the axis acting on the teeth of the horizontal wheel, produce the revolution of the mill-stones, and in the engine a suspended hopper supplying them with grain, in the same revolution the flour is produced.
>
> **Marcus Vitruvius Pollio (ca. 80–15 BCE)**
> *Roman architect and engineer,*
> *in De architectura, translated by Joseph Gwilt, 1826*

There is still considerable skill required to build a water mill from that description, especially since it is lacking any diagrams, but otherwise, a description in a modern textbook would look very similar. The list of ancient writers on manufacturing technology would not be complete without Hero of Alexandria (ca. 10–70 CE). He authored numerous books on technical topics, including *Pneumatica*, detailing hydraulics, pneumatics, and steam power, and *Automata*, a description of mechanical or pneumatic machines.

While many of these descriptions were still rudimentary, and in some instances, simply incorrect, mankind was building up a body of knowledge for the next generations. Books by the ancient writers on science and technology were widely copied and translated and were still read by professionals up to the beginning of the industrial revolution. This significance of the texts for the development of science and technology is also the main reason why these texts survived to the modern period.

4.3 THE FIRST STANDARDIZATION— THE HARAPPAN CULTURE

During the building of Indian Railways in the 1850s, workers discovered large piles of old but well-made bricks along their route. They happily recycled them to build the railway. It was believed that due to the high quality of these bricks, they were only about 200 years old. Well, they were off by a factor of 20. More detailed investigations of these bricks after 1920 showed that these were more than 4000 years old, pretty much doubling the known history of India.* This culture is now known as the Harappan culture, named after the village where the first excavations were made.†

The Harappan culture was at its peak between 2600 and 1900 BCE and, at that time, probably the most advanced culture of the Bronze Age. It covered an area almost the size of Alaska, with more than 1000 settlements found to date (Ackermann et al. 2008, p. 1:213), the largest being Mohenjo-Daro and Harappa, with populations of 40,000 and 20,000 people, respectively (Hart 2007, p. 269). They may have been the first to use wheeled ox carts, they were expert potters, and their weaving technology was unparalleled in the world (Keay 2001, pp. 12–13). They built their cities from high-quality fired bricks and also installed both public and private baths. They built aqueducts for fresh water and a sewage system that outperformed most of the sewage handling in existence in modern-day India. They developed a writing system‡ and had significant long-distance trade.

However, what the Harappan culture stands out for is its extremely high levels of uniformity and standardization. Their hundreds of cities and towns were all laid out after the same plan, using a grid pattern with straight roads running from east to west and from north to south. The cities were divided into residential and commercial areas (Fairservis 1992, p. 133). Street width was standardized, with a width of 1.8 meters for an alley, double width of 3.6 meters for a normal road, and triple and quadruple widths of 5.4 and 7.2 meters for the main crossroads (Kulke and Rothermund 1998, p. 23). The cities always consisted of two parts. First was a working and living district with separate manufacturing, trade, agricultural, and residential areas. Second was a large structure, believed to be a citadel, to the west of the working and living district. The houses were also built highly uniform, with standard houses of different sizes,

* Before the discovery of the Harappan culture, the oldest known Indian civilization was the Maurya Empire between 321 and 185 BCE, and the Harappan culture pushed Indian history earlier by another 2000 years.

† Also known as the Indus Valley culture, as it was located along the Indus and the now mostly dried out Ghaggar–Hakra River in the Indian subcontinent.

‡ Unfortunately, this writing is not yet deciphered, and all we know about the Harappan is based on archeological finds. Deciphering the language turned out to be difficult, since the longest uninterrupted text known is a mere 26 symbols, or less than 1/6th of the length of a Twitter message.

FIGURE 4.2
Harappan weights from around 3000 BCE. The larger one has a side length of about 4 centimeters. (Photo by author, dated April 2010.)

from the two-room worker's house to the houses of the wealthy with flush toilets. Clearly, these were the world's first planned cities.

This uniformity also extended to manufacturing. The large numbers of bricks manufactured across the empire all had the same ratio of 1:2:4, mostly in dimensions of 7.5 × 15 × 30 centimeters. Hence, the width was twice the height, and the length was twice the width. This modularity allowed easier building of doors and windows and furthermore allowed the builder to use bricks from different sources, greatly enhancing flexibility. This modularity was not only applied to street widths and brick sizes. The measurement system for both weight and length was also built on a modular system. Their standard weights, for example, also followed a strict repeating pattern* of 1:2:4:10, and these weights did not change for hundreds of years (Selin 2008, p. 2254). Their measurement for length was the most precise of the Bronze Age, with the smallest division being less than two millimeters. Figure 4.2 shows an example of these weight references.

The pottery also shows a highly standardized and uniform design with few decorations. This design varied only very little both over distance and

* Compare this to the inch/foot/yard/mile system in the United States with a ratio of 1:12:36:63,360, not even including additional systems as the nautical system (fathom, cable, nautical mile) or the survey system (rod, chain, furlong, survey mile).

throughout time. The production of ceramics was on such a large scale that some archaeologists call this a *semi-industrial mass production* (Kulke and Rothermund 1998, p. 21). Even the tool sets of the artisans were highly standardized, consisting of a standard toolbox with stone scrapers, knives, and drills as well as some bronze and copper tools depending on their profession. The standard is so thorough that the type of workshop can be identified from a single example on site (Keay 2001, p. 10). The Harappan culture probably had many more standards throughout their daily lives, for example, in agricultural practices or clothes making, but unfortunately, knowledge of these standards is lost in history.

Nevertheless, this level of standardization is unparalleled for the Bronze Age. Not only were the standards enforced over a large area almost the size of Alaska, but they varied very little throughout the 700 years of the mature Harappan period.* During that time, numerous cities were destroyed by floods, but the rebuilding of new settlements on top of the ruins of the old settlements followed the same grid pattern. Settlements 600 kilometers apart from each other followed the same standard. Figure 4.3 shows a map of the civilization and its major settlements. As the Harappan Empire expanded and (presumably violently) conquered new settlements, the standards were quickly enforced, and the previous local standards and pottery designs disappeared.

Everybody who has ever enforced a standard knows that there must have been significant effort and motivation to enable this consistency. The Harappan culture followed their standards almost like a cult, and it is quite possible that these standards were indeed a cult. Unfortunately, we know next to nothing on how the civilization was organized or what they believed in. Some researchers believe these were a group of independent culturally connected chiefdoms (Duiker and Spielvogel 2006, p. 31), whereas others strongly believe that there was a central authority setting the standard. We also do not know why they developed standards. It could be that some ancient genius saw the clear advantage of standards, but it could equally be that Harappan leadership showed obsessive–compulsive behavior, or possibly a combination of both. We will never know.

Around 1800 BCE, however, the quality of work declined. Pottery styles became sloppier. The formerly tightly controlled standard started to slip, and different standards emerged between the cities. Entire cities were

* Again, in comparison, the modern SI system based on meter, kilogram, etc. is a mere 200 years old.

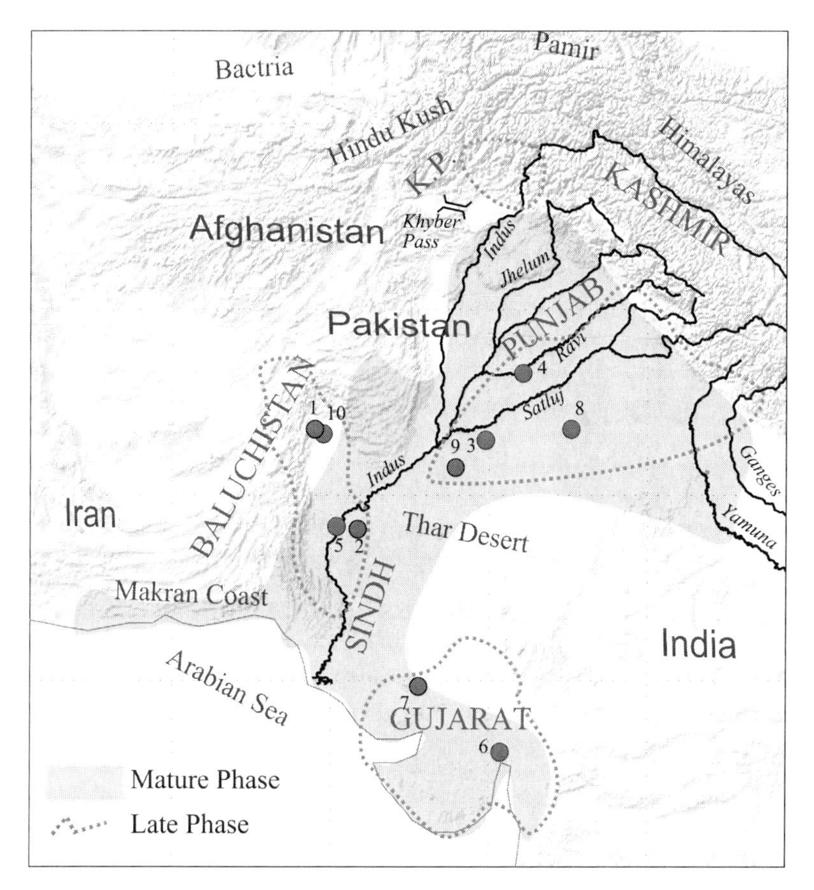

FIGURE 4.3
Map of the Indus Valley culture. The cities are (1) Mehrgarh, (2) Amri, (3) Kotdiji, (4) Harappa, (5) Mohenjo-Daro, (6) Lothal, (7) Dholavira, (8) Kalibangan, (9) Ganweriwala, and (10) Pirak. (Image by Aurangzebwardag and licensed under the Creative Commons Attribution-ShareAlike 3.0 license.)

abandoned (Keay 2001, p. 13). Researchers initially believed that Indo-Aryan tribes invaded and destroyed the Harappan culture. However, modern research shows that the Harappan civilization was in decline even before the Indo-Aryan invasion. Many abandoned cities show no signs of a violent conflict, including the last level of the largest city, Mohenjo-Daro. The decline of the Harappan may have been caused by climate change, where a prolonged dry period led to the drying up of the Ghaggar–Hakra River and possibly a shift of the Indus. The lack of rain also led to soil salination and erosion on the fields (Diamond 2005). This in turn led to a lack of food and malnutrition (Bauer 2007, p. 192ff).

While the Harappan may have had the best standardization of its time, it lacked one essential step that goes hand in hand with standardization in modern manufacturing: continuous improvement! These standards may have been good when they were created, but without an improvement process, they were heavily outdated 600 years later and unable to handle the changed environment. Following these old standards in a new environment may have led to the decline of the Harappan, whose remnants were then mopped up by the Indo-Aryan invasion.

4.4 IMPERIAL CHINA

The history of imperial China spans over 2000 years, alternating periods of peace and stability with periods of warfare and fights for power. Of the periods of stability, the Qin dynasty (221–206 BCE), the Han dynasty (202 BCE–220 CE), the Tang dynasty (618–907 CE), and the Song dynasty (960–1234 CE) stand out in terms of their technical and industrial development. The Qin and Han dynasties developed many cities in northern China into industrial centers, standardized currencies, regulated the economy, and greatly increased trade (Hinsch 2002, p. 28). Especially, the Qin moved the focus of the government away from agriculture and toward industry (Wagner 1993, p. 256). Compared to the Roman Empire, where technical development focused on a few areas, imperial China created a vast body of knowledge and technologies in many different areas. Particularly well known are the so-called *four great inventions* of ancient China: papermaking in the first century,* gunpowder and printing in the ninth century, and the compass in the eleventh century, although there are many more. Most of these technologies were known in Europe only by the end of the Middle Ages.

However, there were also moves backward, where knowledge and technology were intentionally destroyed for political reasons. At the very beginning of imperial China, the first emperor, Qin Shi Huang (260–210 BCE), was influenced by his prime minister, Li Si, to erase all thoughts and ideas except for those suitable to him. Thus, he ordered the burning of books in 213 BCE, and after the protest of Confucian scholars, he also

* There are some indications that the Chinese may have invented paper much earlier, as early as the eighth century BCE.

ordered the execution of 460 of these scholars by burying them alive in 212 BCE (Stearns 2001). This *burning of the books and burial of the scholars* aimed to erase all books except those favorable to Emperor Qin and those about the topics of medicine, divination, agriculture, and forestry. Only a few volumes survived, hidden by scholars or discovered later in tombs and other burial sites.

Nevertheless, imperial China up to the fifteenth century was technologically far ahead of its contemporaries, and numerous inventions originate in China. Many of the inventions and discoveries within imperial China are directly related to manufacturing. Even before the imperial period, the Chinese used wedges, inclined planes, screws, levers, and pulleys, which they knew as the *five labor-saving jewels* in ancient China (Yan 2007, p. 45). Other examples of labor-saving inventions in China are water-powered spinning and weaving machines and the wheelbarrow. Furthermore, printing was done using both woodblock and moveable-type, although the latter was inconvenienced by the large number of characters in the Chinese writing system.

China was also ahead of the rest of the world by 1500 years in the field of metallurgy. From the fifth century BCE, the Chinese used blast furnaces to create pig iron, which was then melted again in cupola furnaces to create cast iron. Hence, the Chinese regularly produced cast iron about 1500 years before the first blast furnace operated in the western world. The produced cast iron was still very brittle. In order to make the iron more flexible, the carbon content needed to be reduced further. The puddling process where molten iron is stirred to burn off excess carbon was invented in China during the third century CE, again producing steel ahead of the rest of the world by 1500 years. Operating both a blast furnace and the puddling process required lots of air to reach the necessary temperatures. The air was pumped with bellows driven by human or animal muscle power. The government official Du Shi is credited with the invention of the water-powered bellow after 30 CE, and historical texts point out the large labor savings and the subsequent widespread adoption of water power to drive bellows. Around the same time, water power was also applied to trip-hammers to crush the ore into smaller pieces so the metal could be extracted more easily.* Archeological evidence, however, finds only few smelters near a river that could have been used as a source of water power,

* Romans also used water-powered trip-hammers around the same time, both for crushing ore and for pounding grain.

indicating that most of the time, the work was still done through muscle power (Wagner 1993, p. 263). From the 11th century onward, the Chinese began to use coke* made from black coal as fuel since the demands on charcoal had led to deforestation. Overall, this led to great productivity improvements, and by 1078 CE, the Chinese Empire produced the large quantity of 114,000 tons of iron per year (Wagner 2001). In comparison, England at the onset of the industrial revolution produced only half that amount, 68,000 tons in 1788.

Due to the significance of iron production for the government, they also aimed to control it, although this differed widely from prefecture to prefecture. For example, the poet Su Shi (1037–1101 CE), who was also governor of the Xuzhou prefecture, wrote a detailed account on the iron industry of Xuzhou. In Xuzhou, the government did not produce iron itself but, rather, issued permits to families to produce iron. Altogether, there were 36 smelters, each run by a wealthy family with hundreds of hired hands working for each smelter, *...for the most part poverty stricken runaways...* Another iron manufacturing center in Dengzhou is described by the official Bao Zheng (999–1062 CE). A number of families in Dengzhou were still on the list of official smelters but could not afford to operate one anymore. Nevertheless, they were still required to deliver iron to the government and therefore had to purchase iron in order to pay their taxes, driving themselves into poverty. Bao Zheng proposed to remove them from the list of smelters and successfully lobbied for the opening of the smeltering business to all families interested, subsequently greatly increasing iron production.

Bao Zheng also reports of the Tongzhou District, writing that of the 700 smelters, around 200 were a sort of joint venture. The government provided the charcoal and the skilled craftsmen, whereas the families provided the unskilled labor (Wagner 2001). Smelters were also completely owned and operated by the government, although this was not consistent and was heavily influenced by lobbyism of different groups. For example, in 1374, there were 13 government smelters, which were all closed by decree in 1385, later opened again, closed again in 1395, opened again, etc., depending on whether the local manufacturers could lobby out the government competition or not. These smelters were usually operated by

* Coke is coal that is heated without oxygen. Hence, it does not burn but merely evaporates undesirable volatile components in the coal. The resulting coke has few impurities and a high carbon content, giving better results in iron production.

soldiers, commoners drafted for this process without pay, and convicts, which naturally led to labor difficulties (Wagner 2005).

Overall, smelters were managed by individual families, by the government, and also by both in cooperation. Up to the eleventh century, these operations required a government permit and therefore were a sort of monopoly, or at least an oligopoly, thus having the potential to create incredible wealth for their owners. These early industrialists had money and power comparable to political rulers. A text from the second century BCE puts it, *...he possessed a thousand slaves. In fields and lakes and in the pleasure of the hunt he equaled the rulers of men* (Wagner 1993, p. 251). However, operating a smelter still required a large starting capital and included the business risk of failure and subsequent poverty. These ironworks were often an almost self-sufficient community in a remote forested location due to the need for large quantities of wood. Most of the work centered on iron production, including the collection and preparation of the ore, the cutting of wood, and the creation of charcoal, but probably also included some agriculture for food. Larger ironworks had easily more than a thousand workers, probably half of them skilled laborers.

Due to the heavy physical demands on the workforce, most of these laborers were probably men employed by the ironworks. Since labor was always in great supply in China, pay seems to have been very low, and most of the hired employees were barely able to make ends meet. Now, having a thousand poor, uneducated, and young men in a remote location with little distraction was a sure way to ask for trouble. Contemporary accounts usually describe these employees in rather unflattering terms, considering *nearly all of them common bandits. [...] Being assembled in deep mountains and remote marshes, engaging in illicit activities and following the powers of factions, their tendency to commit wrongs was a danger* (Wagner 1993, p. 257ff). In some cases, even the owners of the smelter were shady figures. For example, one fellow named Wang Ko hired armed guards to blackmail officials, got into skirmishes with government soldiers, and even attacked an entire city (Eberhard 1957).

Keeping these angry young men in check must have been quite a challenge for the local industrialists. While we do not know their approaches to supervision, there were probably two possibilities. For example, they may have created a chain of command similar government officials supervising their workers or generals supervising their soldiers. But quite possibly, they may also have used a system of subcontracting, where one subcontractor directed a group of five to ten men. The most ancient Chinese management technique was leading by example, an approach well suited to the strong social system of ancient China. This was reinforced through ceremonial etiquette and traditions to increase social integration* (Starbuck and Barnett 2006, p. 425ff). A more detailed approach dating back to at least the first millennium BCE reads almost like a modern leadership book:

> Of the three virtues: The first is correctness and straightforwardness; the second, strong rule; and the third, mild rule. In peace and tranquility, correctness and straightforwardness; in violence and disorder, strong rule; in harmony and order, mild rule. For the reserved and retiring, there should be the strong rule; for the high-minded and distinguished, the mild rule.
>
> **The Great Plan, Shàngshū**
> *Translation by the Chinese Text Project at http://chinese.dsturgeon.net*

Overall, ancient Chinese management used both reward and punishment to manage their subordinates. However, ancient sources also indicate that if there was a choice between the carrot and the stick, the stick was preferred. Prime Minister Li Si (also responsible for the burning of books and burying of scholars described previously) was a strong advocate for punishment in order to achieve his goals. *If a minor offense is punished severely, people can imagine what will be done against a serious offense! Thus, people do not dare to violate the laws* (Starbuck and Barnett 2006, p. 436). In line with his beliefs, Li Si also invented a new form of capital punishment called the *five pains*. The victim is punished by cutting off, in sequence, the nose, a hand, and a foot, followed by castration and, finally, cutting the body in half along the waist. Ironically, this was also the way he was executed in 208 BCE after losing a power struggle related to the succession of the deceased Emperor Qin Shi Huang.

Another large-scale Chinese manufacturing project is the Terracotta Army protecting the mausoleum of the first emperor Qin Shi Huang (259 BE–210 BCE). The Terracotta Army consists of about 8000 life-sized figures, mostly warriors armed with bronze weapons along with hundreds of horses and chariots arranged in battle formation. Each warrior is different from the others, clearly showing different facial features, hairstyle, posture, and equipment. Both in its scale and in its quality, the work is unprecedented. While clay sculptures have been found in other burial

* These traditions exist in many professions all over the world, from the Japanese sword smith observing religious rites when creating a new sword to the *Wal-Mart cheer* chanted at the start of every shift.

sites, there were usually only one or two, none of these life-sized, and the artistic quality was mediocre. While this is a rather unusual manufacturing project, it gives us lots of information on how ancient manufacturing was organized. The mausoleum consisted of not only the thousands of warriors but also a magnificent tomb, reportedly with rivers of mercury and a ceiling with stars made out of diamonds.* Contemporary historian Sima Qian (ca. 145–95 BCE) writes that up to 700,000 people from all parts of China were working on the mausoleum over a time period of about 38 to 39 years (Harman 2002, p. 54). While not all 700,000 people were there at the same time, there may have been up to 50,000 people or more on site at the same time.

The workers were mostly unskilled laborers consisting of prisoners and slaves, many of them sentenced to death, later executed on site and buried in mass graves (Capek 2007, p. 26). Other workers were simple farmers drafted for the project, although there were some concerns that drafting a large number of farmers would reduce the food supply. There was also a smaller group of skilled laborers and artisans from all regions of the country, coming from the provinces of Henan, Hubei, Shandong, Shanxi, and of course, Shaanxi, where the mausoleum is located. Since large numbers of experienced artisans were needed to make the sculptures, the responsible officials accessed the available labor source that was experienced in working with clay. Most of these people previously created clay products for housing like clay water pipes or roof tiles, but there were also some artists working for the government.

Despite the high level of individualization of the sculptures, their production was highly organized mass production. The clay came from local sources and was processed to remove impurities. Quartz and feldspar were added to strengthen the soft clay for molding. Heads, arms, torsos, and legs were created based on molds. The unskilled laborers simply pressed the clay in the molds, removed the clay, and then let it dry. So far, eight basic face shapes have been discovered, indicating the use of at least eight different molds for faces. The production of the legs is very similar to the production of clay pipes, probably owning to the drafting of manufacturers of clay pipes. These body parts were created in different workshops. Each workshop marked their name on their products to ensure quality

* We don't know the details—yet. Archeologists decided not to open the emperor's tomb yet, in order to first gain more experience with conserving their findings before opening the mother lode of the treasures.

control. Some of these workshop names were also found on roof tiles and other clay products dated before or after the construction of the tomb, indicating that commercial potters were recruited to build the Terracotta Army. So far, 87 individual names of master craftsmen have been found on the different statues. This not only shows where these artisans came from but also demonstrates that different master craftsmen where in charge of managing different body parts rather than an entire body and that these master craftsmen also had assistants and workers. It is estimated that at any given time, 1000 workers were working on the modeling and firing of the sculptures, for 12 years (Portal 2007, p. 5).

The final body was assembled using the dried but not yet fired prefabricated clay modules, creating a multitude of different combinations and postures from a small number of standard body parts. Skilled artists then added individual facial features, hairstyles, and other details to the sculpture. The body was then fired in a kiln, requiring rather large kilns to fit an entire life-sized sculpture of a man or horse. Finally, the sculptures were painted individually with multiple layers of paint. The whole process was accompanied by rigid quality controls at multiple points in the process (Portal 2007). Insufficient quality was traced back to the source, and one artisan wrote to his family, *I have to work carefully every day, if I paint the weapons incorrectly, my officer will punish me very severely* (TravelChinaGuide 2011). The creation of the Terracotta Army is one of the earliest, if not the first, known example of mass production using standardized parts and a high level of quality control. Furthermore, such a high level of specialization and division of labor greatly simplified the learning curve and increased productivity.

While these examples focus on large-scale manufacturing, these operations were only large due to either the technical requirements of the process or simply the scale of the order. Nevertheless, the workshops of most artisans and laborers outside of such larger projects consisted of members of one family, possibly with a few hired hands and apprentices. More successful artisans sometimes employed enough workers so that they could focus exclusively on management rather than production. A few artisans were also employed directly by the government, mainly in larger-scale operations, for example, the monopolized iron and salt manufacturing, but also producing some luxury items for the court. Nevertheless, the majority worked for their own profit and at their own risk, and overall, the majority of the manufacturing in imperial China was done in small, family-based workshops. In the social hierarchy, artisans and laborers were considered below the ruling elite and also below farmers, but still above merchants, who were often actively harassed by the government.

The decline of imperial China started during the Ming dynasty (1368–1644 CE), around the end of the fifteenth century. By then, China was the most technologically advanced nation on earth (Landes 1998). The Chinese treasure ships of that time were supposedly the largest wooden ships that ever sailed the earth,* bringing traders and diplomats as far as East Africa. Afterward, however, China curbed any kind of foreign trade. The reports of the travels of the treasure ships were burned in 1479, and the ships themselves were left to rot in the harbor. The construction of any ocean-going vessels was prohibited, and all ocean-going vessels were destroyed in 1525. Furthermore, the land within 20 miles of the coast was declared a forbidden zone, and trespassing was punished by death. While smugglers still managed to evade the ban and selected ports were open for Japan, Indonesia, and the Philippines, trade was reduced to a trickle (Cameron and Neal 2002, p. 83). Knowledge about shipbuilding was lost, and the exchange of knowledge with other nations came to a standstill.

Thus, when foreign powers became interested in China, the Chinese weaponry and technology was very inferior compared with that of the western military. Foreign conquest started in 1839 with the British East India Company defeating a numerically superior Chinese Army during the First Opium War, and ending in 1945 with the Japanese occupation during World War II. Subsequently, this period is known in China as the *century of humiliation*.

While China was, at one point, almost 2000 years ahead of the rest of the world, many historians wonder why the industrial revolution did not happen there.† China had many of the ingredients necessary for the industrial revolution: high levels of literacy; a vast body of technical knowledge; sufficient size; and frequently, a government beneficial to technical development. Nevertheless, China fell behind compared to western nations and therefore was defenseless against Western economic interests. Only within the last 30 years did China start to catch up with the western world.

* According to traditional accounts, these were up to 140 meters long; however, the exact size is disputed, and some scholars argue for only half that length. In comparison, the largest wooden ships in modern times barely exceeded 100 meters length, and most of these were not very seaworthy, due to their size.

† This question is known as Needham's Puzzle, named after Joseph Needham, one of the leading researchers on Chinese history.

Experts have brought forward a number of reasons for stunted development. Some argue that the political system was unsuitable for technical progress, or that the political system was not stable enough. Others argue that there was not enough internal competition within the large unified empire, which was also subject to the whim of a single ruler (Diamond 1999, p. 412). Another plausible explanation is that there was simply no need for labor-saving devices since the population was expanding and labor was cheap. For any work, there were more than enough hands to do it. The increase in manufacturing output was more than matched with the increase in population (Elvin 1973). Additionally, manufacturing was low-prestige work, and the young potentials of ancient China much preferred to aim for a flashy government job rather than a low-prestige manufacturing career. This low prestige of manufacturing also hindered development in another major ancient civilization—the Roman Empire.

4.5 THE ROMAN EMPIRE

The Roman Republic and the subsequent Roman Empire was one of the largest civilizations of its time and is still considered one of the most significant and influential empires ever. The Roman military was the most powerful military in the western world, having 375,000 men under arms at its peak (Gibbon 1998) and controlling most of Europe, Northern Africa, and Arabia. Romans invented numerous devices to aid with warfare and greatly improved artillery techniques. The Roman road network reached every part of the Roman Empire, allowing quick movement of military forces, information, and goods, and hence enabling the successful government of the conquered provinces.

Rome also spent considerable effort to bring Roman civilization to its provinces, partly through Roman achievements in architecture and construction. Numerous aqueducts were built throughout the empire, crossing over valleys and tunneling through mountains to supply the cities in the empire with water. Ruins of Roman public baths can be found almost all over Europe. The Limes and Hadrian's Wall with its supporting military infrastructure are more proof of the technological ability of the Romans. The Pantheon in Rome is still the largest unreinforced concrete dome in the world. The Baths of Caracalla compete in size with many modern water amusement parks. The 50,000-seat capacity of the Coliseum still puts it among the top 50 largest stadiums in the world. The capacity of the Circus Maximus of almost 300,000 people exceeds the largest modern racing venue—the Indianapolis Motor Speedway—which seats only 235,000, although together with the infield around 400,000 people can fit in the venue.

Subsequently, Romans made major advances in agriculture. Archeological remains of animal bones clearly show an increase in the size of domesticated animals, both due to selective breeding and through the improvement in fodder quality. Roman farm animals were of a size that had never been seen before and that would not appear again until the nineteenth century. Similarly, Roman farming also increased in productivity and efficiency, introducing such tools as the long scythe and improved plow design (Margaritis and Jones 2008, p. 170ff). Related food processing technologies also increased, and mills powered by water and animals were used from the second century onward as shown in Figure 4.4 (also see Section 4.1).

FIGURE 4.4
Roman milling scene. Fragment of a sarcophagus from Via Appia, Rome, dated around 225–250 CE. (Photo by author, dated March 2009.)

Food processing operations also emerged on a larger scale. The second-century bakery in Ostia Antica, for example, includes at least 10 animal-powered mills as shown in Figure 4.5 and five animal-powered dough-kneading machines, followed by two large ovens. The size of the bakery clearly indicates a division of labor. Furthermore, the tools are also arranged in the sequence of their use. Flour storage is followed by the milling, then by the kneading, and finally by the baking processes. This is very similar to the flow pattern of a modern production line (Wilson 2008a, p. 406ff). Other examples of large food processing factories are the large fish salting operations near Cotta in Morocco or numerous olive oil processing centers in Northern Africa, the largest at Senam Semana having at least 17 presses. Most of these food processing factories were laid out following the flow of the product through the different processes.

Looking at the numerous innovations and developments in architecture, warfare, logistics, and agriculture, one cannot help but notice the curious absence of innovations related to manufacturing, with the exception

FIGURE 4.5
Large bakery in Ostia Antica, Rome. The room pictured is the milling room. (Photo by author, dated March 2011.)

of food processing. There was some manufacturing-related development, such as the refinement of glassblowing technologies, the invention of the plane in woodworking (although possibly invented in Greece already), or the development of the two-beam vertical loom from the Greek warp-weighted loom. However, all of these innovations pale in comparison to the advancement made in fields outside of manufacturing.

In Roman society, the enterprise with the highest social status was agriculture. Pretty much all senators were also landowners. Farmers also wielded a large political influence and were able to lobby laws and regulations in their favor. Manufacturing, however, did not rank high on Roman social values and subsequently was not in the focus of the ruling elite. Freemen who were working as paid laborers were actually treated with contempt by others and considered not much better than slaves. Being an artisan or workshop owner did not give you any social prestige, and your ability to influence laws or regulations in your favor was nonexistent (Rostovtzeff 1926, p. 167ff).

> Unbecoming to a gentleman, too, and vulgar are the means of livelihood of all hired workmen whom we pay for mere manual labour, [...] for in their case the very wage they receive is a pledge of their slavery. And all mechanics are engaged in vulgar trades; for no workshop can have anything liberal about it.
>
> **Marcus Tullius Cicero (106–43 BCE)**
> *In De Officiis*

Therefore, the unpopular task of manufacturing was primarily delegated to slaves, or in fewer instances, to convicts or poor freemen (Cameron and Neal 2002, pp. 41–43). Often, even the management of manufacturing was delegated to slaves, and—to use modern terminology—there were not only blue-collar but also white-collar slaves (Salmon 1990, p. 71). For example, Roman manufacturing sites in Egypt were supervised not by local free Egyptians but, rather, by educated Greek slaves. Subsequently, there was little incentive to develop labor-saving devices or develop new manufacturing technologies.

From the top down, the ruling class had little interest in common manufacturing, did not understand the process in enough detail to make improvements, and spent little time on organizing or optimizing the manufacturing processes. From the bottom up, slaves did not have opportunities to change their workplace, were never challenged to develop creative thinking to develop new technologies, and did not have the ear of their owners to communicate new ideas upward. Additionally, the ruling class sometimes even opposed labor-saving devices in order to keep employment up. It is said of Emperor Vespasian (9–79 CE) that he forbade the use of advanced technology in transporting a marble column, preferring rather to employ more people instead and saying, *You must let me feed my poor commons* (Suetonius 1997). Similarly, Pliny the Elder reports that Emperor Tiberius (42 BCE–37 CE) executed the inventor of flexible glass in order to preserve the market value of gold and silver (Pliny the Elder 1855, p. 36). Even techniques like waterwheels were barely used until the second century since animal- or human-powered mills were considered sufficient (Harman 2002, pp. 58 and 84). Therefore, manufacturing technology received very little attention and hence developed much less than other fields closer to Roman interests.

The same is true for the working conditions in Roman manufacturing sites, where, again, the slaves had no way of influencing their way of working. The slave was the property of the slave owner, who could do with the slave as he/she wished, including punishing the slave in any way or even executing the slave.* Of course, the treatment of the slave also depended heavily on the owner and the position of the slave. Not all slave owners were cruel, and household slaves were often treated well, whereas working in mines was pretty much a death sentence. In any case, slaves had a pretty weak negotiation position if they wanted to improve their working conditions. Usually, there were only two options available to a slave unhappy with his/her work: First was simply to run away, as they frequently did, hence providing work for a large number of slave catchers. Any slave caught was branded on the forehead with the letter *F*, for *fugitivus*, and was punished severely for his/her crime. Their second option was to revolt in large numbers against their masters.† As there were more slaves in Rome than free citizens, slave revolts were one of the greatest fears of Romans, and these revolts were usually put down with brutal force. The most famous slave revolt under Spartacus (ca. 109–71 BCE) included, at its height, up to 100,000 slaves and was put down only after numerous battles.

* There were improvements to the rights of the slaves later in the Roman Empire, and the owners were, for example, no longer allowed to kill slaves without a legal reason—although this law was not consistently enforced.

† In some cases there was a third option. Exceptionally hardworking slaves also had permission to work partially for their own profit in order to raise money to free themselves. After having paid off their former master, they usually included his name in their own. However, this was not common.

The last 6000 captured slaves were crucified along the Via Appia, and their bodies were left hanging as a warning to other slaves, making it rather unpleasant to travel this road for the following years.

These slave flights and revolts were clearly due to unjust treatment of slaves, but they were primarily aimed at getting away from this unjust treatment rather than to improve the treatment itself. There are only very few reports of organized labor movements in Roman history, and none of these center around slaves. We know of only about a dozen strikes, involving barely organized peasants and quarrymen but also much more structured bankers. With respect to manufacturing, we know of successful strikes of shipbuilders and bakers (MacMullen 1963). Yet overall, organized labor movements had little significance in the Roman Empire. One particular exception was in Asia Minor, where most of the work was done not by slaves but by workers who were not quite free but no longer serfs either. Their demands were supported by organized strikes and demonstrations aimed at changing the social structure (Abbott 1911; Rostovtzeff 1926, p. 169).

As for the manufacturing operations themselves, manufacturing oriented itself on the needs of the customer, and broadly speaking, there were two types of customers. First, there were the ultrarich and powerful elite, able to afford almost any luxury and willing to pay large sums to demonstrate their prestige through materialistic possessions. The second customer group was the poor masses, consisting of the *plebs*, Roman citizens with neither wealth nor blue blood, and of urban and domestic slaves, who by tradition were also allowed to own and spend money. Financially speaking, there was a huge gap between the upper class and the poor masses, with virtually no middle class in between. While the poor masses had limited purchasing power, they vastly outnumbered the financial elite, and the majority of manufacturing was targeted at these poor masses. Hence, there was an enormous demand for functional and cheap products, leading to mass-produced simple and similar-looking products with low artistic value (Rostovtzeff 1926, p. 166).

The majority of the industry was also decentralized and fragmented. Since there were neither labor-saving devices nor any protection of intellectual property, small workshops successfully competed with large workshops. While some richer entrepreneurs owned many workshops, these operated just as separately as if they were owned by different individuals. There were little or no advantages through the size of operations, and subsequently, there was no cost advantage through the economies of scale.

Additionally, since transportation was both expensive and risky,* the customer base was usually the local population, and most goods were transported only over a short distance before they reached the end customer. However, there were a few exceptions where goods were transported over longer distances. For one, Rome depended heavily on food supplies from other parts of the empire, and large quantities of food were shipped to Italy in general and Rome in particular. After unloading the food in Italy, primarily in Ostia, ships returned to their origin to repeat the cycle. Since they had to return anyway, taking some products along was only a small additional effort.

Hence, shipping from Italy to other provinces was less expensive, since the ship was going there anyway in order to get more food. This is comparable to the modern-day situation, were the cost of shipping from China to the United States is much higher than the other way around. This favored manufacturing sites in Italy. Because of this, combined with a desire for Roman goods, a large number of products were exported from Italy to the provinces. One particular example is the oil lamp industry near modern Modena in Italy. While still consisting of many different workshops, this industry initially had almost a monopoly on oil lamps, and oil lamps with the brand names *Strobili*, *Communis*, *Phoetaspi*, *Eucarpi*, and *Fortis* made in Modena can be found everywhere throughout the empire. The *Fortis* brand was of particularly high demand and became synonymous with this type of lamp. The *Fortis* reputation was actually so high that eventually the entire product, including trademark and all, was copied by local potteries throughout the empire† (Rostovtzeff 1926, p. 163). Another example of larger manufacturing centers was products that had a unique local advantage. For example, most paper was manufactured in Egypt, and Asia Minor was the center of carpet manufacturing. Other locations were able to compete with Italy on price, and during the latter centuries of the empire, Gaul was a major producer of glass, certain potteries, woolen products, and bronze. Most items were made to stock. Even customized luxury items were sometimes based on pre-fabricated goods made to stock. See Figure 4.6 for a prefabricated roman sarcophagus, missing only the details of the faces.

---

* Including both natural risks (e.g., weather) and man-made risks (e.g., pirates and robbers).

† Researchers are still debating if these were local branches of *Fortis*, or if it were separate operations making goods without consent of *Fortis*. Due to the effort in controlling long distance operations and to the lack of intellectual property protection it is more likely to be a case of pirated goods.

FIGURE 4.6
Prefabricated Roman sarcophagus from the middle of the third century CE. Note the unfinished faces of the center figures. (Photo by author, dated March 2009.)

Roman wealth was based on conquering other states for financial gain. However, over time, this changed. Rome reached its largest size under Emperor Trajan, after which the expansion stopped and the empire slowly became smaller. Additionally, the provinces became more self-sufficient and less dependent on Roman products (for example, the *Fortis* oil lamps mentioned above or the strong local manufacturing in the Far East). The few regions that did not have a local source for Roman products were violent and difficult to trade with, for example, the barbaric tribes in Germania. These altogether led to a huge trade deficit, making it difficult for the ruling class to finance the empire. In order to still pay the bills, the currency was inflated. The percentage of silver in the *denarius* decreased from almost 100% during the first century CE to trace amounts of 0.2% under emperor Claudius Gothicus (213–270 CE). At the same time, taxation increased, and some freemen even sold themselves into slavery, preferring to be tax-free slaves rather than taxed freemen. Prices were fixed well below their true market value in a vain attempt to maintain stability. Overall, the previously rather free market economy turned into a socialist system (Bartlett 1994).

While historians still debate on what exactly caused the fall of the Roman Empire, most point to the lack of funds and the subsequent collapse of the monetary system. During Roman times, there was no trace of macroeconomic thought, and topics like a balance of trade or other economic theories were completely unknown. The Roman economy burned out, and the spark that started the fire of steam power came only 1500 years later during the industrial revolution.

BIBLIOGRAPHY

Abbott, F.F., 1911. *The Common People of Ancient Rome: Studies of Roman Life and Literature*, Charles Scribner's Sons, New York.

Ackermann, M.E., Schroeder, M.J., Terry, J.J., Upshur, J.-H.L., Whitters, M.F. (eds.), 2008. *Encyclopedia of World History—The Ancient World Prehistoric Eras to 600 c.e.* Facts On File, Inc., New York.

Bartlett, B., 1994. How Excessive Government Killed Ancient Rome. *The Cato Journal* 14 (2), 287–303.

Bauer, S.W., 2007. *The History of the Ancient World: From the Earliest Accounts to the Fall of Rome*. W. W. Norton & Company, New York.

Cameron, R., Neal, L., 2002. *A Concise Economic History of the World: From Paleolithic Times to the Present*, 4th ed. Oxford University Press, New York.

Capek, M., 2007. *Emperor Qin's Terra Cotta Army*. Twenty-First Century Books, CT.

Cuomo, S., 2008. Ancient Written Sources for Engineering and Technology, in: Oleson, J.P. (ed.), *The Oxford Handbook of Engineering and Technology in the Classical World*. Oxford University Press, Oxford, New York.
Diamond, J., 1999. *Guns, Germs, and Steel. The Fates of Human Societies*, 1st ed. W. W. Norton & Company, New York.
Diamond, J., 2005. *Collapse: How Societies Choose to Fail or Succeed*. Penguin (Non-Classics), Viking, New York.
Duiker, W.J., Spielvogel, J.J., 2006. *The Essential World History*. Cengage Learning, Boston.
Eberhard, W., 1957. Wang Ko, an Early Industrialist. *Oriens* 10, 248–252. doi:10.2307/1579637.
Elvin, M., 1973. *The Pattern of the Chinese Past: A Social and Economic Interpretation*. Stanford University Press, Stanford, CA.
Fairservis, W.A., 1992. *The Harappan Civilization and Its Writing*. E.J. Brill, Leiden, Netherlands, New York.
Gibbon, E., 1998. *The Decline and Fall of the Roman Empire*. Wordsworth Editions Ltd, Ware, Hertfordshire, UK.
Harman, C., 2002. *A People's History of the World*, 2nd ed. Bookmarks Publications Ltd, London.
Hart, M.H., 2007. *Understanding Human History: An Analysis Including the Effects of Geography and Differential Evolution*. Washington Summit Publishers, Augusta, GA.
Hinsch, B., 2002. *Women in Early Imperial China*. Rowman & Littlefield Publishers, Inc., Lanham, MD.
Keay, J., 2001. *India: A History*. Grove Press, New York.
Kulke, H., Rothermund, D., 1998. *A History of India*, 3rd ed. Routledge, London, New York.
Landes, D.S., 1998. *The Wealth and Poverty of Nations: Why Some Are So Rich and Some So Poor*, 1st ed. W. W. Norton & Company, New York.
MacMullen, R., 1963. A Note on Roman Strikes. *The Classical Journal* 58, 269–271.
Margaritis, E., Jones, M.K., 2008. Greek and Roman Agriculture, in: Oleson, J.P. (ed.), *The Oxford Handbook of Engineering and Technology in the Classical World*. Oxford University Press, Oxford, New York.
Muljevic, V., 2001. The role of Mediterranean Countries in the History of Science and Technology, in: Proceedings of the 9th IEEE Mediterranean Conference on Control and Automation. Presented at the 9th IEEE Mediterranean Conference on Control and Automation, Dubrovnik, Croatia.
Pliny the Elder, 1855. *Naturalis Historia*. Translated by John Bostock and Henry Thomas Riley. George Bell & Sons, London.
Portal, J., 2007. *The First Emperor: China's Terracotta Army*. Harvard University Press, Cambridge, MA.
Rossi, C., Russo, F., Russo, F., 2009. *Ancient Engineers' Inventions: Precursors of the Present*, 1st ed. Springer, Dordrecht, Netherlands.
Rostovtzeff, M.I., 1926. *The Social and Economic History of the Roman Empire*. Biblo-Moser. Clarendon Press, Oxford, UK.
Salmon, E.T., 1990. *History of the Roman World from 30 B.C. to A.D. 138*, 6th ed. Routledge, London; New York.
Selin, H., 2008. *Encyclopaedia of the History of Science, Technology, and Medicine in Non-Western Cultures*, 2nd ed. Springer, Berlin, Germany; New York.
Starbuck, W.H., Barnett, M.L., 2006. *Organizational Realities: Studies of Strategizing and Organizing*. Oxford University Press, Oxford, New York.
Stearns, P.N., 2001. *The Encyclopedia of World History: Ancient, Medieval and Modern—Chronologically Arranged*. James Clarke & Co Ltd., Houghton Mifflin, Boston.

Suetonius, C.T., 1997. *Lives of the Twelve Caesars*. Wordsworth Editions Ltd., Translated by H. M. Bird. Hertfordshire, UK.

Tapper, R., McLachlan, K.S., (eds.) 2003. *Technology, Tradition and Survival*. Frank Cass, Portland, OR, London.

TravelChinaGuide, 2011. Qin Terra Cotta Artisans [WWW Document]. TravelChinaGuide. Available at http://www.travelchinaguide.com/cityguides/xian/terracotta/artistic.htm (accessed June 9, 2013).

Wagner, D.B., 1993. *Iron and Steel in Ancient China*. E.J. Brill, Leiden, Netherlands; New York.

Wagner, D.B., 2001. The Administration of the Iron Industry in Eleventh-Century China. *Journal of the Economic and Social History of the Orient* 44, 175–197.

Wagner, D.B., 2005. The State Ironworks in Zunhua, Hebei, 1403-158. *Late Imperial China* 26, 68–88.

Wikander, Ö., 2008. Sources of Energy and Exploitation of Power, in: Oleson, J.P. (ed.), *The Oxford Handbook of Engineering and Technology in the Classical World*. Oxford University Press, Oxford, New York.

Wilson, A.I., 2008a. Machines in Greek and Roman Technology, in: Oleson, J.P. (ed.), *The Oxford Handbook of Engineering and Technology in the Classical World*. Oxford University Press, Oxford, New York.

Wilson, A.I., 2008b. Large Scale Manufacturing, Standardization, and Trade, in: Oleson, J.P. (ed.), *The Oxford Handbook of Engineering and Technology in the Classical World*. Oxford University Press, Oxford, New York.

Yan, H.-S., 2007. *Reconstruction Designs of Lost Ancient Chinese Machinery*. Springer, Dordrecht, Netherlands.

5

The Middle Ages in Europe

Stadtluft macht frei nach Jahr und Tag.
(Town air gives freedom after a year and a day.)

Medieval German saying

After the fall of Rome at the end of the fifth century, Europe was in shambles. Toward the end of the Roman Empire, the population in Europe had already declined, and it halved during the sixth and seventh centuries (Russell 1972, p. 25). By the ninth century, most towns in Europe were in ruins and a mere shadow of their former selves. Wars and famine ravaged the land. Public order was practically nonexistent, as were long-distance trade and economy (Harman 2002, p. 102). People were primarily occupied with trying to feed themselves and trying to live another day. There were very few full-time craftsmen. If the Ottoman wanted to invade Europe, they could have easily overrun most of Europe, except that they did not see it being worth the effort (Cipolla 1989, p. 221ff).

Yet, out of this chaos emerged a new society, different from all previous societies, with a new set of values and beliefs. By the end of the Middle Ages, Europe was the economic powerhouse of the world, starting the industrial revolution and leading to the dominance of the European (and hence also the American) culture in the modern world. It was exactly this chaos, this lack of order, that allowed Europe to shed its old values and beliefs and to create something new.

5.1 THE RISE OF THE TOWNS

Due to the weakness or lack of an overarching government, local rulers became more powerful, and the land was divided into many small estates ruled over by a lord, baron, or other nobility that owned it. Peasants were on a status little better than slaves. They had to give almost half of their available work time to the landowner, were unable to leave without the permission of this ruler, and in general had very few rights. As such, people's lives were usually hard and unpleasant. The only general exceptions to this were towns, which by themselves were strong enough to resist the advances of the surrounding landowners. While the power and influence of a landowner was based on the land he (or less commonly she) owned, the power of the town leadership was based on the people of the town. Therefore, the burghers of the towns had a much larger influence in the social and political structure of the town than peasants had on their lives. The German saying *Stadtluft macht Frei*, meaning *town air gives freedom*, expressed the overall feeling of this period. Towns experienced a great influx of people and subsequently a great increase in power between the tenth and thirteenth centuries. From this, towns developed their own social structure and cultural values. The differences after entering a town at that time can be compared to crossing a border into another country nowadays.

The absence of a supra-regional ruling dominance created many small independent towns, where people changed the structure of their society. This fragmentation of Europe into many small townships also meant that there was no central government that defined social standards, set overarching rules, or prohibited things outright. Previously, the leading elite in Rome despised manufacturing and trade, and hence manufacturing and trade had a low social status all over the Roman Empire. Similarly in China, the ruling class looked down on manufacturing and trade, and the commoners followed that example. When the emperor of China ordered the burning of seagoing vessels in 1525, all seagoing vessels in China were burnt, and that was the end of it.

Not so in medieval Europe. Of course, there also were towns that resisted change and stuck to their old social values, but there were many more towns that changed. Moreover, since towns open to change were usually more successful economically, they often expanded and influenced or took over other less successful settlements. For example, German settlements with better technology in agriculture and mining expanded eastward and overtook less advanced Slavic and Polish settlements that were still primarily based on hunting and fishing (Cipolla 1989, p. 205ff). It was a period of rapid social evolution. Not only were the societies small and hence able to change more quickly; there were also many different societies developing in parallel, the most successful ones being able to either influence the less successful ones to either imitate their ways or outright overtake them. As the success of a society is closely related to its economic power, Europe became the economic powerhouse of the medieval world, laying the foundations for the worldwide influence of European society in modern times.

Medieval towns in Europe had a mix of priests, ruling elite, farmers, merchants, and artisans (Figure 5.1), similar to many towns in ancient China or the Roman Empire. In ancient China or the Roman Empire, merchants, artisans, and doctors were in the lowest social levels. In medieval Europe, however, their social standing increased dramatically, and the self-confidence and self-value of merchants and artisans were growing

FIGURE 5.1
Depiction of a smith. (From Hausbuch der Mendelschen Zwölfbrüderstiftung, around 1425.)

significantly. This change in self-respect also led to a change in the society of artisans and merchants. While before a smart and ambitious youngster avoided these *vile mechanical arts* like the plague, now it was a way to rise in society. Manufacturing was in a better position to compete on the labor market for what we nowadays would call high potentials. These new merchants and artisans also strengthened their position in society. For example, generous donations to the church made the church in turn more appreciative of these social groups, and philanthropy increased their status. For the first time in recorded history, you could be proud to be an artisan.

5.2 AGRICULTURAL DEVELOPMENTS

Parallel to the rise of towns, agriculture also changed. In ancient times, a farmer usually made barely enough to feed himself and his family, leaving very little surplus for nonfarmers to live and focus on other tasks. Subsequently, before the Middle Ages, the vast majority of the population was occupied with food production.

However, during the Middle Ages, a number of agricultural techniques were developed that greatly improved food production. From the sixth century onward, a new plow design became commonly used. This heavy plow was able to furrow the heavy fertile soil in Europe. Since more or stronger animals were needed to pull the heavy plow, the amount of manure available to fertilize fields also increased. Around the ninth century, a new collar for horses was introduced in Europe. Previous collars had the flaw that they put pressure on the windpipe of the horse. This chocking greatly limited the horses ability to pull. The new collar put the load on the shoulders and chest, allowing the horse to pull much larger loads than before. Since horses are stronger than oxen, work twice as fast, and have more endurance, the amount of work obtained increased greatly compared to the effort to feed the animal. From the tenth century onward, horseshoes also became common in Europe, reducing the wear and tear on the horses' hooves and hence increasing the ability of the animal to work even further (Cipolla 1972, p. 15).

Another significant change was the three-field rotation system, used from the eighth century onward. Previously, using a two-field rotation system, only half of the fields were farmed each year, while the other half of the fields lay fallow to let the soil recover. This means that half of the fields were unused each year. In the three-field rotation system, a field was laid fallow only every third year, while the other two years, an autumn crop and a spring crop were planted. Not only did this change increase the usage of land by one-third; it also allowed the farmer to spread his work more evenly over the year. In the two-field rotation system, work was concentrated during spring, with plowing and sowing and during autumn with harvesting, leaving the farmer with wasted idle time during summer. In the new three-field system, the sowing and harvesting of the spring and fall crops occurred in different months. The farmer was able to use more of the previously wasted time to work on the spring crop, allowing him to farm even more fields than before (White 1972, pp. 149–150). Finally, since the spring crops were usually legumes, the three-field rotation system also increased the nutritional quality of food.

Another major influence on food production was the change from slavery to serfdom around the tenth century. A slave gained no benefit from working better or harder and subsequently was not interested in the quality and quantity of his/her work. One of the largest problems in slave management was actually to make them work. In serfdom, on the other hand, the farmer had to spend a large part of his time on working for his lord but also part of his time working for his own benefit and profit. This gave the serfs an incentive to work both harder and smarter through the development of new techniques (Harman 2002, p. 143).

The cumulative efforts of these developments increased the productivity of the farmer by over 50%, allowing more and more people to focus on other tasks. While during the Middle Ages 65% to 90% of the population were still farmers (Cipolla 1989, p. 75ff), the increase in food production significantly increased the available manpower for other areas. Hence, the population could not only feed itself but also generate wealth through manufacturing and trade.

Nevertheless, throughout the Middle Ages, the bottleneck to economic growth was still food production. While overall food production increased greatly, it also fluctuated widely. First of all, food production depends heavily on the weather. During normal years, grain yield, for example, was around three to four, i.e., for every grain sown, three to four grains were harvested. In very good years, the grain yield was even six or higher. However, if there was a very dry or wet year, grain yields decreased to below one, meaning more grain was put on the field in spring than harvested in autumn (Cipolla 1989, p. 122ff). This caused starvation and stagnation or recession in both economic and population growth.*

Other major sources of fluctuation were pests and diseases. The Black Death between 1348 and 1350 killed up to half of the population of Europe. This was the largest but, by far, not the only pandemic that ravaged Europe during the Middle Ages, shattering communities and throwing all of Europe back generations. However, while a plague killed people, it usually did not harm livestock or equipment. Furthermore, since many people were killed, labor was in short supply and subsequently more valuable. Overall, survivors were usually better off than before, having more food, more tools, and higher wages. This was quite the opposite of war—the third source of fluctuations. Since Europe was fragmented, there was no shortage of armed conflicts, large or small. While most people were able to survive an army crossing their territory, there was not much left afterward to live from. Food supplies were taken, livestock was killed, and equipment was destroyed or burned down. It usually did not make too much of a difference if the army was hostile or friendly, since the soldiers needed to eat and usually simply took whatever was available (Harman 2002, p. 155). Nevertheless, during the Middle Ages, food production, in general, increased greatly, subsequently feeding many more professions and laying the foundation for economic growth.

5.3 SPREAD OF EARLY LABOR-SAVING TECHNOLOGY

Fueled by the increase in food production and by the rise of the prestige of artisanship, manufacturing also increased during the Middle Ages, both in the number of people involved and in productivity. A large number of labor-saving devices and technical inventions either were invented or became popular during the Middle Ages. First, there was a great change from muscle power (both animal and human) to other sources of energy. While water mills were known at least since around the third century (see Section 4.1), the usage of water mills was not widespread until the Middle Ages, when it became one of the common sources of energy. From the

* For reference, modern U.S. grain yields are around 30. Combined with mechanization and other productivity improvements, nowadays, less than 3% of the population of the United States work in farming.

sixth century onward, water mills became a major labor-saving device in food processing. Later, other water-powered techniques were developed to hammer metal, power bellows, crush malts for beer production, full clothes, grind knives, draw wire, grind food and minerals, and many other processes (White 1972, pp. 156–157). For example, the *Domesday Book*, a survey of England agriculture and industry completed in 1086, shows an average of two water mills for each village or hamlet.* Since a sufficient source of water for a water mill was not always available, horizontal windmills also came into common use from the twelfth century onward. Some places even used tidal mills, for example, in Venice in 1044.

Besides the increase in water and wind power, many other technical devices were invented to speed up production. Around the eleventh century, loom technology made major steps forward. This increased not only the quality of the product but also the output of the weaver by a factor of 3 to 5 (Cipolla 1989, p. 174). While before weaving was the bottleneck, a weaver with a vertical loom was able to keep multiple spinners busy, even after the knowledge of the spinning wheel from China spread rapidly all over Europe. Another great labor-saving device was the printing press invented by Johannes Gutenberg around 1450.† While before a book was a luxury product that had to be written by hand by scholars trained in the art of reading and writing, the printing press made printed documents a commodity. Another simple but extremely effective invention was the wheelbarrow,‡ greatly reducing the labor needed for transport.

This is not to say that this was a smooth process. Labor saving usually frees up the workforce for other tasks, overall increasing productivity and prosperity. Nevertheless, in this transition process, parts of the workforce may be free for other tasks before these other tasks are available. In short, people were out of work, and—naturally—this was not popular, often resulting in public unrest (Cipolla 1989, p. 172). Some regions tried to prevent this by forbidding labor-saving devices to prevent unemployment (Cameron and Neal 2002, p. 113). In the long run, however, these

* Of course, the purpose of the book was not economic statistics but rather to help King William I of England to increase taxation and generate cash for his crown.

† Please note that other countries had moveable-type printing before, e.g., China since the eleventh century. However, the development in Europe was influenced by the (re)discovery of printing by Gutenberg.

‡ Probably invented in China, but common usage started in medieval Europe. Romans did not know the wheelbarrow, and many Roman warehouses had a curved entrance unsuitable for animals or wheelbarrows.

labor-saving devices created more work and employment than before (Thrupp 1972, p. 262).

Overall, the period was a time of tinkerers and mechanical scholars, not only in manufacturing but also in mining, the military, architecture, medicine, trade, and agriculture. Science moved from philosophy toward a modern scientific approach using experiments. Clock making became a common pastime for the elite from 1260 onward, and the first usable mechanical clocks appeared around 1330. During the Middle Ages in Europe, brain power replaced muscle power, and machines made people's lives easier. Productivity per person increased by 30% between 1350 and 1500 (Cipolla 1989, p. 216ff).

5.4 THE ROLE OF MEDIEVAL GUILDS

With the rise of manufacturing both in economic significance and in social status, the organization and regulation of manufacturing also evolved. Not only did the local rulers try to enact laws in order to control manufacturing, but the artisans themselves organized into guilds to structure their profession. In fact, in their heyday, some guilds managed to become the de facto rulers of the town. While artisan associations were also active in previous civilizations,* the emergence of medieval guilds gave manufacturing a power base never seen before.

European guilds started out as religious brotherhoods. The members of a profession formed a group to build or support a church or religious tradition, often related to the saint of their profession. One of the earliest such brotherhoods may be the shoemakers of Würzburg, who formed in 1128 with the purpose to donate money to their church annually. These religious brotherhoods then evolved into economic organizations. One of the first guilds is the bed linen weavers of Cologne, who in 1149 convinced the town officials that all bed linen weavers in Cologne must be a member of their guild (Kieser 1989). This elimination of unlicensed competition therefore established a monopoly for the bed linen weavers in Cologne.

As the economy during the Middle Ages became more competitive, many artisans who could barely count, much less do additions and subtractions,

* For example, the *hanghui* in imperial China, the *collegia* during the Roman Empire, the *shreni* in ancient India, or the *koinon* in Ptolemaic Egypt.

were outsmarted by merchants. Forming a guild offered some protection from the market and simplified the playing field. This economic advantage was probably one of the main reasons that by the fourteenth century, usually every craft in a town was organized in a guild (Centre international de synthäese 1930, p. 362).

While productivity increased greatly, most manufacturing was still only on a small scale. Exceptions were only due to necessity, for example, shipbuilding or smeltering, for which a normal artisan did not have the capital and which also required the coordination of multiple persons in order to do. While many large trading operations emerged during the Middle Ages, there were not yet economies of scale for manufacturing.* Manufacturing was on a much smaller scale, with usually less than a dozen people working together. In most workshops, the master craftsman lived together with his journeymen and apprentices.

Rather than having a working contract with salaries and other organizational issues defined, the apprentices, and to a lesser degree, also the journeymen were more part of the family. A boy† joined a master as an apprentice between the ages of 10 to 15. Depending on the prestige of the master or the craft, the parents of the youngster usually had to pay *Lehrgeld* (German for *apprentice's due*) to the master to cover food and lodging. Being a family relative or having parents in the same or a related guild usually helped acceptance. The master usually had only one or two apprentices at the same time. When a master accepted a new apprentice, the apprentice was obliged to work for the master for the duration of, usually, three to five years‡ and to follow his orders (Mayer-Maly 1991, p. 105). This was, in effect, cheap labor for the master. If the apprentice's family was unable to pay the apprentice's due, the duration of the apprenticeship may have been extended to cover the costs. In turn, the master was obliged to provide food and lodging and, above all, to train the apprentice in the craft in order to enable him to become a journeyman (Engel and Jacob 2006, p. 342). It was uncommon but not unheard of for the apprentice to receive a salary.

* There were very few manufacturing companies, for example, the Société des Moulins du Bazacle in Tolouse, France, a joint stock company for milling established around 1250. Even then, these companies operated similarly to a small artisan workshop.

† Most apprentices were male, although female apprentices were common in some crafts as for example tailoring or baking.

‡ Among the shortest one was probably the wool scouring guild in Strasbourg, who in 1400 required only six weeks apprenticeship. The gold smiths of Cologne, on the other hand, required eight years before promotion to journeyman (Mayer-Maly 1991, p. 102).

After the promotion of the apprentice to journeyman, the journeyman could either travel and seek temporary employment at different places or stay with the master. Travel was rare before the fourteenth century but often mandatory after the sixteenth century. The frequently international travels of a journeyman served both the natural curiosity of a 15- to 20-year old and further education in the arts and crafts, helping him to become master. However, it was in no way guaranteed that a journeyman would become a master. To become a master was expensive, time-consuming, and demanding, and many journeymen never made this step but, rather, remained an *ewiger Geselle* (German for *eternal journeyman*), especially during the late Middle Ages. It is estimated that about one-third of the working population in a town were journeymen (Engel and Jacob 2006, p. 347).

While there were many regional differences in Europe, pretty much all guilds aimed to change the market in their favor or—to be more precise—in favor of the established master craftsmen running the guilds. They manipulated the economic environment to their advantage, similar to modern-day cartels. There was widespread price fixing, both for the purchased raw materials and for the finished goods (Kieser 1989). Outside competition was hindered or eliminated through import taxes in order to benefit local artisans (Cameron and Neal 2002, p. 131). Inside competition was controlled by the guild, and usually, all the town's artisans of the profession were required to be guild members and to follow guild rules. Wages and working hours were also fixed for all guild members. Advertising was often prohibited, lest one guild member have an advantage over another guild member.

Production methods were similar throughout the guild. Quality was regulated and checked, and producing flawed products was punished. Through political influence, the size of the market was increased. In Britain, for example, it was required that a deceased be buried in a woolen shroud, greatly benefitting the textile industry, while fishing guilds were able to establish or promote fish eating during Protestant holidays (Cameron and Neal 2002, p. 155). Overall, rather than acting as numerous individual workshops, guild-organized artisans acted as a larger entity. Hence, they increased their market power over the supplier, customer, government, and outside competition. While this was beneficial for the guild in the short term, these cartels were not good for the economy overall. Often, town governments tried to prohibit or reduce price fixing and other guild actions in order to benefit the overall economy (Thrupp 1972, pp. 265–266).

As the rules of the guild were made by the established masters of the guild, they usually favored the wealthy established masters and disadvantaged the less well-off or young artisans. For example, in London in the 1290s, the weavers' guild required that the loom of a deceased member be destroyed. This reduced the number of looms and increased the cost to obtain a (new) loom. Overall, this made it much more expensive for a young weaver to establish his own workshop. Other guilds required an elaborate and expensive masterpiece before promoting an artisan to master status and allowing him to open his own shop (Thrupp 1972, p. 268). Of course, the number of masters in a town was also often limited. Hence, a journeyman looking for a promotion to master often had to undergo considerable expenses and humiliation and had to produce a time-consuming and elaborate masterpiece before becoming a master. Apprentices and journeymen came under guild jurisdiction but lacked membership rights (Epstein 1998). Effectively, journeymen were mostly wandering laborers subject to the mercy of their temporary masters (Rosenband 1999).

As a response to this, from the fourteenth century onward, journeymen also organized to gain more power over the master craftsmen. These *Gesellenvereine* (German: *journeymen's associations*) started first in Germany, followed by Western Europe, with England being a notable exception (Centre international de synthäese 1930, p. 363). Naturally, the masters strongly opposed the organization of workers over whom they believed they had a God-given right to rule. This resulted in often violent labor disputes and riots.* Journeymen threatened to burn down the goods and houses of the masters, smashed their tools, and even kidnapped masters, carrying them around on a pole through the streets (Rosenband 1999).

Depending on the type of work and the location, working life itself was not bad, compared to the other alternatives to employment. While the worker theoretically had to work every day, the church year was ripe with religious holidays, including Sundays. Moreover, it was often common practice to get half a day off in preparation for each holiday. Hence, a workweek started on Monday and ended on Saturday afternoon or earlier. In total, for about 25% of the year, no work was allowed due to

---

* Sometimes, guilds are mistaken to be the precursor of labor unions. However, since they disadvantaged the lower-ranking worker and benefited the master craftsmen, they are more like industry associations rather than labor unions. Instead, the journeymen's associations could be considered an early form of unions.

religious reasons* (Thrupp 1972, p. 253). Working time itself was also often flexible. Since large numbers of workers were working under the putting-out system (see Section 5.5), they were under their own control, working when they wanted and in the way they wanted. Usually, work started at sunrise and ended in the afternoon, but there were many exceptions, and an artisan may have worked during the night at candlelight and sold his goods during the day. Some less fortunate industries, however, had much longer working times, for example 14 hours per day in construction in Amsterdam or 16 hours per day for the Paris bookbinders (Rosenband 1999). This was often also regulated by the guilds and may have also included seasonal aspects based on demand or the length of the day (Thrupp 1972, p. 252).

Since guilds started out as religious brotherhoods, religion was usually a large part of the guild. Subsequently, while on the outside, guilds acted similarly to cartels, on the inside they acted more like cults, full of rituals and symbols. This gave the guild rules a religious aura of justification, not to be doubted. Singing of guild songs and dances were common. Most of the guild rules and rituals were also secret and not to be shared with outsiders. The guild also had legislative and judicative functions on its members on all non-criminal-related issues. Punishments included monetary fines, incarceration, and expulsion from the guild. Expulsion from the guild was the heaviest punishment, since the former member lost all social status and his source of income and was unable to use his craftsman skills anymore to support himself.

Due to their strong religious aspects and due to the guild leaders usually being older men, guilds were mostly very conservative. Innovation and change were not part of their mindset. Since they also strictly controlled internal competition, members were heavily discouraged from inventing and improving, lest they gain an advantage over other guild members. Most guilds even had written rules against innovation, stating, *No man should think or invent something new or use it, but everyman should follow his neighborough in brotherly love.* Innovation was often punished. A smith in Nuremberg was incarcerated for over a week after making a small invention in 1590, and an improved loom was burnt publicly in Hamburg in 1676 (Kieser 1989). Even merely producing a different product was frowned upon. For example, in 1582, merchant Jacob

* In comparison, a modern worker has about 30% to 40% of the year off, including weekends and holidays.

Seyfrid wanted the weavers in Memmingen, Germany, to make a large quantity of a certain type of blended cloth. However, the guild refused to make this different product, even when offered advance money for necessary investments and material. Nevertheless, not all guilds were so restrictive, and Jacob Seyfrid was much more successful with the weavers in nearby Mindelheim (Safley and Rosenband 1993, p. 1). Overall, smaller guilds or guilds in less competitive environments were usually more open to new ideas. Similar, newly established guilds during their formative years were also more innovative. In addition, in some cases, merchants focusing more on economic benefit than on tradition were sometimes able to influence guilds toward new ideas. Yet overall, guilds usually resisted innovation (Kieser 1989).

This conservative behavior and resistance to innovation also led to the decline of the guilds from the sixteenth century onward. This conservatism, combined with the cartel-like behavior, damaged the local economy. Guilds were trapped in a vicious cycle. They responded to economic threats by reinforcing their cartels, which, over time, worsened the economic threats, leading to even stronger cartels (Kieser 1989). For example, guilds responded to a declining market share by making it harder for journeymen to become masters, hence protecting the masters' market share. However, these journeymen still needed to make a living and, in response, often established themselves outside of the sphere of influence of the guild. Journeymen moved to the countryside or worked in secret in their own chamber and hence were mockingly known as *chambrelans* in French (Schmoller 1900, p. 341). Producing goods in the countryside, of course, took away even more of the market share of the guilds, closing the vicious cycle. Soon, a large part of the manufacturing in Europe occurred outside of the control of guilds, despite the guilds' efforts to suppress this competition (Rosenband 1999).

Eventually, local rulers started to see guilds no longer as a solution but, rather, as a problem. Guilds' activities and rights were more and more restricted for the benefit of the economy. The final nail in the coffin of the guilds, however, was that they were no longer able to compete. Less restricted and better-organized manufacturing systems were able to produce faster, better, and cheaper than guilds. Already during the Middle Ages, the putting-out system threatened the monopoly of the guilds, and the efficiency of a guild-regulated workshop was no longer viable by the time of the first factories. Finally, during the industrial revolution, guilds were a mere former shadow of themselves.

5.5 THE PUTTING-OUT SYSTEM

During the early Middle Ages, an artisan usually had a number of different options available to make a living from his skills. A wealthier artisan owned his own workshop, worked for his own profit and at his own risk, and may have employed other artisans or apprentices. These employed artisans were initially apprentices and journeymen living with the master. Payment by piece, day, week, or year was probably the most common way of remuneration in the Middle Ages (Thrupp 1972, p. 265). Only from the fifteenth century onward were there also either salaried employees or workers paid to do a certain amount of work or to complete a number of pieces. Another common way for an artisan to produce was to rent his tools or his entire workshop, which was usually the more capital-intensive part of the fixed cost (Cipolla 1989, p. 105).

For most workers during the Middle Ages, the reason for work was not to become wealthy but, rather, to have enough to live. Many workers preferred to party rather than to amass riches. Or, in the words of Daniel Defoe in 1704, *There's nothing more frequent, than for an Englishman to work till he has got his pocket full of money, and then go and be [...] drunk, till, tis all gone* (Defoe 1704). Merchants, on the other hand, were, by the nature of their business, more profit oriented and therefore, on average, wealthier than the craftsmen. Subsequently, merchants, over time, expanded their control of the production process. Initially, merchants ordered products from the artisans and paid per piece. Since raw material and finished goods were usually the largest part of the artisans' expense, merchants eventually also purchased the raw materials for the artisans to transform into goods. And finally, the merchants provided the artisans with everything they needed to work, including material, tools, and sometimes even the workshop. The system where another party provides funding and/or raw materials for the artisan to process at home and also organizes the sale of the finished goods is called the putting-out system.*

This putting-out system was a common way for merchants to invest from the fifteenth century onward. Due to the better efficiency based on the division of labor and the more profit-oriented approach by the merchants, the putting-out system had significant economic advantages over the traditional guild system. During the sixteenth to eighteenth centuries,

* Other terms for this system are workshop system, domestic system, *Verlagssystem* (German), or cottage industry.

the putting-out system increased significantly, while the guilds declined. Especially for merchants, it was common practice to integrate horizontally by taking control of the manufacturing process through this putting-out system. Often, a merchant had a large group of artisans working for him. In some cases, the merchants even took control of the entire production in a town, for example, in Flanders, the Po Valley, or parts of Spain (Thrupp 1972, pp. 247–249). However, rather than pooling the labor in a large factory, the artisans worked in numerous smaller workshops. Even a wealthy artisan with his own workshop preferred to finance another artisan rather than enlarge his workshop to a size that he no longer could oversee personally. Having direct control of everything in a workshop was important to the owner. There was usually no middle management.

It turned out that direct control was usually a dire necessity. Artisans hated the putting-out system. They no longer had contact with or control over the customer, or the supplier. They were fully dependent on the merchants in their work. Yet, as they needed the work to make a living, they often had no other opportunity but this putting-out system. Employer–employee relations in the Middle Ages were usually strained at best and often turned hostile. The merchant could pay the artisan the price he saw fit for the quantity and quality produced, often shortchanging the artisan. If money was short, the artisan could be paid in kind rather than in cash. Since there was usually no shortage of labor, the artisan had no other choice but to accept. Wages rose only after a plague, as there were not enough workers (Thrupp 1972, p. 268). Overall, the employed artisan was in a much weaker position, of which the merchant took advantage. This was especially true for part-time artisans, who often farmed during farming seasons and worked as artisans during winter. Hence, the loyalty of most employed artisans to their employers was very limited.

Subsequently, the employed artisan often tried to cancel out the disadvantage secretly. It was common practice for an artisan in the putting-out system to secretly put aside some of the received raw material and to produce finished goods for his own profit using the provided workshop, and tools. Hence, the employees often helped themselves to whatever they needed when they were not watched. They even used the masters' tools and materials to produce products for their own profit. Overall, there was often a—fully justified—high level of distrust between employer and employee.

Nevertheless, manufacturing during medieval times greatly increased in productivity due to the rise of the towns, the advances in agriculture, and the invention of labor-saving devices. While the putting-out system, over time, became more prominent and profitable than guild-organized artisans, there was not yet any manufacturing on a larger scale. It was still very much based on individual craftsmen working in small groups.

BIBLIOGRAPHY

Cameron, R., Neal, L., 2002. *A Concise Economic History of the World: From Paleolithic Times to the Present*, 4th ed. Oxford University Press, New York.

Centre international de synthäese, 1930. *L'Encyclopedie et les encyclopedistes*. Burt Franklin, Paris, France.

Cipolla, C.M. (ed.), 1972. *The Fontana Economic History of Europe: The Middle Ages*, 1st ed. Fontana, London.

Cipolla, C.M., 1989. *Before the Industrial Revolution: European Society and Economy, 1000–1700*, 2nd ed. Methuen & Co. Ltd, London.

Defoe, D., 1704. *Giving Alms No Charity*. Booksellers of London and Westminster, London.

Engel, E., Jacob, F.-D., 2006. *Städtisches Leben im Mittelalter: Schriftquellen und Bildzeugnisse*. Böhlau Verlag Wien, Köln, Weimar.

Epstein, S.R., 1998. Craft Guilds, Apprenticeship, and Technological Change in Preindustrial Europe. *The Journal of Economic History* 58, 684–713.

Harman, C., 2002. *A People's History of the World*, 2nd ed. Bookmarks Publications Ltd, London.

Kieser, A., 1989. Organizational, Institutional, and Societal Evolution: Medieval Craft Guilds and the Genesis of Formal Organizations. *Administrative Science Quarterly* 34, 540–564. doi:10.2307/2393566.

Mayer-Maly, T., 1991. *Ausgewählte Schriften zum Arbeitsrecht*. Böhlau Verlag Wien, Köln, Weimar.

Rosenband, L.N., 1999. Social Capital in the Early Industrial Revolution. *The Journal of Interdisciplinary History* 29, 435–457.

Russell, J.C., 1972. Population in Europe 500-1500, in: Cipolla, C.M. (ed.), *The Fontana Economic History of Europe: The Middle Ages*. Fontana, London.

Safley, T.M., Rosenband, L.N. (eds.), 1993. *The Workplace Before the Factory: Artisans and Proletarians, 1500–1800*, 1st ed. Cornell University Press, Ithaca, New York.

Schmoller, G., 1900. *Staats-und sozialwissenschaftliche Forschungen*. Duncker & Humblot, Leipzig, Germany.

Thrupp, S., 1972. Medieval Industry 1000–1500, in: Cipolla, C.M. (ed.), *The Fontana Economic History of Europe: The Middle Ages*. Fontana, London.

White, L., Jr., 1972. The Expansion of Technology 500–1500, in: Cipolla, C.M. (ed.), *The Fontana Economic History of Europe: The Middle Ages*. Fontana, London.

Section II

The Industrial Revolution—Manufacturing Gets Mechanized

6

Early Modern Europe

As in the Arsenal of the Venetians
Boils in the winter the tenacious pitch
To smear their unsound vessels o'er again,

For sail they cannot; and instead thereof
One makes his vessel new, and one recaulks
The ribs of that which many a voyage has made;

One hammers at the prow, one at the stern,
This one makes oars, and that one cordage twists,
Another mends the mainsail and the mizzen;

Thus, not by fire, but by the art divine,
Was boiling down below there a dense pitch
Which upon every side the bank belimed.

Dante Alighieri (1265–1321)
Italian poet, in Divine Comedy, Book 1: Inferno,
translated by H.W. Longfellow

During the early modern period, fundamental changes in society happened. Arrangements of city-states merged into modern centralized governments. The Sun King Louis XIV of France (1638–1715) lived by *L'État, c'est moi* (*I am the state*).* Only a few years later, King Frederick II of Prussia (1712–1786) lived by his motto, *Ich bin der erste Diener meines*

* The quote is often attributed to Louis XIV, but historians disagree if he really said this phrase. In any case, he strongly believed in the divine right of kings.

Staates (I am the first servant of my state).[*] From the sixteenth century onward, the ideas of mercantilism became popular. Governments actively encouraged exports through subsidies while at the same time trying to curb imports through taxes and tariffs.[†]

Modern science emerged, critically analyzing formerly considered true religious beliefs and leading to the scientific revolution. People started to measure things and tried to understand the laws of nature behind this. Copernicus and Galileo laid the modern groundwork for astronomy, while Newton established classical mechanics and the rules of gravity.[‡] King Louis XVI of France even had his own locksmithing workshop. Most previous kings in Europe or Asia would probably rather have been dead than caught with a craftsman tool in their hands, and here we have the king of one of the leading nations in Europe tinkering with locks. King Louis XVI was reportedly rather skilled in making locks and other mechanical gadgets. Some even say he was a much better locksmith than king. Ironically, he met his end through another popular mechanical gadget of the time, the guillotine, in the wake of yet another social upheaval, the French Revolution.

6.1 MANUFACTURING TECHNOLOGY

Parallel to the change in society, manufacturing technology also evolved, leading to the first steps of the industrial revolution.[§] As textiles were still the largest part of the manufacturing economy besides food processing, there were numerous improvements to spinning, weaving, and knitting tools. Yet, the laborers and the government often opposed these labor-saving improvements.

For example, in 1589, the Englishman William Lee invented a knitting machine that was 10 times faster than a human knitter.[¶]

[*] Quite a difference compared to Louis XIV, considering that only three words need to be added in the English translation.

[†] Modern economists believe that mercantilism in its entirety is actually harmful for the economy, and the system of free trade has since mostly replaced mercantilism.

[‡] Of course, early modern science also included a lot of nonsense. For example, Giovan Maria Bonardo calculated that the exact distance from hell to earth is exactly 3785.25 miles.

[§] Some researchers consider the first factories as already part of the industrial revolution, while others see the steam engine as the start of the industrial revolution. Both views are equally valid.

[¶] Reportedly, he invented the knitting machine in order to impress a woman who liked to knit. It is not known if the lady was impressed with being replaced by a machine.

Queen Elizabeth I of England, however, refused his patents numerous times, worrying about the many knitters in the kingdom. Eventually, Lee moved to France, successfully promoting his invention. Another example is the new improved Dutch loom, which appeared in Leiden before 1604. Despite the protests of the established weavers, the invention prospered and spread to Europe. Charles I of England prohibited the loom in 1604 because of it *having taken away the work of 1200 or 2000 'natyve borne subjects'* (Kerridge 1988, p. 172). Yet, due to the great economic benefits of the improved loom, even the prosecution of the king was unable to stop this invention.

These are only two out of many examples of new labor-saving inventions, against many of which the established workers protested. However, these protests usually were either completely unsuccessful or, at best, delayed the spread of the technology.

6.2 WORKING TOGETHER—THE EMERGENCE OF THE MANUFACTORY

More significant than the improvement in manufacturing technology, however, were the changes in manufacturing organization. The most fundamental change in early modern Europe was the change from the putting-out system to the factory system or manufactory (from German *Manufaktur*) at the end of the eighteenth century. Rather than having a small number of workers working in small workshops, these manufactories pooled together hundreds of workers at one location under one management. This pooling of labor increased productivity much more than technical changes and inventions (Johnson 1993, p. 44). Due to increased productivity and economic viability, manufactories were able to undercut the traditional workshop-based manufacturing of the guilds. Because of this in combination with the protective measures of the guilds, guild-organized workshops rapidly went out of business, and by the end of the nineteenth century, the output from guild-based workshops was insignificant compared to manufactories and factories.

Manufactories benefitted from the concentration of labor in multiple ways. First, a higher division of labor was possible, where each worker was responsible only for a single step in the production process. This allowed the use of an untrained and therefore cheap worker with minimal training.

Furthermore, this also reduced the scrap rate since the worker was less likely to forget something and was able to master his process faster compared to an old-style worker who had to learn and master multiple processes.

Secondly, since manufactories had a larger production capacity, they also purchased larger quantities of raw materials. They received a discount for purchasing larger volumes, since the seller also had to spend less time selling his goods. Furthermore, due to the larger sums involved, it also made economic sense to spend more effort on finding and negotiating the lowest-priced raw materials available. Similar benefits also apply to the sale of products to other merchants or organizations.

Third, due to the larger scale of production, it made sense to pay more attention to optimizing production. Additionally, most entrepreneurs establishing a manufactory had very good business sense and were willing to pinch a penny and go through the effort of cost accounting. They tried to understand the costs involved in their production and often aimed to systematically reduce costs. For example, they quickly found out that labor costs of apprentices are cheaper than journeymen. Furthermore, boys were even cheaper than apprentices, and girls cheaper than everything else.* Inversely, girls created the least amount of trouble, while free-spirited journeymen caused the most.

Fourth, manufactories created a material flow passing through the different stages of production. This had the ability to reduce the work in process compared to a guild worker processing *a pile of products* from raw material to finished product. Since there was less work in process, less capital was tied up in materials, subsequently either freeing up money for other expenses or reducing the borrowed cash and, hence, the interest. Additionally, less space was needed to store materials; less effort was needed to handle them; and since materials got used more quickly, they did not get old or, in the worst case, spoil. The benefit of reduced work in progress is probably the least obvious, and many manufactories did not pay attention to this.†

Finally, the factory system allowed tighter control of the worker, giving power to capital over labor and reducing theft by the workers (Jones 1987). Under direct supervision, a worker was less likely to embezzle materials

* Nowadays, child labor is not acceptable. Even in the eighteenth century, child labor was controversial but was frequently used as cheap labor. Additionally, the children's income helped to support their families.

† Even nowadays, many factories do not see the benefit of reducing inventory, much less realize these benefits.

or tools or to spend time not working. While thefts and idleness still happened frequently, it was much less so than in the putting-out system.*

The problem most manufactories and, later, most factories faced was how to deal with the worker. Workers of the time were very independent spirits who were used to coming and going as they liked, working when they wanted, and often were more interested in drinking and partying than in working. An employer already considered himself lucky if the workers did not steal materials or tools for their own purpose. Overall, it was difficult for the employers to make the workers do what they are supposed to do.

Employers and manufactory owners, on the other hand, were also not very skilled at managing people. In most cases, their education, training, and skills focused on the technical tricks of their trade. The skill of managing people was, at best, self-taught. In a small guild-based workshop, social interaction was able to ease many of these shortcomings, but a manufactory with hundreds of employees or more did not allow much social interaction between the boss and the workers. Additionally, society in early modern Europe was still very hierarchical. Aristocracy demanded total obedience from their peasants, and similarly, manufactory owners demanded total obedience from their workers. As the aristocracy *by the grace of God* believed it knew what was best for its people, employers believed they knew what was best for their workers. Thus, the employers set the working times, salary levels, and other rules and regulations as they deemed fit, and saw no need to consult the workers on these matters. Even with benevolent masters, any attempt by the workers to discuss their conditions of work with the master resulted in hard feelings or, more often, in immediate dismissal (Burton 1976, pp. 92–93). Many employers abused these privileges, and the situation of their *white slaves* in Europe was often not better than the situation of *black slaves* in America. In the face of this extremely authoritarian management style,† it is no surprise that the workers lacked loyalty to the enterprise, resulting in idleness, absence, and theft.

Naturally, the first manufactories were as diverse as the entrepreneurs who established them and the products they made. Many of them focused

* Theft and idleness are still problems in modern factories, too. Insiders occasionally steal entire truckloads of goods. For example, there are rumors of car factory workers who stole enough car parts to build a complete car. Or, any factory using cable ties or duct tape is always well advised to buy more than theoretically needed.

† Management has evolved since then, but authoritarian management is still common nowadays, resulting in a large number of unmotivated employees.

their improvement effort on different areas and invented different organizational measures needed to run a factory. In the next chapter, we will look at a few entrepreneurs who pushed forward in the way they worked and whose contributions have not been lost to history in this usually poorly documented preindustrial age.

BIBLIOGRAPHY

Burton, A., 1976. *Josiah Wedgwood*, 1st ed. Andre Deutsch Ltd., London.

Johnson, C.H., 1993. Capitalism and the State: Capital Accumulation and Proletarianization in the Languedocian Woolens Industry, 1700–1789, in: Safley, T.M., Rosenband, L.N. (eds.), *The Workplace Before the Factory: Artisans and Proletarians, 1500–1800*. Cornell University Press, Ithaca, New York.

Jones, S.R.H., 1987. Technology, Transaction Costs, and the Transition to Factory Production in the British Silk Industry, 1700–1870. *The Journal of Economic History* 47, 71–96.

Kerridge, E., 1988. *Textile Manufactures in Early Modern England*. Manchester University Press ND, Manchester, UK.

7

Pioneers of a New Age—The Factory System

We have stepped forward beyond the other manufacturers & we must be content to train up hands to suit our purpose.

Josiah Wedgwood (1730–1795)
Potter industrialist, in a letter to Thomas Bentley, May 19, 1770

The great improvement of manufactories was to pool workers together in the same building, increasing the division of labor but still working with traditional tools and techniques. The next step in the evolution of manufacturing was to use machine-based manufacturing. This mechanization separates manufactories from factories. For a localized artisan working from his home, purchasing a very productive but expensive machine was usually not an option. He lacked the space for the machine, he lacked the money, he lacked the advanced technological expertise, and he did not have the manpower to really make the best use of the machine.

Manufactory-style operations, on the other hand, were very compatible with heavy machinery. However, few manufactories upgraded by installing heavy machinery. The problem with the heavy machines was their need for power to drive them. Muscle power, both animal and human, was too expensive and unable to run continuously. The only continuous power source available that was both strong enough and constantly available was waterpower. Hence, a factory needed access to a steady supply of water, a waterwheel, driveshafts running through the building to distribute the power, and rooms large enough to accommodate the machinery. Because of these prerequisites, few manufactories actually evolved into factories.

Rather, newly built factories, over time, took away business from manufactories, driving them out of existence.

This concept of a factory—large-scale mechanized mass production—is the primary driver of modern manufacturing. There is no agreement among experts on where the first factory was located. Some consider the Arsenal of Venice the first factory; others look at ancient Roman or Chinese production sites. There are even arguments for some early mining operations in South America as the first factory. However, in my view, a factory requires mechanized mass production, with the first factories appearing in the Derwent Valley,* between Birmingham and Manchester, England. However, before we start the curious tale of industrial espionage that developed into the first modern factories in Sections 7.4 and 7.5, we will look in more detail at preceding developments.

7.1 THE ARSENAL OF VENICE—THE LARGEST INDUSTRIAL SITE IN EUROPE OF ITS TIME

The first manufactories were established not in order to become more profitable but, rather, because there was no way to produce a certain product in small workshops. A prime example of this is shipbuilding. A single workshop did not have the labor or monetary resources available to build a ship. Hence, the first manufactories were government-owned shipyards, the most famous being the Arsenal of Venice. The doge of Venice, Ordelafo Faliero, established the arsenal around 1104, initially as storage for ship supplies. After enlargement in 1320, the arsenal was able to produce about six ships per year. However, the organization was still similar to a private shipyard, with a guild master being responsible for part of the work but having no fixed team of workers to work with (Lane 1973, p. 164).

Over time, the arsenal changed from an unorganized small-scale shipyard to the most productive shipyard in the world. At its peak, around the sixteenth century, the arsenal was the largest industrial complex in Europe, employing up to 16,000 workers. The arsenal was able to mount astounding production feats. In 1570, facing the Ottoman–Venetian War,

* Please note that there are at least four Derwent Rivers in Great Britain. In addition to the spinning industry around the Derwent River between Birmingham and Manchester, another Derwent River near Newcastle was the location of Ambrose Crowley's iron workshops.

the arsenal outfitted* and launched 100 ships within two months, its largest production ever (Lane 1973, p. 364). On July 24, 1574, visiting King Henry III of France watched the outfitting of an entire ship while eating his lunch (Davis 1997, pp. 80–81). While the King of France probably did not have a 30-minute power lunch, it is still very impressive to outfit an entire warship within the span of a few hours. In comparison, other shipyards in Europe usually needed months to complete one ship. Of course, in times of peace, the demand for new ships was much lower, and the production in the arsenal decreased significantly. Nevertheless, the Arsenal of Venice was the most able shipyard in the world.

Venice used a number of different ways to achieve this productivity. Technological changes had only a small part in this. For example, most of Europe initially used Roman technology that first built the hull of the ships and then added the frame. The Venetians, on the other hand, first built the frame and then added the hull onto the frame, subsequently requiring less wood and less time.

Nevertheless, the large gains in productivity were through organizational changes. Probably the biggest impact was that the ship came to the worker rather than the worker to the ship. The first step was the building of the frame and the hull, including the caulking of the hull. As soon as the hull was able to float, it was towed to the different stages of the production process. After the construction of the hull, the cabins were built, followed by masts. Afterward, different equipment was brought on board at different stages—guns, sails, ammunition, anchors, ropes, chains, oars, and the like—before at the last stop, provisions were added. This division of labor and specialization allowed much quicker outfitting of the ships than would have been possible before. This system was not planned from scratch but, rather, evolved over centuries in a continuous improvement process (Lane 1973, pp. 362–363). Figures 7.1 and 7.2 show a historic map of the Arsenal as well as a schematic material flow with the different stops.

The arsenal also had a high level of vertical integration, meaning they often controlled the material flow starting from the harvesting and mining of the raw materials. For example, Venice managed its own forest in the nearby province of Treviso to ensure a steady supply of wood cut to the standards of the arsenal. Intermediate stages of processing were also

* Please note that this does not include the construction of the hull; these were available in storage. Some references confuse outfitting of the ship with constructing it from the keel up. However, historic sources are unclear on the details. The outfitting definitely included bringing ropes, sails, guns, and other equipment on board, but it may also have included raising the masts and building the cabins.

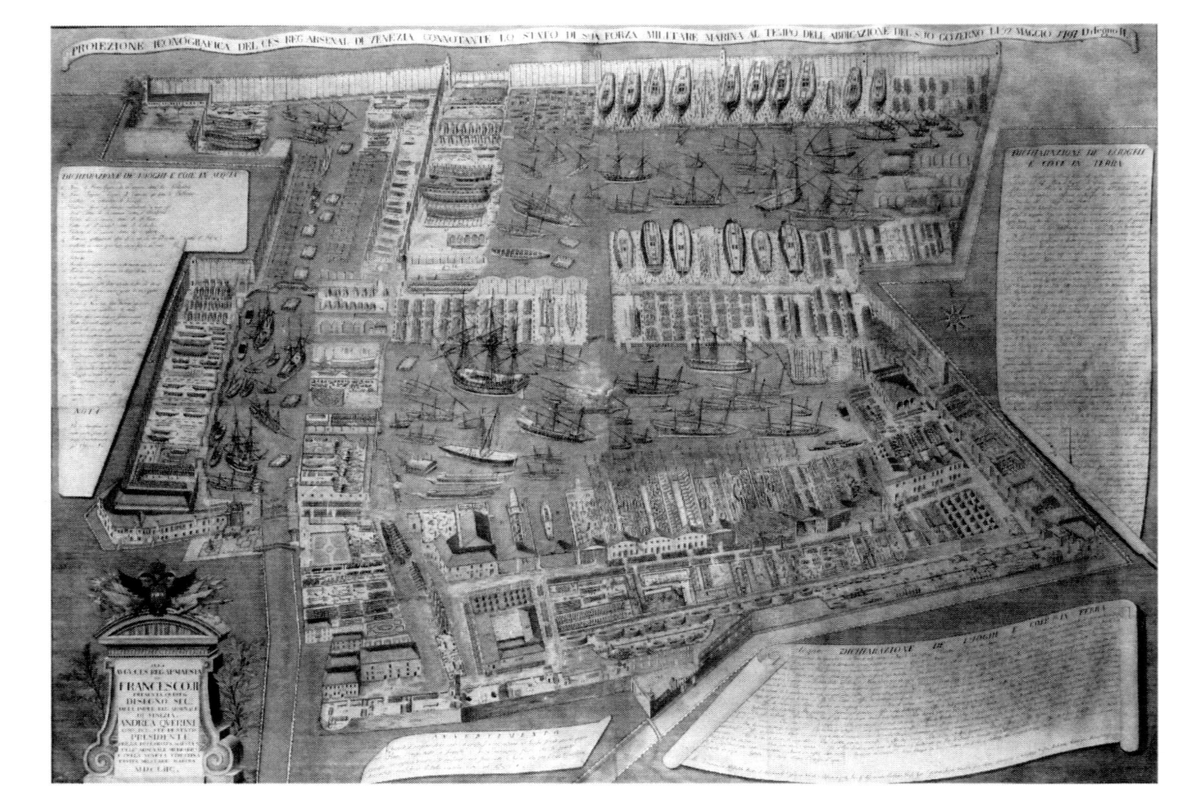

FIGURE 7.1
Map of the Arsenal of Venice in 1797 by Abbot Maffioletti shortly before the fall of Venice, showing in great detail the different workshops in the arsenal.

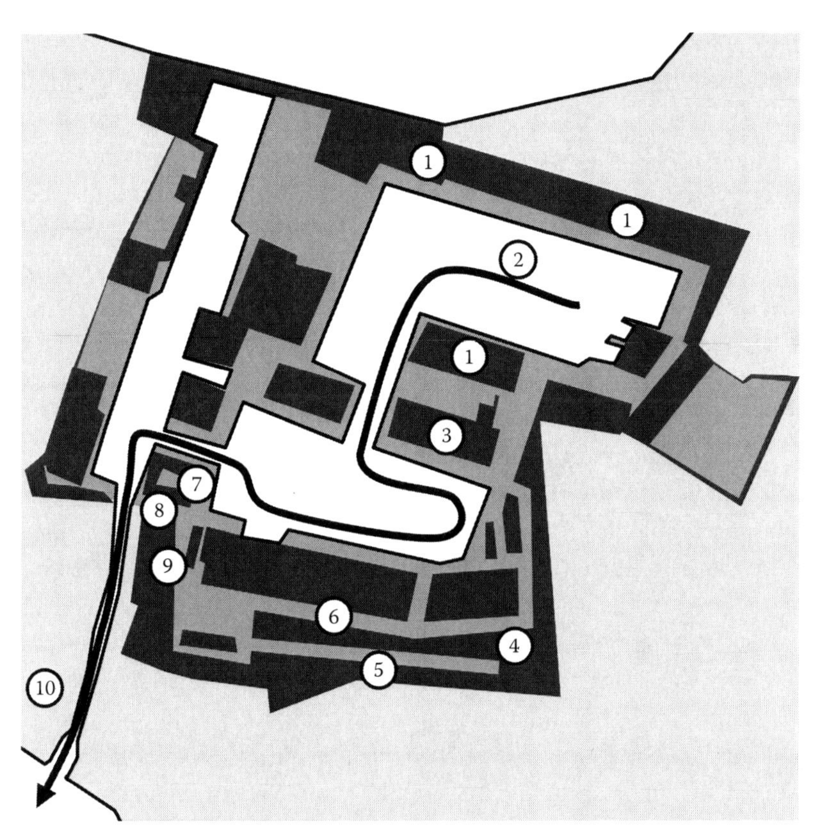

FIGURE 7.2
Schematic of the material flow in the Arsenal of Venice: (1) hull; (2) caulking; (3) masts; (4) ammunition; (5) ropes; (6) cannons; (7) sails; (8) anchors; (9) rudders; and (10) bakery. (Map by author.)

done on site in the arsenal, and besides shipbuilding, the arsenal had for example a rope-making workshop and an armory workshop (Davis 2007, pp. 11–12). Subsequently, most parts were made on site, and these subassemblies were usually located close to the outfitting stops. For example, the weapon making was close to the weapon outfitting, greatly reducing the transport distance of the heavy guns. The bread at the last stop before the Venetian lagoon came fresh from the adjacent bakery. This close linkage of subassemblies and outfitting points made a steady supply of parts possible with less inventory and less waste due to old or outdated products. This is not to say that the arsenal did modern lean manufacturing with few inventories (there were still piles of material everywhere), but for its size and its age, it was one of the most advanced manufacturing systems.

Another major change was the use of simple proto-standardized parts. Due to the large demand, many supplies were mass-produced to conform to the standards of the arsenal. This does not mean these simple standardized parts were interchangeable. Rather, only small adjustments were necessary to customize every part until it fit. For example, when adding masts to the ships, there was no need to adjust the length or the diameter, because a standard mast fit the standardized ship with little or no customization. This also applies to the standard benches for the rowers; the wood was cut to size and merely had to be nailed into place with only few adjustments. The arsenal avoided much fiddling and tinkering that otherwise would have been necessary if the parts were not standardized.*

These changes in material flow probably accounted for most of the extraordinary productivity in the Arsenal of Venice. However, they had numerous problems. Probably the biggest issue they faced was organizing the workers. Initially, they organized their workforce in the way that was standard at the time—different guild masters organizing their part of the work. Workers worked over time for different masters as they pleased. Overall, it was an extremely unstructured process consisting of highly independent and constantly changing subgroups. Generally, a supervisor could supervise between 5 and 25 people depending on the complexity of the tasks, which pretty much defined the group size in the arsenal. Before the fifteenth century, manufacturers simply did not know how to organize a larger workforce of hundreds of workers, let alone the 16,000 workers of the arsenal at its peak. Different organizational structures evolved in the arsenal over time, pushing the boundaries of manufacturing organization for its time but still being miles away from what we are accustomed to nowadays.

Workers received a daily wage in the arsenal, and lifetime employment was the norm. However, this wage was paid for showing up, regardless of the amount of work done by the worker. Hence, the goal of the worker was to show up, and even sick, crippled, or old people dragged themselves to the arsenal every day to receive their wage. Even able-bodied workers often had little interest in working for the arsenal but, rather, spent their time wandering around, sleeping in hidden spots, or enjoying the freely provided wine.† Some were even working for their own profit, often using

* Of course, this is still a far cry from the complex interchangeable mechanical parts shown in Chapter 9.

† In 1630, 1400 workers consumed or stole 650,000 liters of wine, the equivalent of more than 2 liters per worker per day (Davis 2007, p. 33).

arsenal tools and stealing arsenal material, sometimes even constructing entire sheds in the arsenal to run their private business (Davis 1993, p. 200). Additionally, the wages in the arsenal were not competitive, and many workers preferred to work outside the arsenal for better money. For example, in 1559, out of 2183 masters on the payroll, only 960 came to work on a regular basis (Davis 2007, p. 20).

In response to this, Venice, over time, expanded the management of the arsenal. By 1600, there were hundreds of bookkeepers, clerks, supervisors, and managers (Safley and Rosenband 1993, p. 181ff). The focus of management was primarily on materials and much less on workers. A lot of paperwork and bookkeeping were added to keep track of the materials, and arsenal materials were supposed to be marked as arsenal supplies, although these measures had only mixed success at best. A wall was constructed around the arsenal, both to protect secret knowledge and to reduce the loss of material. However, the wall was soon breached by smugglers. In any case, bringing material in and out of the front gate was not a big problem during rush hour. For example, one individual was caught at the gates trying to walk out with a literally staggering 70 kilograms of iron on him (Davis 1993, p. 190).

Besides the focus on materials, there were also some efforts in the organization of workers. Historically, masters and workers were highly independent and followed the traditions of the guilds. The arsenal changed this over time and added more and more organizational overhead, while the masters had less and less freedom to do things when and how they wanted. Because of this change, by mid-1600, these masters were often no longer called *masters* (Italian: *maestri*) but, rather, workers (Italian: *operai*) (Safley and Rosenband 1993, p. 181ff). In 1569, they fixed the group structure, and a master was working with a fixed workforce (Lane 1973, p. 364).

The management of the arsenal furthermore tried to improve attendance and productivity using both the carrot and the stick. As for the carrot, they gave the arsenal workers special privileges such as free food and wine, tax breaks, job security, paid holidays, and shelter in the arsenal during plagues. Working for the arsenal had high social prestige, and the management successfully created what we would nowadays call a corporate identity.

As for the stick, since there was a limited skilled workforce available, management was restricted in the use of punishment and prohibitions; otherwise, workers would have left. There was a police force in the arsenal responsible for keeping order. This force was armed, since their job included a real risk of being attacked by workers unwilling to cooperate. Additionally, there were spies to root out thieves, and an anonymous mail slot existed for whistle-blowers (Davis 1993, p. 190). Workers were required to show up at least six months of the year in order to keep their work. Workers were allowed to attend funerals only of masters in their own guild, whereas before, they could take the day off for any master's funeral, of which there were about 70 per year. It was also common at the time for workers in Europe not to work on Monday (the infamous *Saint Monday*), greatly disrupting work in the shipyard. Hence, management decided that attendance on Monday was required to be paid for the week. However, few of these rules were strictly enforced, and punishment was usually light (Davis 2007, p. 22ff).

Overall, the Arsenal of Venice made great breakthroughs in terms of material flow and vertical integration, establishing the largest and most productive shipyard and probably the largest workforce under the same management of its time. Nevertheless, they faced huge troubles in getting the workforce organized, and organizational measures to improve the productivity of the workforce had limited success. Constant loss of material through the workers was also a never-ending issue. As for the shipyard itself, its significance declined together with the decline and fall of Venice at the end of the eighteenth century, and the arsenal is currently a training location for Italian naval officers.

7.2 IRONMONGER AMBROSE CROWLEY AND THE ECONOMIES OF SCALE

The Arsenal of Venice was established because the magnitude of projects was too big for a single person to handle. Yet, other entrepreneurs also benefitted from the economies of scale for smaller products. One of the early large-scale industrialists was Sir Ambrose Crowley (1658–1713). He was born into a family of blacksmiths in Worchester, England, near Birmingham. After apprenticeship, he started his own business making nails, an industry primarily based on small individual ironmongers located in the greater Birmingham region. The British Navy, however, required enormous quantities of nails for shipbuilding, which so far had been supplied by merchants purchasing them from small workshop owners. This large demand for one product category was just waiting for someone to scale up his business, and Ambrose Crowley did just that.

Very soon in his career, he decided to move his factory to Sunderland, near Newcastle upon Tyne in the north of England. This was a very unusual choice, since at this time, most nails were made in the greater Birmingham region. Yet this decision clearly showed his entrepreneurial sense for profit. Sunderland had both cheap labor and cheap coal. Furthermore, Sunderland being a coastal city, shipping to London was much cheaper than from landlocked Birmingham. After some religious difficulties where his skilled but Catholic Belgian workers were harassed by the Protestant locals in Sunderland, he moved his business to nearby Winlaton, which also provided more space for expansion (Anderson 1973).

Crowley's business prospered, despite the British Navy being both extremely tardy in paying and very liberal in self-applied discounts. For example, a bill from December 1705 was paid over two years later in January 1708, and discounts by the navy of up to 30% of the bill were possible (Flinn 1960). Nevertheless, Crowley apparently priced in these *discounts* beforehand, producing all types of iron products from anchors to tools to 108 types of nails. Through these economies of scale, he was able to supply the navy with consistent quality and standardized products while at the same time still being cheaper than his competition. He eventually supplied 90% of the iron needs of the British Navy (Flinn 1960). His manufactories were soon probably the largest in England, with about 1000 workers (Crouzet 2008) and at least two large waterwheels in different sites around Winlaton.

While Crowley profited from the economies of scale in selling his products and purchasing his raw materials, there was little or no technological change in the production itself. The technology he used was the same every other nail maker in England used. For example, he used two water-powered slitting mills to convert iron bars into rods suitable for making nails, a known and widely used technology. In any case, most of his business was still done by hand (Jones 1987). Even the organization of his workforce was still based on the traditional system, with apprentices, journeymen, and masters. A master together with one or two journeymen processed a nail from rod through four process steps to finished product with almost no division of labor (Flinn 1962). The raw material was supplied by Crowley, and payment was by piece for completed nails (Dobb 1924). In this respect, Crowley's operation was based on more of a putting-out system, except that all workers were located at one site. Similar to the Arsenal of Venice, this manufactory based on traditional methods was a small but crucial improvement in the evolution of manufacturing.

Amazingly, Crowley operated his business from Greenwich, near London, about 400 kilometers away from his production sites in Winlaton. While he still traveled frequently to his manufacturing site (Anderson 1973), his aim was nevertheless to be close to the customer. This means he had to find a way to manage a thousand people, mostly men and boys, while being at the other end of England.

In order to regulate and organize his business, he established a detailed and ever-expanding description of the work rules. Eventually, this *Law Book of the Crowley Ironworks* included more than 100,000 words (Thompson 1967), or almost as much as the book you are reading right now. This law book detailed everyday rules like working times (a mind-boggling 15 hours per day from 5:00 a.m. to 8:00 p.m.), when to have breakfast or lunch (8:00 to 8:30 a.m. and noon to 1:00 p.m., respectively), and rules prohibiting drunkenness and foul language.

However, what made Ambrose Crowley so remarkable in manufacturing was not his technology, or his organizational structure, or the size of his business, but rather, the way he treated his workers. Crowley felt obliged to take care of his workers not only at work but basically from cradle to grave. His benefits included sickness payments, free medical treatment, pension payments, coverage of funeral expenses, food and clothing for the disabled, schooling for the young, and support for widows and orphans (Dobb 1924; Skempton et al. 2002, p. 160). Also unusual was that Crowley paid wages every week on time, different from most other employers, who often paid irregularly. Crowley also solved the management problem by installing different committees and councils to arbitrate disputes. A committee met weekly to settle wage disputes, and other committees handled logistics, transport, production, and supply.

While most council members were supervisors and managers, they also included representatives of the workers. This inclusion of workers into decision making probably helped to ease tensions between workers and management. While the work crew still had quite a rough reputation in the surrounding villages, relations between management and workers were extremely good for its time and the size of the operation. This made Crowley a pioneer not only in the metal industry but also in labor relations.

7.3 POTTER JOSIAH WEDGWOOD—THE SCIENCE OF MANUFACTURING PROCESSES

Another of the major industrial figures in early modern Europe was the potter Josiah Wedgwood (Figure 7.3). Born in 1730 into a potter family, he was fortunate to benefit from many mentors interested in modern science. The modern scientific method evolved during the sixteenth and seventeenth centuries in Europe. This approach uses experiments to gain knowledge, a break from the previous doctrine based mostly on theoretical thoughts by Greek philosophers. While this approach was widely known among scientists in the eighteenth century, it was still uncommon in industry. Josiah Wedgwood was fortunate to have a scientifically skilled schoolteacher, to work as a journeyman for a master who fostered his sense of experimentation,

FIGURE 7.3
Engraving of Josiah Wedgwood by Samuel William Reynolds after portrait by Joshua Reynolds. (Image by Wellcome Images and licensed under the Creative Commons Attribution-ShareAlike 4.0 International license.)

and to become a good friend and partner to the highly educated Thomas Bentley (1730–1780). Through Bentley he befriended many of the who's-who group of modern science, including the discoverer of oxygen Joseph Priestley; the scientist and, later, U.S. statesman Benjamin Franklin*; philosopher Erasmus Darwin†; engineer and pioneer canal builder James Brindley; and contemporary industrialists Richard Arkwright (see Section 7.5) and Matthew Boulton (see Section 8.2) (Burton 1976).

Hence, Wedgwood was good at both sculpting and math and, in addition, had a keen sense for experimentation, all except the first highly unusual for a potter of the eighteenth century. Throughout the years, he kept track of his work in a notebook, writing down the exact ingredients and quantities for ceramics, the duration of firings, an estimate of the temperature, and the color and quality of the resulting products. Later, he invented a thermometer to measure the high temperature in the kiln and was thus able to generate more precise data. Not only did he write down all these details; he studied these data and even conducted experiments to improve his understanding of the relations of material, time, temperature, and quality of his pottery.

Conducting controlled experiments, recording the data, and making a model for predicting the outcome was nothing new. Scientists in the Middle Ages knew this scientific method well. Generating knowledge through controlled experiments was probably first used by Islamic scientists, followed by well-known scientists like Roger Bacon, Galileo Galilei, and Isaac Newton, to name just a few. By the eighteenth century, this method was well established in science. However, there was still a gap between science and technology. There was still a large difference between generating knowledge and the practical application of this knowledge. Wedgwood was probably among the first, if not the first, to use controlled experiments in manufacturing to gain knowledge and then to use this knowledge to improve his products.

While he did not yet understand the chemical reactions in the ceramics, he understood enough to predict the results. Through these experiments, he was able to perfect the production of his pottery, achieving consistent colors and reduced cracks in his ware. This was a critical advantage in the market, especially the high-end market. While the middle classes were happy to have any white ceramics at all, for royalty, it was an issue if their

---

* At a time when scientific reasoning and politics were still compatible.

† Through this friendship, Wedgwood's daughter married Darwin's son, making him also the grandfather of Charles Darwin, originator of the theory of evolution.

complete sets of ceramics contained a single piece with a different shade than the others.

Through his experiments together with his connections, the tastes of the time,* and his friend and partner Bentley's ability to sell, he was soon retailing to the aristocracy in England. Eventually, Wedgwood received the distinguishing title *Potter to Her Majesty* Queen Charlotte. Based on his skills and his reputation, he sold high-quality, highly customized *ornamental ware* to the aristocracy of Europe. However, he also had a second line of products of *useful ware* to the middle class, becoming very wealthy in the process despite the aristocracy's tendency to forgo payments.

Wedgwood ran numerous workshops, always expanding as his business grew. The two final and most important workshops were the *Bell Works*, established in 1762 in Burslem, England, focused on the useful ware, and *Etruria*, opened in 1769 and focusing on the ornamental ware. The name *Bell Works* comes from the installation of a bell to signal the start and end of a shift, while many other contemporary manufactories blew a horn (McKendrick 1961). The name *Etruria* stems from the mistaken belief that the Greek and Roman pottery discovered in Pompeii was of Etruscan origin.

Wedgwood strongly believed in both the division of labor and the separation of processes (McKendrick 1961) to improve productivity. He went as far as to provide different entrances to workers for different processes in the same building. His factory in Etruria was arranged according to a strict division of labor, with all processes located in the sequence of the process flow. The first process was the flint mill grinding the flint and the processing of the raw clay, located next to the canal delivering the raw materials. The next department was the mixing of flint and clay, followed by the throwing and modeling departments. The products were then turned on a lathe, itself also a novelty introduced by Wedgwood. Another process fabricated handles and plates. The products were then prepared for the kiln before being fired by the oven men. The glaze mixing and enameling were then fired again. Finally, the packing and shipping departments were close to the adjacent canal through which the finished goods were transported (Burton 1976, p. 88). Hence, not only was the process flow arranged in a strict logical sequence, but there was also a high division of labor. Wedgwood was aware of the use of division of labor in China through reports by Jesuit priests (Crowther 1972, p. 13). However, this division of labor was new in European pottery. Additionally, Wedgwood also paid

* To follow the quickly changing tastes of the time, both men also heavily consulted their wives.

attention to match the speed of the individual stations in order to avoid both excess inventory and idleness of the workers.

Another skill of Wedgwood was his knack for mathematics, which he used in optimizing his enterprise. While this was not a constant process, Wedgwood investigated his costs in detail during times of economic crisis. Through this cost accounting, he determined the costs of each process and planned his business accordingly. For example, when he initially hired girls due to a shortage of workers, he paid them less because the male workers did not want to work next to girls. Furthermore, he learned that the production cost of an apprentice was only one-third that of a journeyman and that the cost of a boy or girl was even less (McKendrick 1970). Young employees not only had a clear cost advantage, but they were also less likely to create trouble, resulting in a large share of underage workers. For example, in 1790, 25% of his workers were apprentices, mostly girls (McKendrick 1961). Following the threat of a strike in 1772, he analyzed his costs and understood the costs of idle workers and idle machinery. To maximize the use of his fixed assets, he tried to have his potteries run at full capacity whenever possible (Tames 1984, p. 25ff). Working hours covered nearly the entire period of useable daylight. He also tried to increase production by paying a pay-by-piece rate in 1770, but while this resulted in increased production, it also resulted in sloppy work and bad quality (Burton 1976, p. 96).*

Like other employers of his time, Wedgwood had continuous troubles with his employees. Even though he was considered a good employer, he was still thinking in a very hierarchical and authoritarian way. As a master, he was, in his view, superior to the workers, and he believed he should decide for them as a father decides for his children. He said that workers *have like passions with ourselves, & are capable of feeling pain or pleasure, nearly in the same manner as their masters* (emphasis added) (Burton 1976, p. 93). This *nearly* clearly emphasizes that workers for him were not quite the same as masters. His wages also changed very little between 1762 and 1790, while inflation reduced the living standard of the employees (Burton 1976, p. 223). When in 1772 a spokesman of the workers approached Wedgwood to try to negotiate higher wages, the spokesman was fired on the spot.

Unsurprisingly, workers were not overly motivated to contribute to the benefit of Wedgwood. Attendance was a common problem, and workers

* Even nowadays, piece-rate wages tend to have a negative impact on quality.

were absent for the week-long English wake holiday, Saint Monday, fairs, or drinking sprees. While Wedgwood was never able to resolve these absences, he was able to keep track of who was there and who was missing by introducing the first known version of clocking in. In his factory, workers were to drop a preprinted sheet of paper with their name in a box when they arrived and left. A clerk checked the box regularly and investigated irregularities (McKendrick 1961). While Wedgwood was unable to reduce absences, he was able to reduce the idleness of present workers by assigning each worker a fixed task. Out of 178 men and women employed in 1790, only five were designated as *odd men* with no specific tasks, whereas all others were clearly assigned to a certain process (McKendrick 1961). Idleness was further reduced by forbidding walking around between the different processes, and many departments had different entrances to keep the workers separate (Burton 1976). The division of work clearly helped to reduce idleness.

Another problem Wedgwood faced was how to oversee his workers. In 1768, he had 15 workers. This increased to 300 workers in two locations a few kilometers apart by 1790. Even without his frequent travels to London and other locations, it is impossible for one man to oversee so many workers (Burton 1976). Wedgwood installed foremen in all of his departments and provided detailed written work instructions. These not only described the work for all processes but also included details for different products and a final quality checklist (McKendrick 1961). Even measures to reduce health risks were addressed. The grinding of the flint, which could lead to silicosis when inhaled—known as potter rot—was prevented by wet grinding. Lead poisoning due to lead-based coatings was reduced by prohibiting food in the coating room, issuing special protective clothing, setting and enforcing cleaning standards, and using wet sponges instead of dry sponges. Thus, he turned *dilatory drunken, idle, worthless workmen* in 1765 into *a very good sett of hands* by 1775 (McKendrick 1961). However, he never was able to find suitable middle management to run the place when he was absent. His nephew was covering part of these top-level tasks but not to the satisfaction of Wedgwood (McKendrick 1961). The strain of running two workplaces and frequent travels to London at one point even endangered Wedgwood's eyesight, possibly as a result of what is nowadays known as *burnout*. Luckily he recovered, and over time, he was able to transfer the business to his skilled sons. Josiah Wedgwood passed away in January 1795. A private equity company in New York now owns his firm, called Josiah Wedgwood and Sons.

Overall, Josiah Wedgwood was one of the industrial driving forces of his time in England. He transformed his pottery from a small workshop to a large business and conducted scientific experiments to improve his wares. He established cost accounting to find saving potentials and used the first known clocking-in system to track worker attendance. Finally, he applied division of labor and process flow to optimize his production. While he was a good employer for his time, like most contemporaries, he was never able to motivate his workers, which led to constant trouble with absences, idleness, and theft.

7.4 JOHN LOMBE'S SILK MILL IN DERBY—MECHANIZATION AND INDUSTRIAL ESPIONAGE

In 1702,* Thomas Cotchett made the first attempt at mechanized spinning in England (Hills 1970, p. 29). Together with his engineer, George Sorocold, he built a water-powered silk mill in Derby, in the Derwent Valley using Dutch machines to spin the silk. Unfortunately for him, these machines were not well designed, and the resulting threads were of poor quality. Despite much fiddling and tinkering, they were unable to solve the problem and went out of business in 1712.

One employee of Thomas Cotchett was also involved in tinkering with the machines—John Lombe, from a Norfolk wool merchant family. He believed in the abilities of mechanized silk spinning but also knew from the failure of his employer that he did not have the expertise to make it work.

Financed by his half-brother Thomas, Lombe decided to learn more about silk processing. The center of silk spinning in Europe at that time was northern Italy. They offered the best quality and color and already used water-powered machinery since at least 1276, although on a smaller scale (Kuhn 1981). Still in England, Lombe prepared himself by learning Italian, mathematics, and technical drawing before traveling to Livorno, Italy, in 1714. He visited a number of silk plants as a tourist, without gaining much information, before he was able to find employment in a silk mill in Lucca through the help of a (probably bribed) Jesuit priest. Bribing two other workers in the factory enabled him to stay in the workshop at night

* Various sources put the date at different years between 1701 and 1707.

to study the machines used for spinning silk. During these *night shifts*, he also created a drawing of the machine (Knight 1835).

Lombe sent these drawings back to his brother in England, hidden in bales of silk. To protect Italian intellectual property, industrial espionage was punished by death. When the Italians discovered his industrial espionage. Lombe just barely escaped onto an English ship, which was then able to outrun an Italian warship. Unbeknown to When the Italians discovered his industrial espionage Lombe, he could also have avoided all the trouble and just read the widely known book *Novo Teatro di Machine et Edificii* by Vittorio Zonca, published in 1607 in Padua. This work contains detailed technical drawings of silk-processing technology and other machines. His drawings of the silk spinning machines are shown in Figure 7.4.* The book, reprinted in 1621 and 1656, was freely available in the Bodleian Library in Oxford among other places, although merely reading the book would probably not have given him the same understanding as seeing the machine in action (Cipolla 1972).

In any case, both Lombe and the knowledge of mechanized spinning arrived safely in England in 1715. There, Lombe set to work reconstructing and improving the Italian technology, patenting *his* invention in 1718. Again financed by his half-brother Thomas, Lombe also started to build a silk mill next to the former mill of Thomas Cotchett in Derby, in the Derwent Valley. This large five-floor factory included dozens of large spinning machines and employed between 200 and 400 workers. Figure 7.5 shows his plant. Former employees of Thomas Cotchett, including his chief engineer, George Sorocold, aided in this construction. However, John Lombe barely lived to see its completion in 1722. In 1718, a woman from northern Italy came to Derby, befriended John Lombe, and worked in the silk mill. This coincided with a decline in Lombe's health, with him eventually dying painfully from unknown causes on November 20, 1722, at the age of only 29 years. Rumors ran wild. The public opinion was that the woman had been sent from Italy to assassinate Lombe with slow poison in revenge for his espionage. Luckily for the woman, however, the English judicial system was working quite well. An investigation did not find enough evidence, and under the presumption of innocence, the woman was acquitted. Sensibly, she left for Italy before the British could change their minds (Knight 1843).

* In fact, the drawings in the book by Zonca were better than the drawings of John Lombe.

FIGURE 7.4
Vittorio Zonca's drawing of the spinning machine in his 1607 book *Novo Teatro di Machine et Edificii.*

Another brother, Henry Lombe,* took over the business. Yet the mill brought no luck to Henry either. Henry shot himself in the head in June 1723, less than a year after the death of John (Hitchcock 2011). The official cause for the suicide was depression, but this was even more fodder for rumors about Italian treachery. His half-brother Thomas took over the business, and the mill became fully operational in the same year. The mill

* Other sources name the fourth brother, William, as the next owner instead of Henry (Hutton 1957).

FIGURE 7.5
John Lombe's five-floor silk mill in Derby, Derwent Valley. The upper floors burnt down in a fire in 1910. (Photo by author, dated March 2010.)

was a financial success, making Thomas even richer than he already was and leading to his knighthood in 1727. The factory also became a popular tourist attraction and received many visitors interested in seeing the new technology.* For example, Daniel Defoe described it in his book *A Tour Thro' the Whole Island of Great Britain* as *a curiosity in trade worth observing, as being the only one of its kind in England* (Defoe 1727, Letter VIII). Over time, the factory changed ownership numerous times, burned down in 1910, and was rebuilt with only three floors instead of five. The building is now a museum on John Lombe and the industrial revolution.

While the silk mill made Thomas Lombe very wealthy, the work in the silk mill was harsh and often cruel. Hiring and firing was the common approach for adjusting to changes in demand and supply. There were

* The reports usually marveled at the factory but sometimes exaggerated. One quote often misattributed to Daniel Defoe lists 26,586 wheels, 97,746 movements, and 318,504,960 yards of silk per day, a precision that is near impossible to accomplish and most likely exaggerated (Warner 1921, p. 203).

few holidays, and the work hours were long, usually from 6:00 a.m. until 7:00 p.m. In times of peak demand, there was also a smaller night shift. A large part of the workforce were children, often less than six years old. Visitors complained about the heat, noise, and stink, and noted the poor and unhealthy appearance of the children. Beatings were common for even small mistakes. Overall, the first factory was a result of industrial espionage and a marvel in productivity, but to the workers, that time was *the most unhappy of [their] life* (Hutton 1957).

7.5 SIR RICHARD ARKWRIGHT AND THE CROMFORD COTTON MILL—FULL-SCALE MECHANIZATION

Mechanized silk spinning revolutionized the silk industry. Within a few decades, hand spinning was a thing of the past. Yet, then even more than now, silk was a luxury product, and the market was limited. Everyday, ordinary people wore clothes made of much cheaper cotton. Yet cotton was more complicated to process. Silk was already a naturally long fiber, whereas cotton was much shorter and not as easy to spin. After the invention of the flying shuttle in 1733 by John Kay, weaving was much faster than spinning, and demand for cotton yarn increased tremendously. While there were some early machines for spinning, these did not provide adequate quality and were unsuitable for mass manufacturing.

In the 1770s, three major spinning breakthroughs came in rapid succession. Richard Arkwright (Figure 7.6) and John Kay* are attributed with inventing the water frame, patented in 1769, which mechanized the stretching of the raw fibers. In 1770, James Hargreaves invented the spinning jenny,† which mechanized the twisting of the fibers. Finally, in 1780, Samuel Crompton invented the spinning mule, combining the two previous inventions and resulting in a roll of finished yarn (Paz et al. 2010, p. 142).

Using his spinning frame, Arkwright built a successful horse-powered cotton mill in Nottingham, but—probably inspired by the water-powered silk mills in Derwent—set his aim higher and planned a water-powered cotton mill. Looking for the right spot to build a factory, he eventually settled in

* John Kay, inventor of the flying shuttle, and John Kay, coinventor of the spinning frame, are often confused but are two separate persons.

† Yet another disputed invention, as the spinning jenny may have also been invented by Thomas Highs.

FIGURE 7.6
Portrait of Sir Richard Arkwright by Mather Brown in 1790.

Cromford, just 20 kilometers north of the silk mill in Derby. The site was, for his purposes, ideal. A still-working drainage channel from a Roman lead mine provided running water even in the coldest of winter and could be utilized without major construction. The village was far away from everything, including competition and spies. Another benefit was that the trading route passed through Cromford. With the backing of his sponsors, Jedediah Strutt, Samuel Need, and John Smalley, Arkwright signed the lease on a property, including the water rights, on August 1, 1771. He started construction immediately. To speed things up, he bought a house and had it demolished to gain building materials. Within a very fast four months, the Cromford Mill was completed, and workers were hired locally and in Derby. Figure 7.7 shows the factory. In 1772, production of cotton started with multiple water frames, the water-powered version of the spinning frame.

Business was very successful, and by 1776, Arkwright invested in a second mill adjacent downstream to the first mill. This mill was even larger, with seven floors of 325 square meters each. Due to Arkwright's

FIGURE 7.7
Courtyard of Sir Richard Arkwright's Cromford Cotton Mill. The right-hand building in the back was originally five floors high like the building on the left but was partially destroyed by fire in 1930. At the front right edge, you can see the wheel pit of the second mill with the water gushing in. The ruins to the left of the wheel pit (and an equally large ruin on the opposite side of the wheel pit not in the picture) were the second, even bigger mill with seven floors. However, these were completely destroyed by a fire in 1890. (Photo by author, dated March 2010.)

keen business sense, profits were enormous, and he built numerous other mills in the vicinity, including Birkacre Mill at Chorley, Haarlem Mill in Wirksworth, and Masson Mill in Matlock Bath. Arkwright also invested in a large warehouse to buffer the unreliable supply by packhorse. His partner, Jedediah Strutt, also invested in a successful mill of his own in Belper in 1776. Due to legal disputes about who exactly invented the spinning frame,* Arkwright lost his patents in 1785, and subsequently, a bonanza of cotton mills started in England.

The rapid expansion of mechanized cotton-spinning factories of course meant the rapid decline of hand-spun cotton, and subsequently, many spinners were out of work. Often, violent mobs tried to reverse this by smashing cotton mills, and Arkwright's Birkacre mill was burned down in

* Historians are still unsure about who invented what, with possible contenders being James Hargreaves, Thomas Highs, Richard Arkwright, and John Kay.

1779. In return, Arkwright armed Cromford Mill with a cannon and small arms, but the *mill breakers* never reached remote Cromford.

In terms of organizing his factory, Arkwright faced the same problems as most of his contemporary factory and manufactory owners. Most workers were more interested in their independence and their ale than their work (Seabright 2010, p. 210). Arkwright's mills worked on a 12-hour shift system (although the night shift was probably much smaller) six days per week, with Sunday being free. To make sure everybody knew the time, he installed a town clock. While Arkwright was often seen as difficult to deal with by his business partners (Burton 1976, p. 197ff), he was considered a fair employer for the standards of the time. Even though the working hours were as long as in the Derwent silk mill, salaries were comparably high. In order to motivate his workers, Arkwright avoided violence and instead used incentives for families to work for him. While Arkwright employed children from the age of seven, he did not employ single children from parish houses but, rather, aimed to have the whole family employed at his mills. Outstanding workers received special clothes as a reward (McKendrick 1961). He also built houses for his workers and a school for the children, since there would not have been enough dwellings nearby otherwise. Most unusually, in the factory, there was a toilet on every floor, and the buildings were heated in winter. While this was beneficial for the employees, there was also a good business reason for it. Nearby toilets allowed the worker to be absent less, and warmer temperatures were better for cotton processing.

As the Italians tried to protect their intellectual property from John Lombe, the British tried to protect theirs. Similar to Italy, the export of technology from Great Britain was punishable by death. Nevertheless, history repeats itself. A German businessman, Johann Gottfried Brügelmann, worked undercover in the Arkwright mills, studying the machine and stealing spare parts. In 1783, using this stolen technology, he started the first cotton mill outside England in Ratingen, near Düsseldorf, Germany. However, different from Lombe, he gave credit to the true inventor and named his mill Textilfabrik Cromford (German for *Cromford Textile Mill*).*

The technology also traveled to the United States, although this time, it was not a U.S. businessman spying on Arkwright but, rather, Samuel Slater, born in Belper in the Derwent Valley. Slater was employed in the

* Brügelmann never admitted espionage, but neither did he deny it, and the Brügelmann family lore has it that he spied in the Arkwright mills (Gülcher and Gülcher n.d.). Naming his mill after Cromford is, in this sense, a nice way to give credit.

mills of Arkwright's partner Jedediah Strutt. There he learned the intricacies of the water frame and other spinning machineries. Breaking the English law against the emigration of skilled experts, he moved to America in 1789. Offering his services in 1790 to entrepreneurs William Almy and Smith Brown, Slater implemented English technology in their cotton mill in Pawtucket, Rhode Island, establishing the first cotton mill in America (Tucker 1981). Hence, Slater is known as the *father of the American industrial revolution* in the United States but as *Slater the Traitor* in the United Kingdom. I find it most curious how the technology traveled through industrial espionage from Italy to England, Germany, and finally, the United States.*

Overall, Arkwright was an extremely successful businessman, and eventually, he was knighted in 1786. After his death in 1792, his son Richard Arkwright Jr. became the richest man in Europe. The lead mine supplying Cromford with water, however, was deepened; the draining channels were changed; and due to lack of water Cromford Mill never again ran at full capacity. Upgrading the mill with new steam power was not economic, and the mill was eventually closed in 1850. Since then, the buildings have been used as a brewery, a laundry, a warehouse, and a paint shop. A large fire burned the seven-floor second mill to the ground in 1890, and another fire damaged the first mill in 1930, reducing it from five to three floors. This United Nations Educational, Scientific, and Cultural Organization (UNESCO) World Heritage Site is now used to house a small software company, a stop-smoking center, and a small gift shop.

7.6 THE MONTGOLFIER PAPER MILL—PIONEER IN HUMAN RELATIONS

Another fine example of an early factory is the paper mills of the Montgolfier family in France.† The family was producing paper since at least the seventeenth century. They were open to innovation, installing

* It is easy to see analogies. Italian and British industrialists back then were probably as upset about the theft of intellectual property as modern industrialists are about technology theft. Similarly, modern China, on the receiving end of many modern-day transfers, is probably as snug as a bug about this gain as the United States or Germany was back then. In all cases, countermeasures were, at best, only able to slow down the knowledge transfers.

† Two brothers of the Montgolfier family later invented the first hot air balloon, publicly demonstrating its ability to fly in 1783.

numerous machines and labor-saving technologies. They even secured a large government grant to copy superior Dutch machinery in their mills (Rosenband 2000, p. 51). However, the idea of maintenance was still in its very infancy, and—similar to other papermakers of the time—most machines ran until they broke. Repairs were usually nothing more than a quick fix, and the next breakdown was just a matter of time (Reynard 1999). Yet the Montgolfier paper mills are remarkable not for their technical progress but, rather, for their organizational change and their transformation of labor relations.

Up until 1781, the Montgolfier paper mills produced paper in the traditional way of the guilds, based on the familiar system of masters being in charge of a small team of journeymen and apprentices. Traveling journeymen were paid by the piece but otherwise enjoyed a lot of freedom. Maybe it was because papermaking was a very fragile and difficult process, or because craftsmen in this trade considered themselves better than other craftsmen. In any case, they were fiercely independent, probably even more so than other craftsmen of the time (Rosenband 1993, pp. 225–230).

The workers, not only in France but all over Europe, saw themselves more as subcontractors than as employees. They worked when they liked and on what they liked, fiercely protecting their secret knowledge. They also resisted innovation and change. They more or less took over the paper mill where they worked, ignoring the orders of the mill owner. Wages were set more by the journeymen than by the master. In numerous cases, wage disputes led to a strike, where the mill master usually lost out after a few months. The masters then even had to pay the missed-out salary to the striking journeymen before they would work at the mill again (Rosenband 2000, p. 52ff). Traveling craftsmen were generally free to walk in and out of any paper mill. In case they decided to join the mill, all work stopped to have an extensive welcome party for the newcomer. These frequent interruptions and changes made it impossible for the mill masters to plan work or to achieve consistent output and quality. Any paper mill master who tried to bring his workers into line risked being beaten up, having his business burned down, or at the very least, having his workers go on strike (Rosenband 1993, pp. 225–230).

In effect, the traditional system was turned upside down. The master was subject to his workers, the workers made the rules, and the master had very little to say in his own business. Even to outside observers, this disparity was obvious. *The masters are like slaves of the journeymen and workers,* and *the journeymen paper workers form a sort of little republican state in the midst of the monarchy* (Rosenband 1999). Naturally, the masters tried to fight back, establishing a system of credentials where every journeyman had to carry his credentials from his last master in order to show proof of his good behavior when picking up the next job. This would allow the masters to refuse employment to potential troublemakers. Equally naturally, the response by the journeymen was to forge and counterfeit these credentials to their liking.

While the paper mill masters suffered the abuse of the journeymen, there was also significant abuse of the junior workers by the senior journeymen. An apprentice was obliged to pay a substantial apprenticeship fee to the journeymen upon entering the craft. Similarly, payments to the elder journeymen were constantly required for the *honor* to learn from them—regardless of whether the senior journeyman actually spent time teaching the youngster or not. When the youngsters made a career step, for example, moving to a more prestigious and better-paid workplace in the mill, the journeymen required money for their approval. When the junior worker entered a new mill, he had to buy drinks for his colleagues. Likewise, any marriage or birth of a child resulted in free drinks paid for by the newlywed or the father (Rosenband 2000, pp. 54–55). In summary, the junior workers were also disadvantaged from the rules of the established system, which, in some aspects, looked more like the mafia than a workers' organization.

The Montgolfier family suffered from the same problems. In 1781, the conflict escalated. First in their Rives mill and then in their large Vidalonle-Haut mill, the Montgolfiers supported an apprentice who refused to pay his dues to the journeymen. Subsequently, most workers went on strike. The Montgolfier family responded by dismissing all workers. However, more than merely dismissing the workers, they dismissed the traditional ways of the journeyman in its entirety. Rather than rehiring another set of papermaking troublemakers, they decided to look for new sources of labor in the form of untrained peasants and local craftsmen of other industries. Their aim was to have a stable and reliable workforce. By January 1782, this clear cut was completed, and they had a new workforce of five trusted veterans and 20 young workers unconnected to the journeymen's association (Rosenband 2000). While there were still conflicts, these were conflicts with an individual worker and not with an entire organized workforce, and hence, they were much easier for the Montgolfier family to resolve (Rosenband 1993, p. 233ff).

This newly gained freedom allowed the Montgolfiers to run the mill as they wished, with little opposition from the workforce. They established a new system aimed at consistent output and quality. One major change was to set a framework of rules for hiring and firing, payments, bonuses, promotion, and discipline. They set up fixed 13-hour workdays from 4:00 a.m. until 7:15 p.m., seasonally including up to almost seven hours of *candlelight work*. Their work year consisted of 300 workdays, much more than in the old system, with about 200 days (Rosenband 1985). Most importantly, workers were required to follow the orders of the foremen and do only the work they were instructed to do, and were not allowed to do other work for their own profit. While this sounds obvious to us in our modern factories, it was groundbreaking at the time. Another change that appears counterintuitive at first was to base pay on the time worked instead of on the number of pieces produced. Payment by piece is supposed to give an incentive to the worker to produce larger quantities, but the downside is often poor quality. The Montgolfier mills also included bonuses for exceeding production performance and deductions for failure to meet quotations, but their focus was much more on stability than on output (Rosenband 1985).

While the break with the traditional system, at first glance, looks like the workers losing to the masters, this is not entirely the case. Working at the Montgolfier mills was actually very popular. Payment was good, as were the fringe benefits such as food and housing. In addition, there were no elder journeymen freeloading by tradition on the younger journeymen as in the old system, probably to the great relief of the younger workforce. The Montgolfier mill masters also provided cheap housing and food and even gave cash advances to workers in need, such as new workers without working clothes or in the case of a childbirth or death (Rosenband 2000, p. 88). Overall, the duration of employment greatly increased. Most workers chose to stay at the mill rather than travel between mills as in the old system (Rosenband 1993, p. 233ff). To maintain a stable and docile workforce, the Montgolfier family also paid great attention to the hiring process. In many cases, they started with a background check and examined the new worker carefully, including his/her appearance. If the worker appeared to be suitable, they usually hired the person for a probation period of a few months before allowing him/her to become part of the regular workforce (Rosenband 2000, pp. 75–77).

Wages were paid both by piece and for time worked, also unusual for the industry of that time. In the piece rate, the focus was more on consistency than on spectacular performance. For example, if the worker produced above his/her quota, he/she got a small bonus, but if he/she performed below his/her quota, the penalty was about four times the bonus for the same quantity of paper. The Montgolfiers also realized that motivation does not stem from more money but from the possibility of more money. Hence, reliable and skilled workers, over time, received higher wages than new, unskilled, or lazy workers did. If more than one member of a family worked for them, the salary was paid to the head of the household. This was usually the husband, but in the case of some workers who loved wine and tobacco too much, the Montgolfiers wisely paid it to the wives (Rosenband 2000, p. 95ff).

The established workforce that showed up on time and followed orders also generated stability in production. The Montgolfier mills had, on average, an astounding 250 days of regular work per year, significantly more than other mills. Now it was possible to establish a regular maintenance schedule, including daily cleaning of the tools. About 35 days per year were set aside for the annual maintenance and rework of the machines (Rosenband 1985). Overall, the Montgolfier mills were now able to consistently produce over 5000 sheets of paper daily. Creating high-quality paper with no wrinkles, an even texture, and a smooth white color still remained a daunting task. Overall, their goal was to produce 75% good quality, expecting 25% of the products to be unsatisfactory. Since unsatisfactory paper was still sold, the Montgolfier mills offered six different quality types between fine paper and wrapping paper. Combined with different formats, these led to about 90 different products sold (Reynard 1998).

Overall, the Montgolfier mills were probably the first manufacturing operations with labor relations similar to our modern factories. Workers showed up (mostly) on time and did (mostly) what their supervisors told them to do. While normal to us, this was groundbreaking for the time, and it took a long time for the industry to change. In fact, it probably was the exception rather than the norm until the Industrial Revolution came into full swing.

Most other paper mills in France still suffered from abuse by the journeymen. As a side effect, even some technical innovations were created not to cut costs but to reduce dependence on troublesome labor. For example, the continuous paper machine invented by Louis-Nicolas Robert in 1799 was created not to cut cost. Rather, it was built because he had grown *impatient with the irascibility and ill temper of the workers* and aimed at *fabricating paper without their aid* (Rosenband 1999, 2000, p. 148). This papermaking machine increased output in the mills enormously and, over time, ended the dominance of the papermaking journeymen. The Montgolfier mills are now part of the French papermaker Canson, which still produces high-quality paper.

BIBLIOGRAPHY

Anderson, R., 1973. The Winlaton Story: Part 8. Sir Ambrose Crowley. *The Bellman* (Blaydon Urban District Council newsletter).

Burton, A., 1976. *Josiah Wedgwood*, 1st ed. Andre Deutsch Ltd., London.

Cipolla, C.M., 1972. The Diffusion of Innovations in Early Modern Europe. *Comparative Studies in Society and History* 14, 46–52.

Crouzet, F., 2008. *The First Industrialists: The Problem of Origins*, 1 Reissue. Cambridge University Press, Cambridge, UK.

Crowther, J.G., 1972. *Josiah Wedgwood*. Taylor & Francis, Boca Raton, FL.

Davis, R.C., 1993. Arsenal and Arsenalotti: Workplace and Community in Seventeenth Century Venice, in: Safley, T.M., Rosenband, L.N. (eds.), *The Workplace Before the Factory: Artisans and Proletarians, 1500–1800*. Cornell University Press, Ithaca, New York.

Davis, R.C., 1997. Venetian Shipbuilders and the Fountain of Wine. *Past & Present* 156, 55–87.

Davis, R.C., 2007. *Shipbuilders of the Venetian Arsenal: Workers and Workplace in the Preindustrial City*. The Johns Hopkins University Press, Baltimore.

Defoe, D., 1727. *A Tour through the Whole Island of Great Britain*, Volume 3. JM Dent and Co, London.

Dobb, M.H., 1924. The Entrepreneur Myth. *Economica* 66–81. doi:10.2307/2547873.

Flinn, M.W., 1960. Sir Ambrose Crowley and the South Sea Scheme of 1711. *The Journal of Economic History* 20, 51–66.

Flinn, M.W., 1962. *Men of Iron: The Crowleys in the Early Iron Industry*. Edinburgh University Press, Edinburgh, UK.

Gülcher, M., Gülcher, J., n.d. Johann Gottfried Brügelmann. Available at http://www.guelcher-chronik.de/Stichworter/Johann_Gottfried_Brugelmann/johann_gottfried_brugelmann.html (accessed March 8, 2011).

Hills, R.L., 1970. *Power in the Industrial Revolution*. Manchester University Press ND, Manchester, UK.

Hitchcock, S., 2011. Thomas, John and Henry Lombe. Derwent Valley Mills. Available at http://www.derwentvalleymills.org/history/key-figures/108-thomas-john-and-henry-lombe (accessed March 8, 2011).

Hutton, W., 1957. The History of Derby. In *English Historical Documents*, edited by D. B. Horn and Mary Ransome. Eyre and Spottiswoode, Oxford University Press, Oxford, Vol. X, 1714–1783, 458–61.

Jones, S.R.H., 1987. Technology, Transaction Costs, and the Transition to Factory Production in the British Silk Industry, 1700–1870. *The Journal of Economic History* 47, 71–96.

Knight, C. (ed.), 1835. Mr. John Lombe and the Silk-throwing Machinery at Derby. *The Penny Magazine of the Society for the Diffusion of Useful Knowledge* 4, 115–117.

Knight, C. (ed.), 1843. A Day at the Derby Silk Mill. *The Penny Magazine of the Society for the Diffusion of Useful Knowledge* 12, 161–168.

Kuhn, D., 1981. Silk Technology in the Sung Period (960–1278 A.D.). *T'oung Pao, Second Series* 67, 48–90.

Lane, F.C., 1973. *Venice, A Maritime Republic*, illustrated ed. The Johns Hopkins University Press, Baltimore.

McKendrick, N., 1961. Josiah Wedgwood and Factory Discipline. *The Historical Journal* 4, 30–55.

McKendrick, N., 1970. Josiah Wedgwood and Cost Accounting in the Industrial Revolution. *The Economic History Review, New Series* 23, 45–67. doi:10.2307/2594563.

Paz, E.B., Ceccarelli, M., Otero, J.E., Sanz, J.L.M., 2010. *A Brief Illustrated History of Machines and Mechanisms*, 1st ed. Springer, Dordrecht, The Netherlands.

Reynard, P.C., 1998. Manufacturing Strategies in the Eighteenth Century: Subcontracting for Growth among Papermakers in the Auvergne. *The Journal of Economic History* 58, 155–182.

Reynard, P.C., 1999. Unreliable Mills: Maintenance Practices in Early Modern Papermaking. *Technology and Culture* 40, 237–262.

Rosenband, L.N., 1985. Productivity and Labor Discipline in the Montgolfier Paper Mill, 1780 1805. *The Journal of Economic History* 45, 435–443.

Rosenband, L.N., 1993. Hiring and Firing at the Montgolfier Paper Mill, in: Safley, T.M., Rosenband, L.N. (eds.), *The Workplace Before the Factory: Artisans and Proletarians, 1500–1800*. Cornell University Press, Itacha, New York.

Rosenband, L.N., 1999. Social Capital in the Early Industrial Revolution. *The Journal of Interdisciplinary History* 29, 435–457.

Rosenband, L.N., 2000. *Papermaking in Eighteenth-Century France: Management, Labor, and Revolution at the Montgolfier Mill, 1761–1805*. JHU Press. Johns Hopkins University Press, Baltimore; London.

Safley, T.M., Rosenband, L.N. (eds.), 1993. *The Workplace Before the Factory: Artisans and Proletarians, 1500–1800*, 1st ed. Cornell University Press, Ithaca, New York.

Seabright, P., 2010. *The Company of Strangers: A Natural History of Economic Life*, Revised. Princeton University Press, Princeton; Oxford, UK.

Skempton, A.W., Chrimes, M.M., Cox, R.C., Cross-Rudkin, P.S.M., Rennison, R.W., Ruddock, E.C. (eds.), 2002. *A biographical dictionary of civil engineers in Great Britain and Ireland: 1500–1830*. Thomas Telford, London.

Tames, R., 1984. *Josiah Wedgwood: An Illustrated Life of Josiah Wedgwood 1730–1795*. Osprey Publishing, Oxford, UK.

Thompson, E.P., 1967. Time, Work-Discipline, and Industrial Capitalism. *Past & Present* 38, 56–97.

Tucker, B.M., 1981. The Merchant, the Manufacturer, and the Factory Manager: The Case of Samuel Slater. *The Business History Review* 55, 297–313. doi:10.2307/3114126.

Warner, F., Sir, 1921. *The Silk Industry of the United Kingdom. Its Origin and Development*. Drane's, London.

8

Fire Is Stronger than Blood and Water—Steam Power

I sell here what all men desire—Power!

Matthew Boulton (1728–1809)
Industrialist, manufacturer of steam engines

More than any other invention, more than any other device, the industrial revolution is known for the steam engine. While all of the previous developments like the manufactory or the first factories are relevant to the industrial revolution, probably the most significant change is the development of the steam engine. The ancient Greeks were the first to mention steam-powered movement, but these were gadgets with no practical value. Only during the late seventeenth century did steam power become available. This availability of unprecedented power at almost any location changed manufacturing and the world more and faster than any other preceding development in the history of mankind.

8.1 THE DEVELOPMENT OF THE STEAM ENGINE

Thomas Savery developed the first steam engine—or more correctly, steam pump—and patented it in 1698. The technology was still in its infant stages. There were no pistons but merely a set of valves to let steam in the chamber and water in and out of it. Hot steam was filled into a chamber and then cooled down inside it. The condensing steam created a vacuum, sucking up water through an attached pipe. This sucked-up water was then pushed upward with the inflowing steam of the next cycle. Since this engine did not create any mechanical movement but merely pumped water, it was not useable for manufacturing purposes. Additionally, due to the constant heating and cooling of the cylinder and the steam, lots of energy was wasted, and the engine worked very inefficiently. The only use of the engine was to pump up water, for which it was used with limited success to drain water out of mines. Hence, it was also advertised as a *miner's friend*. However, there were numerous problems with the engine. The engine was very inefficient, requiring expensive fuel, and in need of constant maintenance. The engine was also only able to pump water from 9 meters below to ca. 10 meters above the engine. Hence, in order to drain a mine, the engines had to be installed inside of the mine, creating lots of additional problems like providing fresh air and ventilation of the exhaust. Finally, the engine also had a tendency to explode due to weak joints, and therefore, the miners may have been less than excited about their new *miner's friend* deep down in the mine shafts. Overall, only 3% of all steam engines installed in Great Britain during the eighteenth century were Savery steam pumps (Kanefsky and Robey 1980).

A significant step forward was by Thomas Newcomen, who, in 1712, built the first actual steam engine that converted steam into mechanical movement. The condensing steam no longer sucked up water but, rather, sucked in a piston. While the constant heating and cooling of the cylinder still made the engine very inefficient, it was now possible to install the engine on the surface and still bring the movement through shafts and chains to the pumps inside of the mine. The engine was able to raise water over more than 30 meters. The engine worked with less steam pressure than the Savery engine and hence was unlikely to explode, surely a comforting fact to the operators of the engine. Subsequently, the Newcomen engines were very successful, and more than two-thirds of all engines installed in Great Britain during the eighteenth century were Newcomen steam engines. Hence, Newcomen steam engines were the first successful and widely used steam engine (Kanefsky and Robey 1980).

A model of the Newcomen steam engine was brought for repair to the instrument maker James Watt (Figure 8.1) in 1763. The model made only a few turns before stopping. Watt realized that the small size caused an enormous loss of energy. Hence, he started to investigate the principles of the steam engine in his spare time, realizing in 1765 that the efficiency would improve significantly if the steam condensed not inside of the cylinder but externally in a condenser. After additional research on how to

FIGURE 8.1
James Watt, by Henry Howard, ca. 1797.

create a better seal inside of the cylinder, Watt developed an improved and more efficient engine in 1775 (Scherer 1965).* The Watt engine (Figure 8.2) used only about one-quarter of the fuel of a Newcomen engine for the same work, making the former much more preferable over the latter.†

While the major change during the industrial revolution was steam power, most mechanization in manufacturing was still powered by water. Steam power was expensive in installation, required a constant supply

---

* Stating this in such few sentences makes it look obvious, but this does not give justice to Watt, who labored endlessly, calculating and researching uncountable details on the thermodynamics of steam and the principles of the steam engine, and trying out numerous technical approaches and solutions before achieving his breakthrough. Like with most inventors, a great insight is usually preceded by a lot of hard work.

† There were also a large number of other tinkerers working on steam engines like Jonathan Hornblower, Francis Thompson, Richard Trevithick, and many more. While they all contributed to the body of knowledge about steam engines, they were of minor commercial significance.

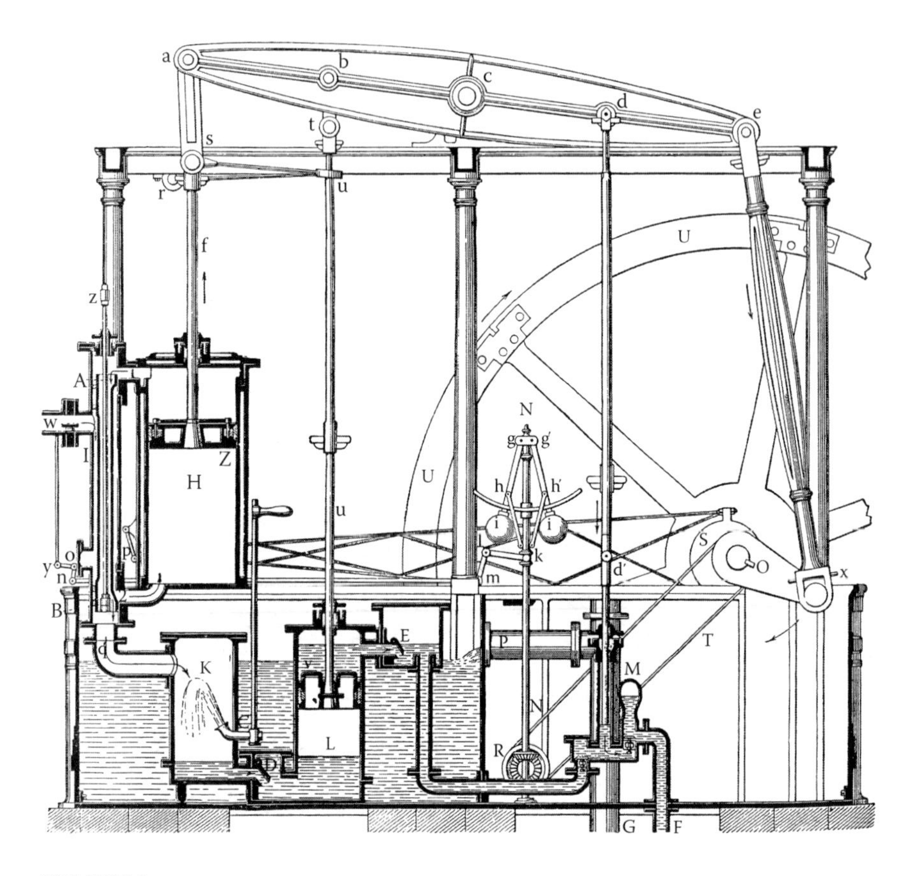

FIGURE 8.2
Steam engine designed by James Watt. (Image from Meyers Konversations-Lexikon 1885–1890.)

of also expensive fuel, needed highly trained experts to run the engine, and nevertheless frequently broke down, even if the risk of explosion was (mostly) under control. The advantage of the steam engine was that it could be installed almost anywhere, whereas waterpower required a constant supply of water. Up to 1780, almost all steam engines were used to pump water out of mines. Since the location of the mine was often not near a readily available water source, steam power was expensive but often the only viable option to pump water out of the mines. Additionally, since the steam engines needed lots of fuel to run, pretty much the only place to run them economically was next to a coal mine. Other locations incurred significant transport cost on coal, making the use of Newcomen steam engines too expensive compared to the use of horses or water. If the mine was near a suitable water source, however, waterpower was usually preferred, even with the risk of flooding the mine. Waterwheels were even installed inside of mines, for example, the Schwazer silver mine in Austria.

Yet, due to its expense, most manufacturing until 1870 was powered by water (Cameron and Neal 2002, p. 195ff), and waterwheels were commonly used until the mid-nineteenth century (Berg 1985, p. 186). In some cases, a steam engine was even installed with the sole purpose to pump water to drive a waterwheel, which provided much smoother motion than any contemporary steam engine. However, the Watt steam engine was initially able to convert only 5% of the heat in movement. Additionally, the waterwheel had an efficiency of less than 50%. Hence, the overall efficiency of the coupled system was probably no more than 2%.

For manufacturing purposes, steam engines were insignificant before 1780. Besides their higher fixed and variable costs and lower reliability, one additional limitation was that they provided only a back-and-forth motion, whereas waterwheels provided a much more useful rotary motion. Only in 1779 did James Pickard modify a steam engine to provide a rotary motion, and from around 1785 onward most steam engines were rotary engines. While many of the rotary engines were also used in mining, from 1780 onward, more and more steam engines were used to drive manufacturing-related machinery. Steam engines were used in textile mills, to turn machinery, to pump air in a blast furnace, to grind corn, and to roll metal. Probably one of the earliest large steam-powered factories was the Albion Flour Mills in London, planned to have 30 millstones and three steam engines. In comparison, the next largest mill in London had four millstones. However, local millwrights were strictly against the new competition, and the mill burned down in 1791 after only a few months of operation (Rosen 2010, p. 187). Overall, about 450 steam engines or about 20% of all steam engines in eighteenth-century Britain were used in manufacturing. Waterpower did generate roughly the same power as steam engines of about 15 to 30 horsepower per installation. Some larger installations both in water and in steam exceeded 100 horsepower (Kanefsky and Robey 1980). Steam engines were mainly used in locations where waterpower was unavailable or unreliable or by early adopters who were interested in new technology despite its risks.

As the technology improved, efficiency and power output went up, and the price went down. More and more steam engines were installed. By 1850, there were about 10,000 steam engines in Great Britain, with efficiencies improving to 15% and an average power output of around 50 horsepower. The number of steam engines was still minor compared to the up to 30,000 waterwheels for the same time and region (Müller, n.d.). The rest of Europe probably had another 10,000 steam engines installed. Waterpower also improved, and around 1860, large wheels provided up to 250 horsepower, but the largest engines quadrupled that with up to 1000 horsepower. Around 1870, steam engines became more prominent than waterpower, and around 1900, the largest steam engine provided 5000 horsepower. Finally, after the invention of the steam turbine by Charles Algernon Parsons in 1884, steam power was able to provide up to 130,000 horsepower (Cameron and Neal 2002, p. 195ff). A single modern steam turbine can provide up to 2,000,000 horsepower, probably more than all medieval waterwheels combined. In fact, most of our electricity is generated with the help of steam turbines, the great-great-grandchild of the Savery, Newcomen, and Watt steam engines.

8.2 THE FIRST ENGINEERING WORKSHOPS—MATTHEW BOULTON AND THE SOHO MANUFACTORY

The development of the steam engine required entrepreneurs and mechanics available to build and repair steam engines, creating a new industry of engineering workshops. One of the first and most well-known of these workshops is the Soho manufactory by Matthew Boulton (Figure 8.3). Born in Birmingham in 1728, Matthew Boulton eventually took over his father's business in making small metal products like shoe buckles,* belt buckles, buttons, hinges, hooks, and other metal trinkets, known at the time as *toy trade.*† As an entrepreneur, Boulton was a whirlwind of activity. Throughout his life, he was active in very different industries, from toy manufacturing to silver mining, from canal building to coin minting. He also delivered metal inserts for Wedgwood and his pottery business (see Chapter 7.3). He even considered entering the pottery business, and Wedgwood had to do considerable arm twisting and even had to threaten his friend Boulton with entering the metal business himself in order to stop Boulton from competing with him in the pottery business.

* Shoe buckles became popular in the mid-seventeenth century but fell out of fashion rapidly at the end of the eighteenth century in favor of shoestrings. This fashion change caused economic ruin for lots of small manufacturers that specialized in shoe buckles.

† Over time, the word *toy* changed in meaning from *small metal parts* to the now familiar *toys for children.*

FIGURE 8.3
Matthew Boulton. (Portrait by Lemuel Francis Abbott, created between 1801 and 1803.)

Boulton first operated from a small factory near his house in Snow Hill. While previously sourcing most of his parts from other artisans, in 1765, he merged all production steps in a new manufactory, naming it the Soho manufactory. Different from many other manufacturing operations of its time, the Soho manufactory, from the beginning, employed full-time workers rather than wandering journeymen. The Soho manufactory was based on individual workshops located in production sequence. Some parts of the Soho manufactory used a high division of labor, while others still utilized the expert artisan in charge of all steps of production. In any case, pretty much all workshops used machines for significant mechanization (Roll 1968, p. 9).

As it turned out, the Soho did not have enough waterpower to power all his machinery. Using horses to drive the machines turned out to be expensive. Hence, he contacted James Watt about the possibility of having a steam engine installed to pump water to his waterwheels or to drive the machinery directly (Scherer 1965). Soon, the never-resting Boulton became interested in manufacturing steam engines. When the financier of James Watt and co-owner of the steam engine condenser patent, John Roebuck, had financial difficulties, Boulton took over Roebuck's share of the patent. Thus, in 1775, started the Boulton and Watt partnership.

Boulton and Watt could not have been more different. Boulton was a whirlwind of business activities, enterprises, negotiation, and risk taking. Watt was the risk-averse tinkerer without any social skills or leadership qualities. He was completely aware of that, and in one of his letters, he clearly stated, *I would rather face a loaded cannon than settle an account or make a bargain. In short I find myself out of my sphere when I have anything to do with mankind.* Watt preferred to tinker alone at home, trying out different approaches to technical problems. He preferred to do things himself, not trusting the abilities of his workers, and unsurprisingly, he was therefore constantly overloaded with work (Scherer 1965). On the rare occasions when he went to Soho, he quickly got angry with the workers. He frequently fired them on the spot—only for Boulton to keep the usually valuable, skilled, and hard-to-replace worker and move the worker out of Watt's sight for a few days until things have calmed down (Pollard 1963).

After Watt finished his invention, he proposed *to cease attempting to invent new things, or to attempt anything which is attended with any risk of not succeeding, or of creating trouble in the execution. Let us go on executing the things we understand, and leave the rest to younger men, who have neither money nor character to lose.* When their partnership ended in 1800, Watt retired at age 64, while Boulton, at age 70, started a new coin minting business (Scherer 1965). However, while they were so completely different characters, they complemented each other very well. While Watt was the technical brain, Boulton was the economic powerhouse.

The development of steam engines proved to be more difficult than initially thought. Especially the sealing of the cylinders was difficult, due to the insufficient quality of the drilling and boring operations of its time. Watt spent significant time trying out different solutions, using tin, copper, wood, and cast-iron cylinders and pistons; experimenting with square, round, and inflatable flexible pistons; using piston disks made from leather, pasteboard cloth, cork, oakum, hemp, asbestos, lead–tin alloys, and copper–lead alloys; and aiding the seal with mercury, oil, graphite, tallow, horse dung, vegetable oil, and other materials. Only after the development of improved boring technologies in 1774 by cast-iron pioneer John

Iron-Mad Wilkinson from nearby Bradley did Boulton and Watt have cylinders with acceptable quality (Scherer 1965).

However, while the Soho manufactory established an engine manufactory division, it never actually produced complete engines but continued to produce *toys* in the form of valves and nozzles for steam engines. The key contribution was to provide blueprints, plans, and drawings and to supervise the construction of the steam engine at the customer site. However, for key parts, for example, the piston, the boiler, and the shafts, the customer was free to use any supplier of his/her choice. Only for the cylinders did they heavily recommend Wilkinson due to its superior quality (Roll 1968, p. 24).

The firm made most of their money from royalties and licensing fees, selling the actual parts at a loss. In fact, the partnership was more of an engineering consultancy than a manufactory. Additionally, most of the many businesses of Boulton were losing money, and the firm was on the constant verge of bankruptcy (Roll 1968, p. 98), losing money every year between 1787 and 1792 (Roll 1968, p. 121). Only after they started to enforce their patents were they able to dominate the industry, creating a near monopoly and putting penalties on manufacturers that used a condenser without their permission (Musson and Robinson 1959). This, of course, also created financial benefits, and Watt and Boulton were finally able to afford the money necessary for a foundry, establishing the Soho foundry in 1796 about a mile from the Soho manufactory.

However, neither Boulton nor Watt was actively involved in the Soho foundry. Watt, who was never active in daily management, retired after the partnership contract ended in 1800 but still tinkered with different inventions until his death in 1819. Boulton focused all his attention on his latest enterprise of coin minting, a business that kept him occupied until his death in 1809. The credit for turning the Soho foundry into a factory far ahead of its time goes to their sons, Matthew Robinson Boulton, James Watt Jr., and Gregory Watt. While the fathers were the builders and tinkerers, the sons were the organizers (Roll 1968, p. 270). Boulton Sr. never spent considerable effort in planning their Soho manufactory nor in improving its organization. His sons, however, created detailed plans for the Soho foundry even before the first stone was set. Based on extensive data mining and statistical analysis of the business up to 1795, they carefully planned the capacity of the Soho foundry to match market demand. This included planned speeds for the machines, completely unusual for its time, where the machinist was normally free to set the speeds as he deemed fit. The plan even included targets for improved speeds for the future (Roll 1968, p. 171).

Material flow was also subject to detailed analysis, with each workshop location carefully decided on based on where the material flows in and where the material flows out. Even information flow was taken under consideration, investigating the information needs of different workshops. Within the workshops, the processes were broken down into small steps, creating a high division of labor. Each worker was responsible only for a small step in the operation, leading to improved quality. Of course, over a number of days, a worker was not only working at one process but covered different processes, but he no longer was doing all processes in sequence (Roll 1968, p. 179). The products themselves were grouped in 14 standard products, easing production, although still without interchangeable parts, i.e., almost all parts had to be fitted in order to be assembled together (Roll 1968, p. 238).

While the quality of the small parts improved, the firm had little experience with casting large metal parts. Subsequently, the quality at the Soho foundry was also—at least initially—inferior to that of their competition. For example, the *Round Foundry* built by the partnership of Fenton, Murray, and Jackson in Leeds provided much better-quality steam engines than Boulton and Watt (Dickinson 2010, p. 171). However, some industrial espionage by Boulton, Watt, and Watt at the *Round Foundry* revealed crucial details on sand casting methods, greatly improving the quality of their own cylinders, pistons, and rods (Roll 1968, p. 166). Additionally, in a shrewd business move Boulton purchased all the land around the *Round Foundry*, prohibiting his competition from expansion and hindering their business (Rosen 2010, p. 207).

The Soho foundry also excelled at cost accounting, and the amount of data gathered on the production processes was enormous. For example, the record keeping for oil and candles was grouped in no less than 23 different consumables (Roll 1968, p. 249). Time studies were undertaken to estimate production ability, and fluctuations in the production times of parts were uncovered. Based on these data, they set performance-based wages for selected operations (Roll 1968, p. 269).

While the company still had workforce problems similar to its contemporaries, for example, heavy drinking and lack of skilled workers, the work at the Soho foundry was good. Besides comparatively high wages, they also had disability and life insurance (established already by Boulton Sr.). The company was also able to motivate its employees toward better quality through benchmarking with the competition (Roll 1968, p. 169).

Overall, the Soho foundry was very well organized and analyzed. Their understanding of the organization of the processes was exceptional. They were clearly anticipating the industrial research and development laboratories of the twentieth century (Scherer 1965). Due to this excellent management far ahead of its time, they were able to generate a profit for every year from 1798 onward (Roll 1968, p. 255). This was especially surprising since their steam engine patent expired in 1800, turning a comfortable near monopoly almost overnight into intense competition.

After the death of the last of the sons, James Watt Jr., in 1848 the firm changed hands, expanding into different product areas as the demand for steam engines decreased. Currently, the site houses the headquarters of Avery Weigh-Tronix, specializing in the manufacture of scales.

8.3 POWER GOES MOBILE—STEAM POWERED LOCOMOTION

The steam engine changed manufacturing by providing a power source that was scalable and independent of local sources like water or wind power. In combination with mechanization, productivity increased multifold. For example, until the end of the seventeenth century, spinning was much more time-consuming than weaving, and one weaver needed multiple spinners to provide him with yarn. This changed with the development of mechanized spinning machines (see Chapter 7, Sections 7.4 and 7.5), now making weaving the bottleneck. This bottleneck was subsequently resolved by the development of the power loom in 1784 by Edmund Cartwright. Now spinning and weaving mills were able to produce enormous quantities of cloth and related products, by far exceeding the local demand of even large cities like Birmingham.

Unfortunately, transport from landlocked Birmingham by road was extremely expensive. The road network in the eighteenth century was still very basic and not suitable for carts of its time. Almost all goods had to be transported on the backs of horses or oxen, a slow and expensive process. This was relieved through the construction of an extensive canal network across Britain, since transport by water was a well-proven technique also suitable for larger loads. The size of the canals, however, limited boats to a width of about 2 meters and a length of about 20 meters. For a number of years, these canals were able to provide slow but cheap transport of goods between cities. Eventually, however, they reached their capacity limit in many places. For example, in 1825, more than 1000 tons of goods were transported from Birmingham to Manchester every day, not to mention raw materials in the other direction, reaching the capacity limitation of the canal network (Rosen 2010, p. 298). It would have been possible to move the factories near harbor cities, allowing for ocean shipping. However, one of the main problems of early industrialists was a stable and reliable labor force, which was difficult to obtain in the rather rowdy environment of a port city (Pollard 1963).

The main problem with land transport was that horses or oxen could bear only a limited load and needed supervision in order to go where they were supposed to go rather than where they wanted to go. A cart was able to carry much higher loads, but the power needed to pull a cart over a usually unpaved road was significant, not to mention the possibility of rain turning the road into mud. Key to developing a suitable road network was to smoothen the road in order to make it easier to pull carts. And, rather than smoothing the entire road, it is more efficient to just smoothen the road where the wheels are in contact with it. The easiest way to do this is with railway tracks, an idea so simple that it was in use since antiquity. Such wagonways powered by horses or humans were, for example, used to carry ships across the Isthmus of Corinth in Greece in the seventh or sixth century BCE. Railway-like cart ruts found in Malta may even be older, possibly dating back to 2000 BCE, although there are still many open questions as to by whom, when, and why these were made. In any case, during the eighteenth century, many coal mines used railways to transport their coal from the mine to a nearby canal. These railways were usually either horse powered or built on a downhill slope so that the horses only had to pull the empty carts back up and were able to ride themselves down in a so-called dandy wagon (Veitch 1930).

The idea of applying steam power to motion was first brought up in 1736, when Jonathan Hulls patented the idea of a steamboat using a Newcomen engine. However, it is unclear if he actually ever built one. The first steamboat was actually built in France by Count Joseph d'Auxiron in 1772. There was significant opposition by the local boatmen worrying about their jobs, and despite military protection, d'Auxiron found his vessel at the bottom of the Seine before it could be tested. Other Frenchmen developed a working version, which they were able to test successfully (Flexner 1992, p. 40). Research on steam-powered boats was well underway in France when, during the French Revolution, the French put all their most ingenious heads in a basket—quite literally—with the help of the guillotine.* Those inventors and tinkerers who escaped made sure they did not stick their heads out afterward.

Hence, the idea of the steamboat continued in Great Britain and America. Oliver Evans claimed to have built an amphibious vehicle in Philadelphia, but this claim is highly dubious (Lubar 2006). What is not dubious, however, was his great idea to put fire inside of the boiler, rather than underneath. While the steam engines of Newcomen and Watt had a fire underneath of the boiler, similar to a teakettle, Evans put the fire inside of a combustion chamber. This increased the efficiency of the heat transfer, thereby reducing coal consumption. Publications about this reached Richard Trevithick, who subsequently made significant improvements on high-pressure steam engines. Most steam engines before worked by condensing steam and thereby sucking the piston in; hence, they were called atmospheric steam engines. However, high-pressure engines actually pushed the cylinder outward, creating much better efficiency at the cost of…well…a high risk of exploding. Luckily, through the incorporation of some safety devices, Trevithick was able to reduce this risk significantly.

The improved efficiency was crucial for the development of the steam train. While a large-enough atmospheric engine of Watt could have easily moved a wagon, it was not strong enough to move itself, let alone the coal and water needed for a longer trip. The high-pressure engines had a much better power-to-weight ratio, consumed less fuel, and therefore were much better suited to drive a train. After different advances and a number of coal-mine railways, the first public railway, the Stockton and Darlington Railway, opened in 1825. The engines for the initially 40-kilometer-long railway were built by George Stephenson. The railway was still operated similarly to a canal, as the tracks were owned by the Stockton and Darlington Railway, and anyone with a steam engine could ride it for a fee anytime he/she wanted to. In the beginning, most of the tracks were single tracks, and due to the lack of a schedule, engines were frequently facing each other on the single track, resulting in quite some friction between the engine operators.

These problems were resolved in the next large rail project, the Liverpool and Manchester Railway. With a length of 50 kilometers, this railway was the first intercity railway, having double tracks over its entire length and

* Including, for example, the *father of modern chemistry* and discoverer of oxygen, Antoine Lavoisier, and the astronomer Jean Sylvain Bailly. Overall, the French Revolution is one of the possible reasons why the industrial revolution took place not in France but in Great Britain.

a fixed timetable for the trains. A large competition with significant prize money was held to determine the best way to power the trains, and numerous inventors submitted ideas, from stationary steam engines along the tracks to horses on treadmills. The winner was decided in a race, which was also used as a large PR event. The clear winner was George and Robert Stephenson with their *Rocket*, being both faster and stronger than the competition. The railway was opened on September 15, 1830, and soon, large quantities of raw material and finished goods were hauled between Liverpool and Manchester.* The trip was both significantly faster and significantly cheaper than transport by canal boat, much to the dismay of the canal owners, who, until then, made a fortune with their canals. To their surprise, the operators soon found out that goods were not the only things worth transporting, and passenger cars were added shortly thereafter.

Due to the success of the railway, many other railways were built, and by 1860, there were more than 15,000 kilometers of railway track in Britain (Puffert 2009, p. 44). The railway significantly improved both the speed and the cost of transport, both for goods and for passengers. Since transport cost depended on the distance of the transport, lower transport cost allowed entrepreneurs to sell their goods to more parts of Britain, in turn allowing them to build bigger factories and benefit from economies of scale.† This development of rail transport greatly changed the size of factories, allowing larger factories than ever before.‡

8.4 STEEL—A NEW INDUSTRY BASED ON FIRE AND IRON

The development of the steam engine was the kickoff for a whole set of new industries. It was not a bubble, it was not a temporary fad. Instead, it

* The opening of the railway also had the world's first death from a steam-powered train. William Huskisson, a member of the parliament, was run over by *Rocket* when he wanted to cross the tracks, and he died in a hospital shortly thereafter.

† This applies even today. For example, cardboard has little value compared to its volume, and transport costs are a large part of the total costs. Hence, large cardboard factories cannot compete in further-away markets, and you will find a small cardboard factory every 100 to 200 kilometers all over Europe and America. In comparison, cars have quite a high value for their volume, and transport cost has little influence on their overall price. Subsequently, car factories are rather large and often serve the entire world market.

‡ Steam ships also improved significantly, but since sailing ships were already able to transport large quantities of goods, the impact on manufacturing was less radical.

fundamentally changed the face of the manufacturing industry and therefore also the world. Steam engines were initially designed to pump water out of mines and to power cotton mills. Yet, a steam engine needs coal to run, therefore increasing the demand for coal. Water and steam power also increased the production quantities. The larger quantities manufactured then required railways for transport to reach the larger markets. This, in turn, again increased the demand for coal. The steam engines, the machines, and the railways required large quantities of iron, for which—you guessed it—more coal for the smelters and steam engines for processing was needed, not to mention more railways to transport the coal. This, in turn, led to a whole lot of other minor and major inventions and discoveries, for example, typewriters, sewing machines, and bicycles, to name just a few. Not only did all these technologies foster and support other technologies; they also created enormous savings in labor, which freed up the manpower to develop and build these technologies. This self-accelerating system started with the steam engine but is still accelerating nowadays with new products and technologies appearing at an astonishing speed.

One of the key materials for this enormous increase in productivity is iron. Iron is the sixth most common element in the universe.* Iron is also the most common element on Earth, with Earth's core consisting mostly of iron. Iron is also the fourth most common element of Earth's crust (after oxygen, silicon, and aluminum), although very little iron in Earth's crust is pure; rather, it is in different forms of iron oxide, better known as rust. Iron itself is a very versatile material. While pure iron is relatively soft, iron alloys can have a wide range of properties depending on the mix of elements. Iron is easy to shape and cut but holds its shape very well. Through heat treatment, it is possible to adjust the properties of the iron alloy without changing the alloy itself. Overall, this makes iron a very versatile material and one of the key ingredients of the industrial revolution, leading to a whole new industry.

The key problems of ironmasters in the industrial revolution were (1) how to generate the heat necessary to extract the iron and (2) how to control the traces of additional elements responsible for the properties of the alloy. Especially, achieving the desired content of carbon needed for

* Although this will increase over time since iron is also the final end product of both radioactive decay of higher elements and nuclear fusion of lower elements. Hence, over time, all material in the universe will turn into iron. Don't sell your steel stocks yet, however, since it will take about 10^{1500} years (a 1 with 1500 zeros, or roughly a complete page of this book filled with zeros) before the universe consists of giant spheres of pure hot iron (Dyson 1979).

steel and removing sulfur to avoid brittleness were difficult. Iron with a carbon content between 0.2% and 2.1% turns into hard steel. If the carbon content is below 0.2%, the iron will be too soft. If the carbon content is above 2.1%, the iron will be too brittle. Early ironmasters had little knowledge of these relations, and making steel was very difficult and error prone.

The blast furnace was first invented by the ancient Chinese empire (for details, see Chapter 4.4). The technology was known in Europe already during the Middle Ages. However, these blast furnaces were very inefficient, creating low-quality iron and consuming large quantities of wood or coal. A significant step forward was made through the invention of coke, which occurred independently in ancient China and, a few hundred years later, in England. Coke is coal that has been heated without oxygen to drive off unwanted components, for example, tar and water, while at the same time improving the quality of the remaining carbon. The process is similar to the production of charcoal out of wood, although coke requires much higher temperatures. Coke burns hotter and more efficiently than coal, charcoal, or wood, allowing more efficient production of iron. As the cost of iron was reduced, iron was used for more and more products that were previously built from other, cheaper materials. This applies not only to technical products but also to architecture. One of the first uses of iron as the structural element in architecture was the Iron Bridge, built between 1777 and 1781 across the River Severn at the Severn Gorge in England, which has since been renamed Ironbridge Gorge. Over time, numerous improvements and developments were made to the smelting process. For example, the exhaust heat of the furnace was used to preheat fresh air to improve fuel efficiency, or the pig iron was refined through a finery forge. While this produced better iron, the much-harder steel was still an expensive product.

The first more-or-less reliable way to make larger quantities of steel was through puddling, invented by Henry Cort and patented in 1783. In this process, puddlers stirred the molten iron, allowing oxygen to burn off excess carbon. The puddler visually identified pieces of low-carbon steel in the molten mix, and formed a ball of steel, which was eventually lifted out of the molten iron for further processing.* Still, the process was expensive, yielding an inconsistent quality of steel.

---

* Probably the most demanding voluntary manufacturing job in history. Imagine working in a 100°C sauna vigorously stirring a blindingly bright pool of 2800°C liquid iron with an iron bar, which you also use to lift 40 kg steel balls out of the molten iron, while at the same time wearing heavy and hot protective clothing (Dennis 1967, p. 15).

The next significant improvement was invented by Henry Bessemer, who patented his Bessemer process* in 1855. Rather than having workers stirring the mix to bring oxygen into the iron in order to burn off impurities, the Bessemer process blew air through holes in the bottom of the furnace. This allowed the reliable production of steel on an industrial scale, generating significant cost savings, not to mention the buckets of sweat of the puddlers. The price of steel dropped to a fraction of its former price, and steel was now available for almost the same price as iron. Since steel is, for most applications, more suitable than iron, the usage of steel exploded, and many products that were before made out of wood or iron were now made out of steel. While in 1880 only 1 ton of steel was produced for every 10 tons of iron, it was just the opposite 10 years later, with 10 tons of steel produced for every ton of iron. The total quantity produced also increased significantly (Landes 1969, p. 260).

Until this point, the industrial revolution was driven primarily by Britain, with other countries being far behind. However, having an extensive cotton industry and a working iron industry, Britain started to fall behind. From 1900 onward, Germany was the leading industrial powerhouse in Europe, second in the world only to the United States. The adoption of the improved Siemens–Martin process was very rapid in Germany, but Britain's entrepreneurs were hesitant or unable to change their existing investments and buildings and mostly stuck to the Bessemer process. Hence, German steel outperformed British steel both in cost and in quality. While in 1870, Britain produced twice the quantity iron of Germany, by 1910, Germany produced twice the quantity of iron of Britain (Landes 1969, p. 269).

8.5 THE INVISIBLE INDUSTRY—CHEMICALS

Starting in the nineteenth century, another major technology emerged—the chemical industry. This industry is generally underappreciated in its significance, since the general consumer purchases very few chemical products directly and less than 1% of manufacturing employees in the United States work in the chemical industry. In general, the consumer notices the chemical industry only in a negative context, i.e., if something

* Also independently invented by William Kelly in the U.S., and possibly others.

goes wrong. Even when we see a chemical plant, we usually see only pipes and vats, or if we are lucky, an undistinguishable liquid product. This is all much less interesting to the general public than other forms of manufacturing with sparks and chips flying and identifiable parts coming together to a recognizable final product.

However, almost every product we buy required some chemicals during its manufacturing process. When you see a colored or black product, you also see a chemical dye. If the product is white, then it was probably bleached through other chemicals. Anything made out of plastic required polymer resin. Anything moving probably has some lubricants or coatings. Anything that should not move may have glue or adhesive tape. In any case, it is probably treated against corrosion. Even foods are most likely grown with the aid of chemical fertilizers, insecticides, and pesticides. In short, products from the chemical industry are all around us even if we do not see them.

Furthermore, while less than 1% of manufacturing-related employees in the United States work in the chemical industry, its revenue is second only to the automotive industry, with 20 times the number of employees. If you include the petroleum industry in the chemical industry, then its revenues by far exceed the automotive industry, even though it has only about 10% the employees (U.S. Census Bureau 2010). This makes the chemical industry one of the strongest drivers of manufacturing in the United States, which due to its low percentage of labor is also quite robust against outsourcing.

Before the nineteenth century, the chemical industry was almost nonexistent. There were very few chemical products, and most of them were animal- or plant-based organic chemicals. For example, one of the most common bleaching agents for clothes and wool was urine, or, to be more precise, weeks-old stale urine that has an increased ammonia content. Human urine was commonly used due to the difficulties involved in making an animal pee in a bucket. For tanning leather, the animal hide was allowed to rot in stale urine before being kneaded in animal dung and treated with tree bark.* Fertilization was done using animal excrement, if at all. Early machines, for example, waterwheels and windmills, were lubricated using animal fat. Most dyes were based on animal and plant products, and fancier colors often required prohibitively expensive processes. The most expensive purple or red colors involved dried beetles or

* Unsurprisingly, most tanneries were located outside and, if possible, downwind of towns.

processed sea snails and hence were used mainly by royalty and high-ranking clergy.

As can be seen, most of these products were animal or plant based, and only very few were mineral based; for example, lime and its derivative products like quicklime were used. Collecting these chemicals was time-consuming, and large-scale production was not done, due to the decentralized nature of the collection process. Most chemical products before the eighteenth century were made in a workshop only slightly larger than a household kitchen. The resulting chemicals also had widely different properties, resulting in inconsistent performance during subsequent processes.

Scientists started to ramp up the theoretical knowledge in the seventeenth century and especially the eighteenth century, building up the theoretical foundations for the chemical industry based on experiments. The first chemical element to be discovered through science was phosphorus in 1669. While it took almost 100 years to identify the next element, cobalt, this was followed by a rapid series of discoveries. By the end of the eighteenth century, 34 elements had been identified, and by the end of the nineteenth century, 84 out of roughly 114 elements* were known. These 84 elements include the majority of the industrially significant elements. The first major discovery on their interaction was the effect of oxygen in combustion in 1778, with more following in the decades afterward. The underlying periodic table was also discovered around 1870. However, from knowing the parts, it still was a big step to the knowledge of their use, and from there, it again was a significant step out of the laboratory toward industrial-sized implementation. In any case, during the early phases of the chemical industry, entrepreneurs did not understand much of the scientific background but, rather, discovered new processes by trial and error.

In the early nineteenth century, inorganic chemicals slowly started to replace organic chemicals. Inorganic chemicals had the advantage that in many cases, they were not only cheaper to obtain but also available in large quantities on one site, enabling economies of scale for the chemical industry. Furthermore, due to the more consistent raw materials, the final products were also more consistent. Possibly the first key product was

* The scientific process is ongoing, and more elements are discovered every few years. Most of the current discoveries are highly radioactive, and the few atoms created decay almost immediately. Both the quantity of a few atoms and the half-life measured in milliseconds make any industrial use unlikely.

sulfuric acid, an important chemical for the treatment of metals. The first industrial-sized process started in 1736, when Joshua Ward from London developed a large-scale method to produce sulfuric acid. Further developments reduced the cost even more, and by 1800, the price per ton was only £30, compared to £280 around 1740 (Aftalion 2001, p. 10ff).

Another major product was alkalis for bleaching, dyeing, and the production of soap and glass. The key process to produce sodium carbonate, a type of alkali, was discovered and patented by the Frenchman Nicolas Leblanc in 1791. Sponsored by the Duke of Orleans, he built a large processing plant. Unfortunately for Leblanc, during the French Revolution, the duke lost his head in the guillotine. The plant was repossessed, the patent annulled, and the financially ruined Leblanc ended his own life with a bullet to the head in a poorhouse in 1806. His Leblanc process, however, survived, and many entrepreneurs in France and even more in England made a fortune using his process (Aftalion 2001, p. 11ff).

Yet another key product of the early chemical industry was potassium chloride used for bleaching textiles. Initially, textiles were bleached with urine or buttermilk in combination with sunlight, a time-consuming and expensive process. Most critical, however, was the use of large areas of land to spread out the sheets in order to expose them to sunlight, a process that can easily take months. With the development of cotton mills, large quantities of textiles needed to be bleached, overwhelming the available bleaching capacity. Sulfuric acid helped to speed up the bleaching process and to lower the cost, but it had the disadvantage that it also started to dissolve the textiles. Yet, many people preferred cheap but slightly degraded textiles over the more expensive variant. The real breakthrough came with the development of potassium chloride by Claude Louis Berthollet from France and the subsequent development of a bleaching powder by Charles Tennant in 1797 in Scotland. This bleach reduced the time for bleaching from months spread out in the sun to hours in a vat, with a corresponding reduction in cost. This chemical was finally able to handle the onslaught of unbleached yarn and woven textiles due to the boom of the cotton mills (Wolff 1974).

Another often-overlooked key ingredient in modern manufacturing are lubricants (or alternatively) is lubrication. Many products with moving parts contain lubricants, but more significantly, almost all processing machines in manufacturing are lubed. Lubricants reduce friction. This allows more of the kinetic energy to reach its destination. Additionally, this reduces the wear and tear on the moving parts and also reduces the heat generated from friction. On top of that, many lubricants also reduced corrosion. Until 1850, pretty much all lubricants were based on animal fat or vegetable oil. These have very inconsistent properties, are more susceptible to changes in temperature or humidity, age quicker, and all of that for a relatively high price. Most animal or vegetable oils also had a tendency to thicken with oxidation, requiring frequent replacement.

One of the notable exceptions is olive oil, which has very good lubricating properties and ages only very slowly (Muendel 1995). However, with olive oil being unavailable in large parts of Europe, animal fat was still frequently used. The quality of the fat was also highly significant. For example, an ironmaster in 1823 reported that he preferred first-rate animal fat from Russia since it gave him 10 more revolutions per minute on his waterwheel. The benefit of this increased speed exceeded the considerable expense for the fat. The issue became more pressing with the introduction of steam power and the resulting fast and continuous movements. From the 1850s onward, mineral oils started to replace animal fat as grease for lubrication, often also in a mix of organic and inorganic lubricants with additional additives to improve performance. The machines were also adapted to make sure the moving surfaces stayed lubricated, by constant immersion in a lubricant, gravity feed, or pumped lubricants. The development of ball bearings also reduced friction but increased demand for quality lubrication (Landes 1969, p. 298f). Nowadays, most lubricants are synthetic or mineral oil–based lubricants, although organic lubricants are still used in areas where pollution is a concern, for example, in chainsaws or food processing.

Environmental concerns were nonexistent for the nineteenth-century chemical industry, and pollution of the environment was heavy. Waste was dumped in the water, on land, and in the air. Some plants used very high chimneys to disperse the toxic fumes, which eased the direct effect around the plant by spreading it to a wider area. Nevertheless, for many people, wealth was more important than health, and there was a lot of money to be made. In addition, over time, these entrepreneurs developed new technologies to create other products based on these wastes, and eventually, public pressure moved the government to establish laws reducing pollution.

While initially France was chemically more advanced than Britain, the French Revolution slowed down development significantly. While the surviving French scientists and entrepreneurs in the aftermath were very innovative in finding substitute products due to war shortages, the British chemical industry overtook the French industry to become the leading chemical producer of the first half of the nineteenth century. The largest chemical plant of the world in 1830 was St. Rollox in Glasgow, owned by Charles Tennant and producing bleaching powder (Landes 1969, p. 112). Started in 1799 with a total production of 52 tons, it produced 13,000 tons all over Britain by 1852 (Aftalion 2001, p. 13).

While Britain was the largest chemical producer in the early nineteenth century, Germany was catching up quickly. Germany had, in general, a higher educated labor force. Furthermore, while Britain had a significant industrial boom due to the textile mills, this boom was slowing down. Germany, on the other hand, did not yet have such a period of extensive growth and was eager to catch up. Hence, from 1850s onward, Germany started to become the leading chemical producer in Europe, if not the world. By 1895, Germany was the leading nation in the production of artificial dyes, and by 1913, also in the production of sulfuric acid (Landes 1969, p. 186ff). New products emerged frequently, changing the manufacturing landscape. These included artificial fibers and leather, fertilizers, paper, dyes and paints, etc. In many cases, the chemical industry enabled new products that were impossible with the old ways, for example, plastics and rubbers, photography, explosives, insecticides, pesticides, and many others.

In the chemical industry, the processing environment has to be tightly controlled, as are the ingredients going into the process. Yet, in terms of manufacturing, chemicals are less complex products. A product as simple as a ballpoint pen has many more specifications than the plastic it is made of. The geometric dimensions alone are orders of magnitude more complex than all manufacturing parameters of the plastic. This does not even include processing parameters for machining and injection molding. For the most part of history, producing mechanical products included lots of time-consuming tinkering and fitting, making the final product expensive. This changed only slowly starting around 1800, as we will see in the next chapter.

BIBLIOGRAPHY

Aftalion, F., 2001. *A History of the International Chemical Industry*. Chemical Heritage Press, Philadelphia, PA.

Berg, M., 1985. *The Age of Manufactures: Industry, Innovation and Work in Britain, 1700–1820*. Fontana, London.

Cameron, R., Neal, L., 2002. *A Concise Economic History of the World: From Paleolithic Times to the Present*, 4th ed. Oxford University Press, New York.

Dennis, W.H., 1967. *Foundations of Iron and Steel Metallurgy*. Elsevier Publishing Co. Ltd., Barking, Essex, UK.

Dickinson, H.W., 2010. *Matthew Boulton*. Cambridge University Press, New York.

Dyson, F.J., 1979. Time without End: Physics and Biology in an Open Universe. *Reviews of Modern Physics* 51, 447 doi:10.1103/RevModPhys.51.447.

Flexner, J.T., 1992. *Steamboats Come True: American Inventors in Action*. Fordham University Press, New York.

Kanefsky, J., Robey, J., 1980. Steam Engines in 18th-Century Britain: A Quantitative Assessment. *Technology and Culture* 21, 161–186 doi:10.2307/3103337.

Landes, D.S., 1969. *The Unbound Prometheus: Technological Change and Industrial Development in Western Europe from 1750 to the Present*, 1st ed. Cambridge University Press, New York.

Lubar, S., 2006. Was This America's First Steamboat, Locomotive, and Car? *Invention and Technology Magazine* 21.

Muendel, J., 1995. Friction and Lubrication in Medieval Europe: The Emergence of Olive Oil as a Superior Agent. *Isis* 86, 373–393.

Müller, G., n.d. Water Wheels as a Power Source.

Musson, A.E., Robinson, E., 1959. The Early Growth of Steam Power. *The Economic History Review, New Series* 11, 418–439 doi:10.2307/2591464.

Pollard, S., 1963. Factory Discipline in the Industrial Revolution. *The Economic History Review, New Series* 16, 254–271 doi:10.2307/2598639.

Puffert, D.J., 2009. *Tracks across Continents, Paths through History: The Economic Dynamics of Standardization in Railway Gauge*. University of Chicago Press, Chicago; London.

Roll, E., 1968. *An Early Experiment in Industrial Organization: History of the Firm of Boulton and Watt, 1775–1805*, New issue of 1930 ed. A. M. Kelley, New York.

Rosen, W., 2010. *The Most Powerful Idea in the World: A Story of Steam, Industry, and Invention*. Random House, New York.

Scherer, F.M., 1965. Invention and Innovation in the Watt-Boulton Steam-Engine Venture. *Technology and Culture* 6, 165–187 doi:10.2307/3101072.

U.S. Census Bureau, 2010. 2010 Economic Census. United States Census Bureau.

Veitch, G.S., 1930. *The Struggle for the Liverpool and Manchester Railway*. Daily Post Printers, Liverpool, UK.

Wolff, K.H., 1974. Textile Bleaching and the Birth of the Chemical Industry. *The Business History Review* 48, 143–163 doi:10.2307/3112839.

9

Interchangeable Parts—The End of Filing in Assembly

There is nothing that cannot be produced by Machinery.

Samuel Colt (1814–1862)
Inventor and industrialist

As the price of iron and steel dropped, more and more products were made out of metal. Steel was especially suitable to mechanical devices with moving parts, for example, manufacturing machines, winches, clocks, and guns. However, the technology of shaping metal was still in its infancy. Metal workshops of that time were able to make similar, but not quite identical, parts. This, of course, created problems during final assembly. To assemble two or more metal parts, they have to fit into each other. As the parts were similar, but not identical, they usually did not quite fit. Hence, rather than merely assemble the parts, the worker had to make them fit. Therefore, the assembly of parts was a time-consuming task requiring highly skilled workers. Their most important tool was a file to make parts fit that did not fit before. The use of these fitters was widely used until the early twentieth century, and even Henry Ford relied on teams of fitters in the early years of the Ford Motor Company (Hounshell 1985, p. 220).

There were numerous reasons why fitters were a key part of most metal workshops. First of all, there was no precise way to describe how a part should look. Technical drawings during the early industrial revolution used the same methods already known in antiquity, either a top-down *shadow* plan* or a—more or less skilled—perspective drawing as shown

* One of the earliest such drawings is the floor plan of a house in the scale of 1:360 from Mesopotamia dating back to around 2000 BCE (Feldhaus 1953, p. 9).

in Figure 9.1.* Some of the best surviving drawings are from Leonardo da Vinci (1452–1519), whose works give a general idea of how the machine should look like, but without any measurements or details, making an exact reproduction impossible. Eventually, some drawings started to include selected measurements. For example, the length of a rectangular beam was given, but no width or height, leaving much to the imagination of the builder. The description of complex geometries that are not based on either a line or a circle was also unknown. So were the technique of showing the inside structure of parts, for example, holes using dashed lines, and the combination of front, side, and top-down views used by modern technical drawings. All of these originated only in the nineteenth century. Hence, describing the part precisely to the worker using a technical drawing was not yet possible.

Furthermore, even with a detailed drawing, the worker was unable to absolutely measure the dimensions of his finished part precisely enough. They simply did not have the tools to measure dimensions with the required accuracy. And even if they could measure it, they would have to deal with a bewildering mix of measurement units, including feet, inches, yards, and lines, which also differed in different locations. The first precise measurement tool, the micrometer, was created by Henry Maudslay in the early eighteenth century, measuring with a precision of at least 25 micrometers. Due to is outstanding precision, Maudslay nicknamed it *The Lord Chancellor* (Rosen 2010, p. 196), but this was a one-of-a-kind achievement. Only from the 1870s did similar tools become widely available (Roe 1926, p. 211ff).

Finally, metalworking technology was unable to reproduce parts exactly. Machines were as imprecise as the parts, and the same settings on one machine may have produced completely different results on another machine. Even on the same machine, it was difficult to produce identical parts. This problem was compounded if the part had to be hardened, in which case the heat treatment often warped the part. Even if the parts fit together before, they would no longer fit after hardening.

The best approach in the early days of the industrial revolution to reproduce a part was to give the worker a reference part from which he could take

---

* The perspective was often distorted. For example, a wheel as seen from the front would correctly be drawn merely as a rectangle. To make things clearer, the artist therefore simply flipped the wheel sideways to show it as a round wheel, or sometimes an elliptical wheel-a-like. This gave an effect similar to Egyptian drawings, where the head is seen from the side, the torso from the front, and the arms and legs again from the side (Feldhaus 1953, p. 21).

FIGURE 9.1
Water-raising machine with different gears from Ismail Al-Jazari's *Book of Knowledge of Ingenious Mechanical Devices*, 1206.

relative measurements. Using simple tools, for example, a ruler and a compass, the worker tried to reproduce the part. The use of simple gauges was also known, where each part had to fit through a standard-width gauge. For example, since medieval times, cannonballs were tested with a round gauge to ensure that they fit into the cannons. However, there was no test for if they were too small or if they were not round but elliptical (Alder 1998).

Possibly one of the earliest uses of interchangeable parts were by Swedish inventor Christopher Polhem. Polhem was active in many different businesses, including clock making and lockmaking. Around 1720, he reportedly achieved interchangeable parts for his clocks through the use of gauges. However, his approach did not spread beyond the walls of his factory (Woodbury 1960).

In summary, during the early modern age, it was common for numerous highly qualified and highly paid fitters to painstakingly make things fit. Every final product was unique. Parts could not be swapped with other parts. This also made it difficult to ship spare parts for repairs or replacements, or—particularly significant for rifles during war time—to make one working product out of two broken ones. Yet, the first interchangeable parts were created for a completely different reason, as we will see in the following section.

9.1 HONORÉ BLANC AND FRENCH MUSKET PRODUCTION

The French Army, until the beginning of the eighteenth century, sourced their weapons from a large number of master gunmakers, who produced or purchased the parts and fitted them together individually before delivering the weapons. Initially, each colonel bought the weapons for his regiment separately, leading to a very fractured market. The weapons used in the field differed widely not only between regiments but also within the regiments, not only due to the different sources but also due to the habit of the soldiers of customizing their weapons. Naturally, this led to a wide variety of different and incompatible weapons, making it difficult to both do field repairs and supply ammunition of the correct caliber. At the beginning of the eighteenth century, the purchasing of all weapons was centralized with the military engineers of the artillery service, in order to give the army more purchasing power (Alder 1998).*

In the 1730s, Florent-Jean de Vallière, director general of the Battalions and Schools of the Artillery, made another logical step and standardized artillery sizes to five classes of cannons: 4-, 8-, 12-, 16-, and 24-pounders (Corvisier and Childs 1994, p. 837). In 1750, the idea was brought up to also standardize the artillery carriages that carried the cannons in battle.

* The details of French gun making in this subchapter are mostly based on the research of Ken Alder, professor of history at Northwestern University.

At that time, the idea was more for aesthetic reasons of officers who disliked individuality and favored uniformity. This idea was included in reforms starting in 1763 under the leadership of General Jean-Baptiste de Gribeauval and his *Gribeauval system*. To achieve this pleasant uniformity of their carriages, they created detailed technical drawings and provided standard gauges for the construction of the artillery carriages. By 1780, these standards had progressed well enough that—much to the surprise of the officers—these wooden parts were interchangeable, and did not need to be fitted for the assembly of the carriage. This resulted in both easier assembly during construction and easier field repairs (Alder 1997).

Besides interchangeable artillery carriages, Gribeauval also initiated strict standards in the production of cannons and cannonballs. While the upper tolerance of cannon balls had been checked using a simple ring gauge since medieval times, this did not check if the ball was too small. The team around Gribeauval was probably among the first to use a second gauge to check if a cannon ball was too small. If the ball also passed through the second, smaller gauge, then it was too small. A cannon ball was only acceptable if it passed through the larger but not the smaller gauge. Through this combination of a *go gauge* and a *no-go gauge*, the French were able to set both an upper and a lower tolerance limit for the cannon balls. Since this still allowed oblong or football-shaped balls to pass, they further refined their gauges, and a ball was required to roll through a series of angled tubes to ensure roundness (Alder 1998). These military engineers were probably the first to use the term *tolerance* and among the first to work with both upper and lower tolerance limits (Alder 2010, p. 148).

Similar precision checks were now also taken of the cannon itself. The inner diameter of all new cannons was measured with a precision of ±0.025 millimeters using a special gauge called *etoile mobile*, followed by an visual inspection using a mirror, a hook test checking for cavities, and wax impression. The whole procedure was repeated after two test shots (Alder 1998). Gribeauval also initiated a redesign of the cannons and set standards for their use. Overall, this made the French artillery probably the best artillery of its time (Chartrand 2003, p. 1).

Through this approach using gauges and checks, interchangeable parts for cannons and their carriages were achieved. However, the technical complexity of these products was rather low. As for the cannon, the only critical dimension was that the ball had to fit the cannon. In addition, the dimensions of the round hole and the round ball were very easy to describe and reasonably easy to measure. While the carriage had more critical dimensions, these were also easy to use. Most nonmoving parts were made of wood and hence had rather wide tolerance limits. As for the moving parts, for example, the axle, these were also rather easy to describe using simple geometry and equally easy to gauge.

The Gribeauvalists, having significantly improved cannons through rigorous use of standards, also wanted to apply the same approach to muskets. However, the matter was very different for muskets. The firing mechanism of the musket—the gunlock—consisted of a number of metal parts and required tight tolerances to make the mechanism work. Probably the most complex part of the entire lock was the tumbler (Figure 9.2). This *brain* of the lock transferred the force from the spring to the flint to create a spark and to ignite the gunpowder. Not only did this part have tight tolerances; it also had an irregular geometric shape. It was sort of—but not quite—round with a number of protrusions, a shape that was difficult to describe with the geometry available at that time (Alder 2010, p. 197).

Nevertheless, the possibility of interchangeable parts for muskets had already been demonstrated. Since 1704, armorer Guillaume Deschamps had experimented with interchangeable parts for gunlocks. In 1723, he demonstrated his achievements before King Louis XV by disassembling

FIGURE 9.2
Tumbler for U.S. muskets around 1850. (Photo by Schönwälder.)

50 locks, mixing up the parts, and reassembling them again. By 1727, Deschamps' factory in Toulon had built 660 such locks. Unfortunately, each lock cost up to seven times as much as a noninterchangeable lock. Despite Deschamps' efforts to promote his goods, he was unable to convince General Vallière that the benefit was worth the cost. However, he did convince General Vallière that there should be only one type of lock for the army, not a multitude of different designs. As such, the army started to use a standard lock, although without interchangeable parts (Peaucelle 2005).

In 1777, the army introduced a new and improved musket designed by master gunmaker Honoré Blanc, who had been in charge of inspecting the final quality of guns since 1763. As part of these reforms, starting in 1763, the French Army set prices and standards not for the complete new musket but, rather, for its parts. Additionally, the prices for these parts were not negotiated but, rather, calculated by the army based on the theoretical work needed to make them. Naturally, these prices were way off and no longer profitable for the gunmakers. On top of that, resulting from their success with standard gun carriages, the army set rigorous standards in many dimensions and enforced them with gauges. As a result of these increased quality requirements, a substantial number of guns were rejected, at the cost of the artisan (Gillispie and Alder 1998). Hence, the French gunmakers preferred to supply guns to the American Revolution for profit rather than to deal with the scrupulous French Army at a loss (Alder 1997).

As the supply of guns for the French Army dwindled, the officers dug up an old law stating that every gunmaker is—technically speaking—also a soldier and simply ordered them to make guns for the French Army. Numerous gunmakers were thrown into jail for failing to follow the commands of their *superior officers*. Now, gunmakers were highly skilled and respected artisans of their time, and naturally, they fought back using their political influence. Very soon, in August 1781, this political tug-of-war ended with a clear victory for the gunmakers. While disrespecting the officers and failing standards still had consequences, selling the guns to other customers had none (Alder 1997).

The army was in a fix. Their prized armory was running out of guns and by 1785 could no longer supply arms. In their desperation, they even considered starting musket production themselves but rejected this approach. Financial incentives also were not attractive enough for the gunmakers to produce for the army. Drawing from their experience in making artillery carriages with interchangeable parts, they aimed for interchangeable parts in their muskets in order to control production and to lower prices (Alder 1997). The reason for using interchangeable parts had nothing to do with cost and very little with quality and field repairs, but it was all about controlling the supply and replacing skilled and stubborn master gunsmiths with hopefully more docile unskilled labor.

Honoré Blanc was in charge of musket quality since 1763, and he instructed numerous gun makers in the use of gauges, whom he believed to be *our guides and ought to be our laws.* Additionally, since 1782, he experimented with the production of interchangeable parts for his musket (Alder 1997). On July 8, 1785, Honoré Blanc set up a stunning demonstration in the courtyard of the Chateau de Vincennes. Taking parts for 25 locks at random from 50 disassembled locks, he assembled 25 working locks without any filing or fitting. Each randomly selected part fit into the lock as if it were made for it. The officers, representatives, and officials were amazed. Among those highly impressed dignitaries was the U.S. politician and future president Thomas Jefferson, who even assembled a number of locks himself (Peaucelle 2005). Jefferson tried (unsuccessfully) to convince Blanc to move to the United States. However, Jefferson also wrote letters about the event to the American secretary of war and to U.S. Chief Justice John Jay. He also sent a number of locks along (Alder 2010). Back in the United States, Jefferson also wrote to the future president James Monroe. Through this promotion of Honoré Blanc's work, Jefferson set the basis for the manufacture of interchangeable parts in the U.S. Army (see Section 9.4).

Based on Blanc's experience, the French Army decided to set up an experimental workshop for the production of interchangeable parts for locks. While Honoré Blanc had his workshop in Saint-Étienne, his achievements in interchangeable parts made him very unpopular with his fellow gunsmiths in Saint-Étienne. Hence, the French Army decided to relocate his workshop to a safer location, selecting a most unusual location for a manufactory—the dungeons of the Château de Vincennes near Paris (Figure 9.3). These dungeons have been home to many famous prisoners in the past, including philosopher Denis Diderot, politician de Mirabeau, writer Jean Henri Latude, and the Marquis de Sade.* Blanc's workshop was actually in the very cells that housed the Marquis de Sade. However, even the fortress Château de Vincennes was unable to protect the workshop of

* If you don't know who Marquis de Sade is, let's just say that the term *sadism* bears his name.

FIGURE 9.3
Even the walls of the Château de Vincennes near Paris were not able to protect Blanc's workshop located in its dungeon. (Photo by author, dated September 2011.)

Blanc when a revolutionary crowd entered the dungeon to free prisoners in 1791. While they were at it, they also destroyed some of Blanc's machinery before order was restored (Alder 2010, p. 246).

Despite these interruptions, Blanc was able to successfully produce locks with interchangeable parts. To ensure interchangeability, he used filing jigs, machine tools, technical drawings, and—above all—elaborate sets of go and no-go gauges. He also changed the hardening process. The rapid and uneven changes in temperature often warped the parts out of interchangeability. Blanc wrapped the parts in a mixture of clay and horse manure and slowly heated them over four hours to avoid warping (Peaucelle 2005). While he was able to reduce cost compared to the efforts of Guillaume Deschamps in 1723, Blanc's interchangeable locks still needed between 30% and 100% more time for production than the regular-style locks from the other manufacturing locations. However, the workshop in Vincennes was of a much smaller scale than the other gun workshops, and a larger-scale production probably would have reduced the cost even more (Alder 2010, p. 245).

Blanc's main supporter, General Gribeauval, died in 1789, the year the French Revolution started. His successor, François Marie d'Aboville, was also supportive of Blanc's ideas, but political opposition was mounting. Hence, Blanc demonstrated his achievements again on November 20, 1790, in the Hotel des Invalides, assembling working locks in front of military dignitaries and officials from individual parts of 500 locks selected at random. He repeated this demonstration in 1791. A military committee also confirmed after a 15-month investigation that these locks were perfectly interchangeable (Alder 2010, p. 238f).

However, the large contract and the full-scale manufacturing of interchangeable locks Blanc hoped for never came. Besides the cost disadvantage, the main reason for the hesitation of the government was their fear of social unrest due to the use of unskilled labor rather than skilled gunsmiths. Blanc realized that there would be no government workshop with interchangeable parts and eventually set up his own factory in Roanne in March 1794 (Alder 2010, p. 247). There, he successfully produced large quantities of interchangeable locks, eventually achieving about 10,000 locks per year, or about 5% of the entire production for the French Army. While his machines were multipurpose, Blanc dedicated the machines to one task only in order to reduce setup time and to increase accuracy. Furthermore, he established a strict division of labor, dividing production into 156 distinct steps (Alder 1997). Nevertheless, his interchangeable locks were always about 10% more expensive than conventional locks, his factory depended on state support, and he never made a profit (Peaucelle 2005).

Honoré Blanc died impoverished in 1801, and the support for interchangeable parts was dwindling. Their strongest opponent was General Jean Jacques Basilien Gassendi, protégé of Napoleon and director of the influential Division of Artillery. General Gassendi's focus was on cost and—even more so—on social peace, both of which he saw to be endangered by the system of interchangeable parts. Yet, for political reasons, General Gassendi could not simply withdraw support for the factory in Roanne, since General d'Aboville and others still supported the Gribeauval system. What followed was a prime example of political maneuvering and, once more, proof not to believe any statistics unless you have doctored it yourself.

While Blanc was on his deathbed, General d'Aboville demonstrated in 1801 for the Central Committee of the Artillery that 492 out of 500 locks from Blanc were interchangeable. Only a week later, the team around

General Gassendi made a countertest and found only 152 of roughly 500 guns useable. General d'Aboville was able to establish his eldest son as the official inspector of the gun factories in 1802, receiving favorable reports. But in 1803, General Gassendi was able to replace the inspector with one of his people, who, unsurprisingly, found many flaws in the Roanne factory. Another test by General Gassendi found only 3 of 100 locks as interchangeable, yet General d'Aboville's supporters demonstrated, in yet another test, very good interchangeability. Again, in 1804, General Gassendi's inspector tested the locks, finding them inadequate, while General d'Aboville's controller found them interchangeable. General Gassendi was able to replace this controller with one of his own men, who discovered significant flaws with the locks (and was rewarded with a sizeable sum by General Gassendi for his efforts). The last-ditch effort of General d'Aboville, a public exhibition at the 1806 World Fair Exhibition in Paris, went pretty much unnoticed (Alder 2010, p. 325f).

General Gassendi won. The conclusion in the French military was that interchangeable parts had been tried, and found unsuitable, even though the failure was more for economical and social reasons, not technical. Blanc's factory in Roanne closed in 1807 (Alder 1997). As a side effect, the quality of French firearms decreased considerably, and by 1811, French muskets were the worst of Europe (Alder 2010, p. 339). By 1850, the French military had forgotten everything it ever knew about interchangeable parts, even the fact that they produced them successfully. It was up to America to pick up the effort, started by Thomas Jefferson and supported by many French officers and engineers emigrating to the United States. However, before this story, another, independent, manufacture of interchangeable parts has to be told.

9.2 BLOCK PRODUCTION AT PORTSMOUTH

Independently of the developments of Honoré Blanc in France, interchangeable parts were also developed in the block production of the Royal Navy Dockyards in Portsmouth. As with Honoré Blanc, the interchangeability was more of an accidental side effect of other goals.

A block is a mostly wooden pulley with metal axles. Blocks are needed on ships to lift heavy weights or to tighten ropes. As such, they were frequently used on sailing ships of the navy. To equip a new ship, about 1500 blocks of different sizes were needed. Additionally, these blocks were often under heavy loads for prolonged periods of time, not to mention the peak load during a storm, and wore out quickly. Overall, the British Navy needed about 100,000 blocks every year to maintain its naval superiority (Rosen 2010, p. 200).

Up to the end of the eighteenth century, most of these blocks were supplied by either Bartholomew Dunsterville in Plymouth or Walter Taylor in Southampton. Both suppliers relied primarily on manual labor of skilled craftsmen, with very few machines used, although Taylor used a horse mill and a water mill for sawing and drilling, and also experimented with a steam engine. However, these blocks were expensive, and production was slow. Quality from the primary suppliers was inconsistent, even more so from the few smaller suppliers. Due to the inconsistent quality, blocks often broke, snagged lines, or caught fire due to friction on rough surfaces. All these failures were unpleasant in peacetime but could quickly become critical while under heavy weather or in combat (Cooper 1984).

Following the significant changes in industry due to the industrial revolution, a new department was added to the British Navy in April 1795 with the task to modernize warship construction. The head of this department was Brigadier General Sir Samuel Bentham, Inspector General of Naval Works. Bentham was an excellent choice for the head of the shipbuilding operations of the British Navy. He was a trained shipwright, had combat experience as an officer, invented different woodworking machines, and was famous for his engineering skills (Coad 2005, p. 21ff).

While working in Russia together with his brother, he designed a new type of prison architecture for his brother Jeremy. This new design was called Panopticon and was a round prison with a central overseer in the middle being able to monitor all prisoners any time, reducing the manpower needed for supervision. Interestingly enough, Bentham also considered the same design for a factory, where a single foreman would be able to supervise many workers. While Bentham never built such a *prison-factory* himself, the idea was used for other factories, for example, the round mill in Belper, a cotton mill built by William Strutt, son of Jedediah Strutt, the partner of Richard Arkwright (see also Section 7.5).

The naval works of the time were the usual chaotic mess, similar to larger workshops of the Middle Ages or the Arsenal of Venice. Workers did more or less whatever they wanted whenever they wanted, each of them in their own ways, while taking a generous share of material, tools, and time to work for their own profit rather than for the products the British Navy paid them for. Bentham thoroughly changed these old ingrained work habits, despite resistance from the workers and from conservative members of the British Navy board and the admiralty. Bentham was a firm believer of visiting the location and seeing with his own eyes, rather than merely receiving reports from subordinates. He stated, *There seems reason to believe that those works might go on better still were I to go there myself* (Coad 2005, p. 34).

Historically, the tasks in the navy shipyards were divided according to product, as was common in the Middle Ages. Bentham rearranged the workers according to their tasks, not their end products, aiming to use skilled workers primarily for jobs requiring skill, and unskilled workers for unskilled tasks (Ashworth 1998). Bentham also established an elaborate bookkeeping system to keep track of the materials (Hume 1970).

Since 1634, the workers of some British shipyards had the right to carry off chips of wood three times a day. Mind you, a *chip* was anything less than one meter in length. Additionally, the workers interpreted this rule as permission for the *entire family* to carry away any piece of wood less than one meter in any dimension *three times a day*. In case of need, workers frequently shortened longer and useful pieces to make them into *chips*. As can be expected, almost all wooden products in the surrounding towns like stairs, doors, windows, cupboards, and so on were just barely below the one-meter restriction. It is estimated that five-sixths of the wood brought into the yard was carried out as chips (Rosenband 1999). Bentham himself watched one night in secret how the workers carried out bundles and bundles of wood to sell them on the market. Up to then, no administration had the guts to restrict this practice, and quite possibly, they did not even know how much wood vanished as chips. For the workers, however, this was part of their salary, making up to 50% of their wages, and was essential for survival, since the navy was sometimes up to 15 months late in paying wages (Ashworth 1998). Bentham thoroughly changed the system. By introducing steam power, he made the cutting of wood a machine task, not a worker's task, and hence made it more difficult to cut wood into chips. In 1801, he forbade the taking of chips and, as compensation, introduced *chip money*, an additional payment on top of daily wages.* Remaining smaller pieces of wood were now intended for a new block mill in the Royal Navy Dockyards in Portsmouth (Ashworth 1998).

---

* While chips were initially only for wood workers, now everybody wanted a piece of the pie, and even sailmakers petitioned to receive chip money (Ashworth 1998).

Up to the end of the eighteenth century, most manufacturing breakthroughs were the result of one mastermind putting things together, possibly with a sidekick for financial support or sales (Arkwright, Watt, Crowley, Wedgwood, Montgolfier, Boulton, etc.). If there were aides or other inventors on the team, they were glossed over, as historians of the time also tended to idealize their subjects, making them larger and more perfect than they really were. But by 1800, manufacturing projects had started to reach a complexity where a single person was no longer really able to do everything. Almost all breakthroughs after 1800 were a team effort, even though the credit was not always equally shared. Additionally, historians at the beginning of the nineteenth century started to include these other contributors of ideas in their writings. As for Bentham, he would have never achieved what he did without the help of engineers Marc Isambard Brunel* and Henry Maudslay.

Marc Isambard Brunel (1769–1849) was born in France, trained initially as a priest and later as a cabinet maker. During the French Revolution, he made the mistake of talking ill of Robespierre, after which he quickly emigrated to America in order to escape the Reign of Terror and to literally save his neck. After receiving American citizenship, he was appointed chief engineer of New York City. During a dinner with founding father Alexander Hamilton, he overheard a conversation about the expense in block making in England. Seeing the potential to use his engineering and woodworking skill, he moved to England.

Besides his entrepreneurial interest, there may have been another reason for him to move to England. During his escape from France, he had to leave behind his sweetheart, Sophia Kingdom, who was arrested as an English spy and escaped execution only when Robespierre was executed before her. Afterward, she moved back to England. They finally met again in London, and within less than one year, they were married. Conveniently enough, Sophia's brother was an undersecretary to the navy board, giving Brunel access to the network inside of the navy (Cooper 1984). Brunel contacted the primary block supplier, Taylor, with his ideas about mechanizing block production. However, Taylor rejected the offer of the untried machines, a mistake that put him out of business within five years (Coad 2005, p. 51).

* Not to be confused with his even more famous son, Isambard Kingdom Brunel, genius engineer and builder of bridges, ships, and railways in Great Britain.

While investigating different business opportunities, Brunel was also introduced to British machine toolmaker genius Henry Maudslay (1771–1831). Maudslay set up his own shop after his former employer refused a wage increase, and already employed over 80 men only two years later (Rosen 2010, p. 196). Maudslay also made himself a name in the industry by fundamentally changing the way a tool is used on a lathe, and also as the first to use standard screw sizes within his own workshop. He may even have made navigational instruments with interchangeable parts, although only in small quantities (Rosen 2010, p. 192). He was probably the very best machine maker of his time and is considered to be the father of machine tool technology.

Brunel and Maudslay worked together to build some scale models of machines for block making in order to propose the idea to Bentham, who accepted the proposal. With Bentham providing the leadership and financing, Brunel designing the machines, and Maudslay building the machines, they set to work. This is not to say that there was no friction among them; in fact, they disagreed a lot, but together, they managed to push manufacturing into a new era of specialized machine tools.

Before he met Brunel and Maudslay, Bentham already installed the first steam engine in 1799 to pump dry docks dry during the night and to power machine tools during the day. He also installed the first reciprocating power saws powered by the steam engines. Bentham's aim was to build skill into the machines, rather than painstakingly train skilled workers (Coad 2005, p. 63). In most cases, it was no longer the worker who determined the final shape of the product but the machine. For example, the holes created during the first boring of the shell were used as reference points for all subsequent machines. Using these holes, it was possible to clamp every shell in exactly the same position, allowing for accurate manufacturing in the subsequent steps. Similarly, the internal hole in the sheave was also used to position the part in a machine. Most machines stopped automatically once the product was finished. All the worker had to do was to clamp the part, start the machine using a clutch, and maybe push the part through the machine until a fixed stop was reached. After unloading, the cycle repeated. The precision of the part was now independent of the skill of the worker. This also allowed Bentham to use cheaper low-skilled labor, for example, untrained boys rather than highly skilled and strong-willed artisans (Cooper 1984).

Probably the most advanced machines were the mortising machine making the slots for the sheaves in the shell and the shaping machine for rounding the cornered shells. The shaping machine was able to make 10 blocks

FIGURE 9.4
Portsmouth block mill shaping machine. (Image from the Edinburgh Encyclopædia, 1832.)

at the same time, a marvel of early-nineteenth-century technology (Figure 9.4). While Brunel designed all machines with a wooden frame, Maudslay built them using iron frames, a technology he used before for his lathes. While wooden frames move slightly while a machine is in use, a metal frame is much sturdier and hence allows for much higher precision. All machines in the Portsmouth block mill had iron frames, with the exception of the largest saws. Also, each machine was a single-purpose specialized machine. Except for an early sheave cutting machine that was later replaced, there were no general-purpose drills or lathes in the workshop; each machine was designed to do one task and one task only (Cooper 1984).

It has to be pointed out that there were not yet suitable metalworking machines to produce the machinery. Rather, all the machines in the workshop were made by hand. With the exception of round parts, for which lathes were available, all other parts had to be cast or forged into a rough shape, with more important surfaces filed by hand and the most important flat surfaces even scraped.*

* Even nowadays, for the highest-precision flat surfaces, scraping is industry practice; milling or planing machines have less precision than the human hand for extremely precise flat surfaces.

Overall, there were three sets of machines, one for making small blocks from 4 to 7 inches, one for medium-sized blocks from 7 to 10 inches, and one for large blocks from 10 to 18 inches. The first machines to be ordered were for the most commonly used medium-sized blocks, ordered in 1802 and installed in 1803, when the Portsmouth block mill started production. In 1804, the small-block machinery was installed, and in 1805, the large-block machinery (Coad 2005, p. 52). With the installation of the last set of machines, there were a total of 45 machines of 22 different types, and the mill was making enough blocks to cancel all contracts with outside suppliers. By 1807, all the staff were fully trained, and from 1808, the mill was running at full capacity of 130,000 blocks per year (Coad 2005, p. 75).

Besides the novel highly mechanized and automatic machinery, Bentham also spent considerable time organizing the workshop. All machines in the workshop were set up in sequence. The machines for the small blocks were aligned along the south wall, the medium-sized machines along the longer west wall, and the machines for the largest blocks were aligned along the north and east walls. Some machines processing parts for all types of blocks were set up in the middle of the room (Cooper 1984). This greatly reduced the transport distances of the parts, although there was probably still a batch production in place, where the operator made a lager quantity of parts before moving them to the next location.

The machines also allowed for a fixed production speed. No longer did the worker determine the speed of production; rather, the speed was given by the machines. The machines were also designed to be roughly the same speed, or, in some cases, two machines at half the speed. While not perfectly equal, this system allowed the production of approximately one block every 75 seconds. Furthermore, every worker was trained on two machines. Hence, absences and illnesses did not lead to production delays. The workers on slightly faster machines could work on different machines while these slower machines caught up with the faster ones in production. Except for the absence of a conveyor belt, the production in the Portsmouth block mills was mass production, using special-purpose machinery aligned in sequence and timed to similar speeds (Cooper 1984).

Economically, the mill was a great success. The investment paid for itself within four years, and the prices of blocks went down dramatically. What before was done by 110 skilled (and expensive) craftsmen was now done by 10 unskilled (and cheap) workers using the new machinery.

The quality of the parts also increased dramatically. The new blocks had less friction, broke less often, snagged lines less, and overall had a much longer life at a cheaper cost than the old blocks (Cooper 1984). Almost by accident, these parts were also interchangeable. The pins for the sheaves fit any metal insert in the sheaves without modification. The sheaves also fit any shell of the same type without adjustments (Alder 2010, p. 237).*

The project was such a success that a second set of machines was constructed in 1815–1817 as spare machinery in case the workshop burned down. These were stored in a fireproof room at the Chatham Dockyard, but they were never used, and this *fireproof* room burned down during the nineteenth century (Coad 2005, p. 99). The mill was also famous for its technology beyond the military and was mentioned in most guidebooks. The *Encyclopedia Britannica* even had its own article about the mill in six consecutive editions from 1801 until 1890. Brunel even complained about the frequent interruptions by visitors and wanted to put a fence around the mill, but Bentham was actually promoting this publicity (Cooper 1984).

The Portsmouth block mill was producing pretty much all blocks for the navy after 1805. The mill closed around 1960 but not because the machines were worn out or the system did no longer work. The reason for its closure was that there were simply no more sailing ships to be equipped with blocks. The first steam ship of the navy was the HMS Devastation in 1870, and the last sailing battleship was produced in 1903. After that, almost no more blocks were needed, and modern blocks made of metal and plastic have since replaced wooden blocks. The machinery designed by Brunel and built by Maudslay is still functioning as intended. While most of it is now in a museum, some is still used nowadays, although for different products (Cooper 1984).

The ideas behind woodworking machinery, however, did not gain hold in England, probably because there was simply not enough wood. Additionally, wood was about 20 times more expensive there than in Germany and the United States (Cooper 1984). Furthermore, the British consumer was not interested in mass-produced goods but preferred custom-made items. Brunel, for example, built a factory for the mass production of shoes in 1812, supplying to the military. When the Napoleonic

* Keep in mind, however, that in terms of complexity, these blocks were much simpler than a gunlock.

wars ended quicker than expected in 1815, Brunel was left with a large number of unsold shoes. While these were durable and cheap, the general public preferred custom-made shoes rather than standard shoe sizes,* and Brunel even had to spend a few weeks in a debtors' prison (Cooper 1984).

The technology, however, did find its application in the metal industry, where numerous trained mechanics and engineers from the Portsmouth block mill greatly contributed to the development of machine technology. However, the idea of mass production did not gain hold in England. Despite the Portsmouth block mills in England being the first example of mass production with automated machinery, and despite the French musket production using interchangeable parts long before the United States, the system gained its biggest popularity in the United States. Hence, it is known as the American system of manufacturing. In contrast, the system of fitting parts by hand is known as the English system of manufacturing.

9.3 THE UNWILLING ENTREPRENEUR—ELI TERRY'S WOODEN CLOCKS

The first production of interchangeable parts in America was the result of ingenious clockmaker Eli Terry (1772–1852). Clockmakers started making clocks in the United States around the eighteenth century. Since metal was very expensive, only the most crucial parts were made of metal. The rest of the clock was made out of wood. Of course, a wooden clock was much less accurate than a metal clock and also needed to be rewound much more, often due to its increased friction. Despite the mostly inexpensive material, a wooden clock was still a luxury piece of mechanical art, and most clocks were made to order. All gears and parts were made by hand and filed to match each other during the tedious and time-consuming process of assembly (Murphy 1966).

It was in this industry that a young clockmaker, Eli Terry, started his own business in 1792 in Connecticut. Although Terry apprenticed as both a wooden clockmaker and a brass clockmaker, he focused on wooden

* Interestingly enough, arguably the most famous car companies in Britain are Rolls Royce, Bently, and Aston Martin, all specializing in high-end, customized, almost handmade cars. Compare this to the large car companies in America, continental Europe, and Asia, which made their fortune through mass production. Even the small British car brand Mini distinguishes itself by selling unique rather than cheap cars.

clocks, cutting each tooth of every gear by hand, just like every other clockmaker in the world did. Shortly after 1800, however, he invented a simple machine to mechanize the cutting of gears. Rather than shaping each wooden gear by hand, he was now able to cut multiple wooden disks into gears at the same time using a circular saw. The machine did not only the cutting but also the positioning of the parts for the next tooth of the gear (Raber et al. 1989, p. 464). As wooden parts require much less precision than metal parts, almost by accident, these gears were interchangeable. This allowed him to produce significantly more than the 10 clocks a common clockmaker was able to produce per year (Murphy 1966).

In January 1806, he expanded his business by establishing a small factory with a supply of water power. In the same year, he also made contract with two merchants for an enormous number of clocks. Terry promised to build 4000 clocks within three years, a multiple of what a conventional clockmaker could achieve in his lifetime (Hoke 1990, p. 52). However, this meant that his new factory was too small only a few months after establishment. In June 1806, using the capital provided by the merchants, he started to build a much larger factory. Even though it took him two years to build and fine-tune the factory and the machines, the factory started producing clocks in very large numbers (Muir 2002). Terry was able to produce 3000 clocks per year, not only a hundred times faster than any other clockmaker but also with interchangeable parts and significantly cheaper.

Having fulfilled his contractual obligation to make 4000 clocks, rather than continuing to make lots of money, Terry sold his factory to his two assistants in 1810 for $6000—about $100,000 nowadays. Aged only 38, he retired with a modest fortune to do what he really loved—to tinker with his clocks. During his tinkering, he developed a radically new pillar-and-scroll shelf clock, which was much easier to build and to adjust than any other clock before. Adjusting the critical escapement that determined the speed of the clock was now possible from the outside after the clock was assembled (Hoke 1990, p. 69). It was now no longer necessary to disassemble the clock, adjust the escapement, and reassemble it until the adjustment matched. In 1812, he came out of retirement and started another clock factory to produce his clock designed for ease of assembly and low cost, using interchangeable, machine-produced parts.

Terry patented his inventions, but like many other inventors, he had to find out the hard way that having a patent and enforcing it are two very different things. Hundreds of workers in dozens of clock factories in Connecticut copied both his machines and his clock, creating a boom in cheap wooden clocks. Some of them even advertised their copies as *patent clocks* (Muir 2002). Terry and his three largest competitors produced 15,000 wooden clocks per year in 1820 (Murphy 1966). Before Terry, a wooden clock including the case was around $20 to $40 (the equivalent of $350 to $700 in 2016). Through mechanized production, interchangeable parts, and easy-to-assemble clocks, the price dropped to around $10. A wooden clock turned from a luxury item for the rich to a standard household item in 1830 (Hoke 1990, p. 56ff). While in 1800 every clock was handmade, there were virtually no small clockmakers left in the business in 1830.

However, Terry was interested in making clocks, not money. After having made enough for the rest of his life, he sold his factory in 1820 and retired (again). For the rest of his life, he did what he forced so many others to quit—he produced clocks the old-fashioned way by hand until his death in 1852. His workshop was one of the last of many traditional craftsman clockmaker workshops.

9.4 SPRINGFIELD AND HARPERS FERRY ARMORIES—THE AMERICAN SYSTEM OF MANUFACTURING

The achievements of Honoré Blanc were soon forgotten in France. The interchangeability of block components was for a product with only very limited complexity, and wooden clocks were only marginally more complex. It was the armories of the American government that put interchangeable parts firmly and permanently on the agenda of manufacturing. Hence, production using interchangeable parts and heavy mechanization would eventually become known as the American system of manufacturing, or armory practice. But in the eighteenth century, America was still a technological backwater. Most manufacturing was little different from that during medieval times, with most products manufactured either at home for family use or in small, often seasonal workshops. Great Britain treated America, like most other colonies, as a source of raw materials and a market for finished goods. Most metal and many other finished products were imported from Great Britain.

This started to change only with the American War of Independence in 1775–1783 and the Declaration of Independence in 1776. Despite England's strict laws prohibiting export of technology and skilled technicians, the first cotton mill was established in the United States in 1790 using British technology with the help of the—from the British point of view—illegally emigrated Englishman Samuel Slater (see Section 7.5).

Already during the war, the U.S. Army was created. Outfitting the troops with small arms was done through small purchases from individual gunmakers, both domestic and abroad.* To reduce dependence on foreign suppliers and to take control of the production of weapons, the U.S. Congress decided in 1794 to establish three to four armories. After quite some political wrangling about possible locations, two armories were eventually built. The first one was the Springfield Armory in Massachusetts, built in 1794. The site was already an arsenal during the war and was selected for its good accessibility, being connected to three rivers and four major highways. The second location, Harpers Ferry Armory, established in 1798 and located in Virginia (now West Virginia), was much more remote. President Washington, who happened to grow up in Virginia, had to personally push for this location on account of it being well protected and far inland.

The idea of interchangeable parts, however, was quickly and firmly rooted in the U.S. military. Thomas Jefferson promoted the idea of interchangeability, which he observed in France (see Section 9.1). Furthermore, French officers that emigrated to the United States also brought with them technological experience with interchangeable parts (Raber et al. 1989, p. 122). This vague idea was the key topic of a meeting in 1815, where the chief of ordnance Wadsworth discussed the *uniformity system* with the superintendents of both armories, Roswell Lee of Springfield and James Stubblefield of Harpers Ferry. As a result, interchangeable parts were the vision and the direction of the U.S. armories, although probably none of the participants imagined that it would take 35 years to come into effect.† However, interchangeability was still very expensive. Any firm subject to the market forces of supply and demand would not have been able to produce interchangeable parts. Only a government had the resources to push the boundaries of manufacturing.

* As you may remember, due to the demand for arms by the Americans, French gunmakers preferred to sell to America than to the French military. This gave the French military problems equipping their troops for their wars, setting in motion a process that led to interchangeable parts in France and—as we will see—eventually also in the United States.

† I find it amazing that an organization was able to keep up the momentum for this vision for 35 years. Nowadays, many visions don't even survive the current CEO.

Initially, both armories used traditional techniques and organization. Workers had a high level of independence, working on a wide number of different tasks to produce guns. Assembly still was a trial-and-error process, and most parts had to be filed extensively until they fit into this particular weapon, but no other. Even the screws were filed by hand. Unsurprisingly, the quality of the work was unsatisfactory, and many weapons failed. Working times were more of suggestions, and employees came and went as they pleased. Drinking alcohol, gambling, singing, chewing tobacco, and quarrels were widespread. Everybody had access to the sites, and with lots of merchants, butchers, beggars, and peddlers visiting, it looked more like a market than a manufacturing location. Even when a high-ranking official visited the armory, workers did not even bother to look up from their newspapers* (Raber et al. 1989, p. 277f). It can be safely assumed that lots of the materials intended for government use were pilfered by craftsmen, either as raw materials or as finished products.

The evolution of the two armories differed significantly. Springfield pushed heavily for structure and organization. One of its biggest achievements was a rigid division of labor. While in 1810 there were 20 distinct jobs in making a weapon, this increased to 34 tasks in 1815, 86 tasks in 1820, 100 tasks in 1825, and 400 tasks in 1855 (Hindle and Lubar 1986, p. 228; Smith 1980, p. 82f). Through this finely detailed division of labor, it was much easier to train a worker in his task, and he could become much more skilled and faster in his particular task. At that time, Springfield had the finest division of labor in the world. As expected, the quality of the products improved significantly, while cost dropped.

Combined with the division of labor, Superintendent Lee also aimed for a consistent workforce. In 1812, he changed the majority of the work from a daily wage to a piece-rate wage, combined with a quality-based payment. This greatly improved product quality. To limit access to the site, he put a fence around it. Fighting and drinking were forbidden. An effort to introduce a fixed 10-hour workday, however, failed miserably. For decades, workers were able to resist losing control of their time (Raber et al. 1989, p. 25). Lee also established superior accounting practices for both materials and quality by tracking them. The materials a worker received had to equal the sum of products, rejected parts, and wasted material. The worker had to pay any difference out of his own pocket (Chandler 1977, p. 73ff). Overall, this made the Springfield Armory, with 250 employees, not only

* Apparently, the official was not pleased, as reading newspapers was forbidden shortly thereafter.

the largest but also the best-managed factory of the first half of the nineteenth century.

The other armory at Harpers Ferry, however, did much worse. Being far inland, it was believed to be better protected from external enemies than Springfield. However, exactly that location was the cause of most of the problems in Harpers Ferry Armory, including its violent end. The location was so remote that it took materials and visitors between 3 and 13 days to reach the armory, depending on the weather. There were a shortage of labor, frequent flooding, seasonal lack of water for the mills, summer heat, and diseases (Smith 1980, p. 33ff).

Due to its labor shortages, workers had lots of power, and due to its remoteness, there was a lack of administrative oversight. The Harpers Ferry Armory quickly became a swamp of corruption and bribery. Work was highly unregulated. A skilled worker was able to produce his monthly quota in half a month, spending the other half on private business (Smith 1980, p. 65f). Partying, drinking, and gambling were actively supported by the superintendent, James Stubblefield. This was reportedly to support the morale of the workers but quite likely also because Stubblefield owned the local brewery. Both management and workers were highly skilled in lining their own pockets at government expense. Relatives of arsenal management soon established businesses providing the arsenal with overpriced materials and workers with overpriced food and goods, which the workers paid for with overpriced wages, courtesy of the U.S. government (Smith 1980, p. 140ff).

The government made some moves to get control of the situation. It was proposed that Superintendents Lee from Springfield Armory and Stubblefield from Harpers Ferry change positions temporarily. While Lee reluctantly accepted this bag of problems without being able to change anything, Stubblefield just went home to Virginia, called in sick, and then did not respond at all until he went back to Harpers Ferry. A phony investigation of the situation found no faults with Stubblefield, even though he refused to implement wage cuts, as ordered by his supervisors. Eventually, he was forced to resign in 1829 after 22 years of mostly mismanagement and corruption, largely influenced by an even more corrupt master armorer, Armistead Beckham (Smith 1980, p. 171ff). His successor, Thomas B. Dunn, tried to establish proper business procedures. However, in doing so, he made himself very unpopular in Harpers Ferry. He was shot to death in 1830 after less than six months in office by a worker discharged for *worthlessness* (Torp 1830). The worker was executed for his crime but turned into a folk hero in Harpers Ferry (Smith 1980, p. 256). Another superintendent in 1841 tried again to establish proper procedures, resulting in a large strike and a committee of workers traveling to Washington to complain personally to President Tyler about the situation. Overall, Harpers Ferry was *one of the most troublesome and disorganized labor forces in the country* (Smith 1980, p. 270).

Hence, it is most curious that in this swamp of corruption, the first weapons with interchangeable parts outside of France should emerge. Directly responsible for this outstanding act of manufacturing is John Hancock Hall (1781–1841). Hall was an experienced gunmaker and invented a new breech-loading rifle, where the bullet and the charge are loaded in the rear of the barrel rather than being pushed down the front. His model M1819 rifle was the first breech-loading rifle produced in large numbers, greatly increasing firing speed. In 1816, he proposed producing his gun for the U.S. Army, using interchangeable parts, with

> ...every similar part of every gun so much alike that it will suit every gun, e.g., so that every bayonet will suit every barrel, so that every barrel will suit every stock, every stock or receiver will suit every barrel, and so that if a thousand were taken apart and the limbs thrown promiscuously together in one heap, they may be taken promiscuously from the heap, and will all come right.

The U.S. Army readily accepted and set him up with his own workshop in Harpers Ferry, giving him lots of independence. This was, of course, a major problem for Superintendent Stubblefield, who ran the Armory like his own kingdom. He did not like it at all that there was a workshop in his armory but out of his control and run by a strong-minded private outsider. It did not help that Hall had a curt and outspoken personality. Hence, Stubblefield and most of his successors tried to slow down Hall wherever they could. Hall had constant problems getting materials, men, and machines for his work. The workshop he was given was too small, and he constantly had to move his machines around to make space for the current production process (Hounshell 1985, p. 43).

However, despite these obstacles, Hall succeeded. He soon realized that interchangeability was all about precision. While interchangeable wooden parts like Eli Terry's clocks could be produced with ±2-millimeter accuracy, for metal parts, the accuracy needed to be closer to ±0.2 millimeters to work and almost ±0.02 millimeters for interchangeability. Hence, Hall focused intensely on precision. Back then, there were no reliable rulers or micrometers that could measure with the required precision (Gordon 1988b). Hall therefore developed an extensive system of gauges to determine if a part was within tolerance, too large, or too small. More than 63 gauges were needed to determine if the rifle met specifications.

To further increase accuracy, he created what he called a bearing point. Measuring different dimensions with different gauges creates the risks of errors adding up. If the distance from one hole to the next is measured, and then the distance from the second hole to the third, any error in measurement is cumulative and will sum up to a larger error. Hall's bearing point was the point against which all dimensions were verified, avoiding accumulation of errors and increasing accuracy (Hounshell 1985, p. 41).

Hall also created high-precision machinery. For starters, he had two types of machines, high-precision machine-making *primary machines* that were used to make other *secondary machines*, which actually produced the parts of the gun. All of his machines also had a cast-iron frame, much sturdier than other contemporary wooden-frame machines. These solid machines were able to work with greater accuracy than ever before. He also balanced the driveshafts that delivered the power from the water mill to his machines, therefore significantly reducing vibration (Smith 1980, p. 228ff). Through his focus on machine design, he greatly increased the accuracy of his parts.

Additionally, he also used new machine technology. His machines needed no skilled operator but usually only a boy, who put the part in the machine, started the process, and removed the part after the machine had stopped automatically. He invented drop forging for metal parts (Hounshell 1985, p. 41) and utilized and improved the milling machine invented by Simeon North in 1816 (Hindle and Lubar 1986, p. 182). Thus, he was able to mill a part directly to the correct dimensions where Springfield still filed away large amounts of metal by hand.* Even after hardening—which usually created problems in Springfield—Hall's parts were still accurate. Hence, by 1824, his rifles were fully interchangeable. Any part would fit any other part to form a working rifle.

After producing fully interchangeable rifles in his workshop in Harpers Ferry, Hall worked together with the aforementioned Simeon North, who was producing the rifle invented by Hall under license. North was an

* It is estimated that for every musket produced, one file was worn out. Springfield, for example, showed 360,000 files in its inventory in 1856. Roughly half of all metal working tools were files (Raber et al. 1989, p. 234ff).

independent arms manufacturer, who looked into interchangeable parts before but dropped the idea because it was too expensive. However, something like a big price tag does not stop a government from doing what they think is right. Hall worked with North to have their weapons interchangeable not only with other weapons from the same workshop but also across workshops. This required a centralized set of master gauges, with secondary gauges being distributed to the different workshops. Starting in 1828, it took a six-year effort, but in 1834, the weapons produced by Hall and by North were completely interchangeable despite being produced in different workshops. This was the first time in the world that interchangeability across different sites for a complex product was realized. Hall, however, died impoverished, probably from tuberculosis, in 1841 (Smith 1980, p. 223).

The workshop of John Hall was the most accurate metal workshop of its time, far ahead of Springfield and light-years ahead of the mess of the rest of Harpers Ferry Armory. Springfield did not yet have interchangeable parts and achieved interchangeability only in 1849. Before that, weapons from Springfield were not interchangeable, and every part in the final assembly had to be filed manually to match another part. Their goal, nevertheless, was still interchangeability, as envisioned in the 1815 meeting, although they tried to achieve it mainly through the use of gauges. Even noncritical parts were measured with gauges. For example, the trigger guard, having very little requirements on accuracy, was still filed down from a 2-kilogram block of metal to a 200-gram final part in 1917 (Gordon 1988b).

Where Springfield excelled, however, was—besides administrational efficiency—mechanization of production. From the beginning, Springfield had a small water shop. However, due to geographic conditions, it was away from the main workshops, where basic milling, grinding, hammering, and polishing of barrels was the routine in 1815. During the next decades, Springfield made numerous advancements in machine technology. While not achieving the precision of Hall's machines, they greatly reduced the cost of production. The mastermind behind these machines was Thomas Blanchard (1788–1864). Throughout his lifetime, he created numerous inventions, from a tack-making machine at age 18 to steam cars and boats to a machine for folding envelopes and many more. From the 1820s onward, he worked as an independent contractor at the Springfield Armory. The armory provided space, materials, and power, while Blanchard hired his own men to build and operate the machines.

In Springfield, one of his main achievements was the automatic gun barrel lathe (Figure 9.5), where the gun barrel was completed using one machine for both the round parts and the octagonal barrel. More significant, however, than the gun barrel lathe was the invention of a copying lathe for manufacturing gunstocks as shown in Figure 9.5. The gunstock was a very irregularly shaped piece of wood that, until then, was produced mainly by hand, first by sawing and later by filing the part to shape. Naturally, the differences between one stock and the other were significant, and even for the looser tolerances of wood, interchangeability

FIGURE 9.5
Thomas Blanchard's copying lathe. (Photo by James Langone at the Springfield Armory National Historic Site. With permission.)

was impossible. Blanchard developed a lathe that was able to copy the irregular shape from a model and form the workpiece in its likeness. The principle was similar to a modern key-copying machine, where a roller moved across the rotation surface of the model and, by moving up or down, moved the cutting tool across the workpiece accordingly. What was a time-consuming and highly variable process done by a skilled craftsman became a routine operation for the machine. An unskilled operator merely loaded the part, started the machine, and unloaded after the machine stopped operation upon completion.

However, there were still many steps left to complete the gunstock; a number of holes and cavities had to be shaped; and also, the thin end of the gunstock had to be formed. This prompted Blanchard to continue his drive for mechanization and automation. By 1827, he mechanized almost the entire process, using 14 different machines, from automatically sawing off stock material to its proper length to forming the bed for the lock plate and boring ping holes. With these machines, 17 men were able to do the work of 75 men before, and with much higher accuracy and precision (Cooper 1988). This made the Springfield Armory the most automated workshop in the United States and probably the most automated workshop since the block production at Portsmouth.

Even though the workshop was—and still is (Raber et al. 1989, p. 24)—praised as a fully automated production line, there was significant manual work to be done. Not all of the machines were fully automated; for example, the bedding machine making the groove for the barrel still required a skilled operator to control two independent movements, one in each hand. In addition, out of the 17 people working in the workshops, only 10 operated machines; the other 7 did manual work on the stock. Even after Blanchard built new machines and redesigned existing ones, there were still about half a dozen people working by hand on the parts (Cooper 1988).

Blanchard was also skilled in protecting his copyrights. He went after each and every violator like a bloodhound. Another inventor, Azariah Woolworth, invented a similar machine independently. Woolworth's machine used a back-and-forward motion to make wooden shoe lasts, while Blanchard's machine rotated to make gunstocks. Woolworth challenged Blanchard's patent, and despite the machines operating on different principles, Blanchard won after many years in court (Cooper 1991). After his 1820 patent was extended for another 14 years in 1834, he lobbied Congress for an unprecedented second extension of another 14 years in 1848. Knowing how to flatter congressmen, he copied their marble busts using his machine, claiming that this was a new use and, hence, he deserved another extension. He also claimed that he did not make any money, due to legal fees suing other violators. Unbelievably, he got his patent extended again until 1862 (Cooper 1991).

With this extended patent, he now sued some of his previous customers that correctly paid the royalty for his machine, claiming that he now deserved more money since the patent now applied to a longer period. Fortunately, this idea did not fly in court. He was more relaxed, however, with himself using the intellectual property of others. For example, he utilized another invention to use scale models instead of full-sized models in his own patents without blinking an eye (Cooper 1991). Overall, his inventions and his skills made him wealthy. He died in 1864 in Springfield.

Together, the Springfield and Harpers Ferry Armories—guided by the vision of Wadsworth, the organizational skill of Lee, and the technical ingenuity of Hall, Blanchard, and North—created the most advanced and influential manufacturing site of its time.

The Springfield Armory continued to be a cornerstone of the U.S. Army, supplying weapons during World War I, World War II, and the Vietnam War until production ended in 1968 after over 170 years. The site, however, continues to exist as the Springfield Armory National Historic Site.*

Harpers Ferry, however, was not so lucky. The beginning of the end was in 1859, when abolitionist John Brown tried to start an armed slave revolt by raiding the Harpers Ferry Armory.† This resulted in numerous deaths among both the abolitionists and the townspeople. Two years later, in 1861, the Civil War started. Harpers Ferry was very close to the Mason–Dixon line, the cultural border between the North and the South. Union soldiers tried to burn down the Armory to prevent it from falling into Confederate hands, but the townspeople were able to extinguish the fires and to reduce the damage. The technologically inferior Confederates soon moved the materials and machines to the south. During the Civil War, Harpers Ferry changed hands multiple times, with the bridge being burned down and rebuilt at least nine times. Most of the town was destroyed, and the armory

* Interestingly, the name Springfield Armory was licensed to a weapons company in Geneseo, Illinois, which—although with no connection to Springfield Armory whatsoever except for the name—refers heavily to its *founding date* in 1794.

† Curiously enough, John Brown lived for many years in Springfield, Massachusetts, and gained most of his radical views while living next to the Springfield Armory.

was never rebuilt. While the site was selected far inland to protect it from enemies outside, it fell victim to the enemies within.

The idea of interchangeable parts, however, was to stay. While the French achieved interchangeable parts already during the end of the eighteenth century, this effort was forgotten in France only 30 years later. Yet, the spark of the idea made it to America. Similarly, while the British achieved interchangeability using automatic machines around 1800—albeit for a much less complex part—this never spread much outside of the British Navy. Only in America did the idea of interchangeable parts and automatic machines spread outside of military workshops. Numerous technicians and workers who learned their crafts in one of the armories moved to other industries, bringing their ideas about interchangeability with them. The production of sewing machines, typewriters, commercial firearms, reapers, bicycles, and eventually the automobile all benefitted from the developments at the Springfield and Harpers Ferry Armories. Hence, this system using interchangeable parts is known as the American system of manufacturing, compared to the older British system of manufacturing, where parts had to be fitted by hand during assembly.

Through this effort, the technological backwater America of 1800 overtook the rest of the world only 50 years later. During the Crystal Palace World Fair in London in 1851, most countries presented products so heavily decorated and artistically enhanced to the point of being no longer useable. American exhibits on the other hand were no frills, no thrills—but they worked perfectly. While at first American products were smiled upon for their lack of artistic refinement, this turned to stunning disbelief after they won award after award. The United States received awards for rifles, presses, pianos, colts, textile machines, fireproof safes, meat-processing rubber products, and artificial legs, to name just a few of the 159 awards won (Hindle and Lubar 1986, p. 250ff). In some cases, the U.S. product was even the only one that worked at all. For example, during a test of automatic reapers in bad weather, the McCormick Reaper from the United States was the only one to actually work (Pursell 1995, p. 186). A particular embarrassment to the British was that American locksmith Alfred Charles Hobbs picked the *unpickable* British locks, while the British were unable to pick his American lock, and that America won the yachting race, now known as America's Cup. Overall, one out of four U.S. products won an award, and the mocking of American products soon turned into talk about the American superiority and the American system of manufacturing (Hindle and Lubar 1986, p. 250ff).

The American system of manufacturing was also known as the American plan, American principle, or armory system. However, it was never really defined what exactly constitutes an American system of manufacturing. Some see it only as the use of a customized automated machine that requires little operator input; others see it only as heavy mechanization. The use of interchangeable parts, precision machining, machine arrangement in the sequence of production, stringent organization of the labor force, and high division of labor is also often seen as part of the American system. In short, everybody in industry was talking about it, but the understanding of the concept was very vague. Everybody brought in a new idea or approach, and in the minds of some, it eventually encompassed all that is good in manufacturing.* In any case, due to the superiority of American technology, these technologies—although not all invented in the United States—became known as the American system of manufacturing (Hounshell 1985, p. 331ff).

9.5 WHAT ABOUT ELI WHITNEY?

At this time, if you are well versed in industrial history, you may wonder: what about Eli Whitney? Shouldn't he be in the story here? After all, Eli Whitney (1765–1825) is hailed as the inventor of the cotton gin, the milling machine, and interchangeable parts in the U.S. armories and almost single-handedly invented the American system of manufacturing. Well… unfortunately, more recent research shows that most of this is not true, and the rest is highly doubtful. He was, however, most certainly a very gifted self-promoter.

His famous invention of the cotton gin, a machine to separate cotton fibers from its seeds, was most likely no more than a minor improvement to existing technology rather than a fundamental breakthrough (Lakwete 2005). In protecting *his invention*, Whitney applied for a patent in 1793. However, the patent was contested and was not validated until 1807, after which there was only one year of patent protection left. Due to the expensive legal proceedings, and problems with his cotton mill, Whitney faced bankruptcy and urgently needed money.

* Curiously, we will see something similar 100 years later. Everybody nowadays talks about the Toyota Production System or Lean manufacturing, but nobody ever defined what exactly this includes and—even less so—what it does not.

Hearing about the new idea of interchangeable parts, he proposed to the government to make interchangeable weapons, as long as he got a cash advance. In 1798, he got a contract to produce 10,000 muskets within only two years for the extremely low price of $13.40 per musket, despite having no experience in the metal industry, never having produced a single gun, having no experience with interchangeability, and having no suitable workshop or manufacturing location. Nevertheless, the enormous cash advance of $5000 immediately solved his financial problems. It did not help, however, in producing the weapons. By September 1800 rather than having delivered 10,000 weapons, he had delivered exactly none. The first 500 weapons were delivered one year late in September 1801, and not until 10 years later in 1809 did he complete the 10,000 weapons (Woodbury 1960). He did, however, deliver frequent excuses regarding the delay, emphasizing his honest effort. He also asked for more cash to keep his operation going and continued to milk money out of the government (Hounshell 1985, p. 29).

Surprisingly, in 1801, Whitney demonstrated the interchangeability of his guns for President John Adams and Vice President Thomas Jefferson by disassembling and reassembling weapons at random. However, in all likelihood, this was a carefully staged dog-and-pony show. The critical point of interchangeability was to swap the high-tolerance metal parts of the complex lock mechanism. Whitney, however, did not disassemble the complex lock mechanism at all but merely interchanged the locks among different guns. Hence, the locks only had to match the groove in the stock, which had a much more relaxed tolerance than the internal components of the lock (Hounshell 1985, p. 31). He also used only 10 guns as examples, which were carefully selected beforehand and hand-shaped to make the groove in the stock match every lock of the 10 guns.

None of his weapons were anywhere near interchangeable. Detailed modern investigation of his weapons showed all signs of batch production and filing to fit (Battison 1973), and the weapons were not even close to being interchangeable (Woodbury 1960). His parts were actually the most sloppily manufactured parts of all U.S. gun manufacturing sites (Gordon 1988a). Even taking into consideration that the idea of interchangeable parts was still ill defined in 1801, Whitney successfully deceived the president of the United States and the rest of the audience* about the abilities of his products, allowing him to extract even more money from the government.

---

* And also a large number of historians and writers afterward.

He also claimed to have invented a barrel-turning lathe in 1808, but there is no evidence other than his claim (Smith 1980, p. 117). A milling machine discovered in 1912 was quickly attributed to the then still legendary Eli Whitney by his grandson as the first milling machine ever, but the evidence was very sketchy (Cooper 2003; Woodbury 1960). That particular machine was later dated to 1827, two years after Whitney's death (Battison 1973). Overall, pretty much all of Whitney's achievements have been debunked by modern historians. He did, however, greatly promote the idea of interchangeable parts, helping the government to keep up momentum toward this vision, although his motives were probably much more financial than ideological.

BIBLIOGRAPHY

Alder, K., 1997. Innovation and Amnesia: Engineering Rationality and the Fate of Interchangeable Parts Manufacturing in France. *Technology and Culture* 38, 273–311. doi:10.2307/3107124.

Alder, K., 1998. Making Things the Same: Representation, Tolerance and the End of the Ancient Regime in France. *Social Studies of Science* 28, 499–545.

Alder, K., 2010. *Engineering the Revolution: Arms and Enlightenment in France, 1763–1815.* University of Chicago Press, Chicago; London.

Ashworth, W.J., 1998. "System of Terror": Samuel Bentham, Accountability and Dockyard Reform during the Napoleonic Wars. *Social History* 23, 63–79.

Battison, E.A., 1973. A New Look at the "Whitney" Milling Machine. *Technology and Culture* 14, 592–598. doi:10.2307/3102445.

Chandler, A., 1977. *The Visible Hand: Managerial Revolution in American Business, new edition.* Belknap Press of Harvard University Press, Cambridge, MA.

Chartrand, R., 2003. *Napoleon's Guns 1792–1815 (1): Field Artillery: 66*, illustrated edition. Osprey Publishing, Oxford, UK.

Coad, J., 2005. *The Portsmouth Block Mills: Bentham, Brunel and the Start of the Royal Navy's Industrial Revolution.* English Heritage, Swindon, UK.

Cooper, C.C., 1984. The Portsmouth System of Manufacture. *Technology and Culture* 25, 182–225. doi:10.2307/3104712.

Cooper, C.C., 1988. "A Whole Battalion of Stackers": Thomas Blanchard's Production Line and Hand Labor at Springfield Armory. IA. *The Journal of the Society for Industrial Archeology* 14, 36–58.

Cooper, C.C., 1991. Social Construction of Invention through Patent Management: Thomas Blanchard's Woodworking Machinery. *Technology and Culture* 32, 960–998. doi:10.2307/3106158.

Cooper, C.C., 2003. Myth, Rumor, and History: The Yankee Whittling Boy as Hero and Villain. *Technology and Culture* 44, 82–96. doi:10.1353/tech.2003.0009.

Corvisier, A., Childs, J., 1994. *A Dictionary of Military History and the Art of War.* Wiley-Blackwell, Oxford, UK; Cambridge, MA.

Feldhaus, F.M., 1953. *Geschichte des technischen Zeichnens.* Kuhlmann K.G.

Gillispie, C.C., Alder, K., 1998. Engineering the Revolution. *Technology and Culture* 39, 733–754. doi:10.2307/1215848.

Gordon, R.B., 1988a. Who Turned the Mechanical Ideal into Mechanical Reality? *Technology and Culture* 29, 744–778. doi:10.2307/3105044.

Gordon, R.B., 1988b. Material Evidence of the Manufacturing Methods Used in "Armory Practice." IA. *The Journal of the Society for Industrial Archeology* 14, 23–35.

Hindle, B., Lubar, S., 1986. *Engines of Change: The American Industrial Revolution, 1790–1860*. Smithsonian Institution Press, Washington, DC.

Hoke, D.R., 1990. *Ingenious Yankees: The Rise of the American System of Manufactures in the Private Sector*. Columbia University Press, New York.

Hounshell, D.A., 1985. *From the American System to Mass Production, 1800–1932: The Development of Manufacturing Technology in the United States*, Reprint. Johns Hopkins University Press, Baltimore.

Hume, L.J., 1970. The Development of Industrial Accounting: The Benthams' Contribution. *Journal of Accounting Research* 8, 21–33. doi:10.2307/2674710.

Lakwete, A., 2005. *Inventing the Cotton Gin: Machine and Myth in Antebellum America*, New edition. Johns Hopkins University Press, Baltimore.

Muir, D., 2002. *Reflections in Bullough's Pond: Economy and Ecosystem in New England.* University Press of New England, Hanover, NH.

Murphy, J.J., 1966. Entrepreneurship in the Establishment of the American Clock Industry. *The Journal of Economic History* 26, 169–186.

Peaucelle, J.-L., 2005. Du Concept d'interchangeabilité à sa Réalisation: Le Fusil Des XVIII et XIX Siècles (the Concept of Interchangeabilioty and Its Realization). Les Annales Des Mines: Gérer Et Comprendre 80.

Pursell, C., 1995. *The Machine in America: A Social History of Technology*, Illustrated edition. Johns Hopkins University Press, Baltimore.

Raber, M.S., Malone, P.M., Gordon, R.B., Cooper, C.C., 1989. *Conservative Innovators and Military Small Arms: An Industrial History of the Springfield Armory, 1794–1968*. U.S. Department of the Interior.

Roe, J.W., 1926. *English and American Tool Builders*. McGraw-Hill, New York.

Rosen, W., 2010. *The Most Powerful Idea in the World: A Story of Steam, Industry, and Invention*. Random House, New York.

Rosenband, L.N., 1999. Social Capital in the Early Industrial Revolution. *The Journal of Interdisciplinary History* 29, 435–457.

Smith, M.R., 1980. *Harpers Ferry Armory and the New Technology: The Challenge of Change.* Cornell University Press, Ithaca, New York; London.

Torp, K., 1830. Assassination of Col. Thomas B. Dunn, *The Baltimore Patriot*.

Woodbury, R.S., 1960. The Legend of Eli Whitney and Interchangeable Parts. *Technology and Culture* 1, 235–253. doi:10.2307/3101392.

10

Social Conflict

Any man who has stood at twelve o'clock at the single narrow door way which serves as the place of exit for the hands employed in the great cotton-mills, must acknowledge, that an uglier set of men and women, of boys and girls, taking them in the mass, it would be impossible to congregate in a smaller compass. Their complexion is sallow and pallid—with a peculiar flatness of feature, caused by the want of a proper quantity of adipose substance to cushion out the cheeks. Their stature low—the average height of four hundred men, measured at different times, and different places, being five feet six inches. Their limbs slender, and playing badly and ungracefully. A very general bowing of the legs. Great numbers of girls and women walking lamely or awkwardly, with raised chests and spinal flexures. Nearly all have flat feet, accompanied with a down-tread, differing very widely from the elasticity of action in the foot and ankle, attendant upon perfect formation. Hair thin and straight—many of the men having but little beard, and that in patches of a few hairs, much resembling its growth among the red men of America. A spiritless and dejected air, a sprawling and wide action of the legs, and an appearance, taken as a whole, giving the world but "little assurance of a man," or if so, "most sadly cheated of his fair proportions..."

Peter Gaskell
In The Manufacturing Population of England, 1833

The industrial revolution introduced a whole number of disruptive technologies, including steam engines, railroads, and ships; automated spinning; automated weaving; and many others. These technologies were disruptive since they obliterated previous technologies, and often also the enterprises that used them. Formerly high-priced goods and services became inexpensive, creating many business opportunities for those who had access to these new technologies. On the other hand, many others saw their fortunes drop. A disruptive technology creates both winners and losers. The industrial revolution disrupted markets on a very large scale, creating some winners but also a large number of losers. As we will see in this chapter, the losers were not about to go down quietly.

10.1 EFFECT ON SOCIETY

With the advent of factories, factory owners were able to profit from economies of scale. Before, a craftsman worked independently. Larger quantities merely meant more craftsmen working independently, producing the same goods. At first, benefits from larger-scale operations were generated through larger-scale purchasing and sales, for example, in the putting-out system. However, centralized manufacturing using machines and mechanization also yielded large advantages due to economies of scale. At one point, half of the cotton used in the world was processed in Britain. It was cheaper to ship the raw materials from India to Britain, process it using mechanized processing, and ship it back to India, rather than using ineffective manual production in the originating country.

These economies of scale meant that the production was centralized in factories. This also meant that the manual labor had to be centralized, too. While there were only two major cities with more than 50,000 people in Britain in 1750—London and Edinburgh—there were 8 by 1801 and 29 by 1851. By then, most people lived in cities (Hobsbawm 1999, p. 64). This was a massive shift in population away from the countryside toward the cities. However, neither were the cities prepared for this massive population increase, nor were the poor farmers moving to the city able to afford adequate housing. They were able to afford only the cheapest rooms, often living with the whole family in one or two rooms. Houses were built cheaply and quickly. There was often no clean water supply, nor was there any sort of garbage disposal, and raw sewage was simply dropped on the narrow streets. In these slums, the death rate from disease and accidents exceeded the birth rates (Cameron and Neal 2002, p. 185f). In Manchester in 1842, life expectancy at birth was only 18 years for common laborers, whereas the middle and upper class could expect to grow 40 years old (Sale 1995, p. 48). Life was often miserable.

However, through the large economic benefits from factories, the overall wealth in Britain grew. The average Englishman was usually better off than the average Frenchman. Workers in Britain were able to afford meat almost every day. French workers and farmhands, on the other hand, ate meat much less frequently, while at the same time paying more for food. In Britain, leather shoes were also quickly becoming the standard, while in Europe, cheap self-made wooden clogs were still causing blisters (Landes 1969, p. 47). The British upper class even lamented that it was no longer possible to determine ranking in society through clothes, as even the lower middle class started to buy second-hand but well-made coats and top hats to wear on Sundays. The way of living changed fundamentally.

Before the industrial revolution, farmers or craftsmen could survive with little or no money, combining farming with manufacturing. They tended their fields or gardens and maybe kept some animals, generating most of their food themselves. As for tools, both farmer and craftsman were able to produce most tools themselves. They generated only a small surplus on goods for sale, but then, they did not need much that had to be purchased. Probably the biggest expense was for clothes, and maybe some metal tools that the average nonmetal craftsman could not produce himself. The people were mostly self-sufficient. This changed radically after the industrial revolution. Living in a city, the people no longer had gardens or farms but, rather, had to purchase food and rent their housing. Life became expensive, and people needed money. Losing their source of income meant a rapid decline down the social ladder (Berg 1985, p. 98).

Scholars are still discussing if the industrial revolution improved the standard of living in the eighteenth and early nineteenth centuries or if the standard went down. In any case, there is general agreement that from 1850 onward, the standard of living improved (Cameron and Neal 2002, p. 185f). Yet, until 1850, there were definitely some groups whose standard of living decreased sharply.

The common artisan was completely unprepared for the events of the industrial revolution. Their world changed rapidly, and it was not to their advantage. Most affected initially were textile-processing artisans, since textiles was one of the biggest markets and hence also the starting point of the individual revolution. Before mechanization, a weaver was a well-respected and well-paid member of society. He worked on his own time, usually drinking and partying during the beginning of the week before working hard to finish the products by Saturday (Landes 1998). His work was frequently interrupted for other household chores, gardening, or feeding livestock (Sale 1995, p. 36).

This all rapidly changed with the industrial revolution. Between 1790 and 1810, due to the mechanized competition, artisan wages were reduced by half, while living costs increased. Within only 20 years, the weaver turned from a well-paid specialist to a poor producer of inferior wares, matching neither the quantity nor the quality of machines (Rosen 2010, p. 241).

Similar events happened for croppers. A cropper used big shears to cut off excess fiber from the surface of the finished textile, and significant skill was needed to make the surface look even without cutting the textile itself. In 1814, there were almost 2000 croppers in Leeds, earning a respectable wage of around 36 to 40 shillings per week. Only 15 years later in 1830, there were only 100 croppers left, earning a measly wage of 10 to 14 shillings per week. The entire industry was turned upside down. In one factory, the work of 27 skilled croppers was now done by five unskilled men or children (Sale 1995, p. 23).

In short, workers lost their income. Their jobs went to the machines. Journeymen were used to losing their jobs, and they just went traveling until they found a new one. But now there were no jobs left in their trade anywhere. It was even worse for established masters, who may have had invested in tools and a workplace. In general, as wealthy members of society masters had much more to lose. They no longer had the means to support their middle-class lifestyle and their families and rapidly fell down the social ladder.

Depending on the craft, this even happened early in the twentieth century. A wheelwright, for example, was prestigious even among master artisans. His wooden wheels had to be completely symmetric, made out of different woods depending on the use of the wheel, and combined with iron parts for the rim and parts of the axle. Significant training was needed before the artisan was able to produce a proper wheel. An apprentice in wheel making was bound for success in life. However, this changed rapidly with the introduction of mass-produced wheels with rubber tires. A former wheelwright apprentice in Germany told me how he and his master went out during the early twentieth century to deliver a painstakingly crafted brand-new wheel ordered by a farmer. This farmer had just learned about the new metal wheels and rubber tires, which were not only cheaper but also better. The farmer, while visibly upset, had to pay for the wooden wheel. But afterward, in his anger, the farmer took an axe and smashed the brand-new wheel to bits in front of the shocked wheelwright. As the wood was breaking and the chips were flying, these wheelwrights realized that their craft had about as much future as this wheel. None!*

However, they lost more than that. Before, they were respected members of the community, proud of their skills and traditions. The new message, however, was crystal clear: Your skills are no longer needed. You are no longer needed. Your craft goes on the garbage pile of history.

10.2 DESTROY WHAT DESTROYS YOU—LUDDITE FRAME BREAKING

The industrial revolution turned many of these crafts into history, replacing years of artisan experience with a machine. Naturally, the artisans fought back. Since they were unable to compete on economic grounds, they fought back with violence. Workers revolted and destroyed machines in Colchester in 1715; in Barking in 1759; in Shepton in 1776; in Chorley in 1778; in Saint-Etienne, France; in 1789; in Bradford upon Avon in 1791; and in Trowbridge in 1785 (Alder 2010, p. 215; Berg 1985, p. 120; Burton 1976, p. 159; Rosen 2010, p. 240). The list could be extended significantly, but in short, violence against machines and factories was a constant threat to factory owners. In response, many of these factory owners sought to protect their enterprise through both defensive and offensive means (Pollard 1964). Arkwright's Cromford Mill, for example, was built more like a castle, with high walls, a gate, and few outside windows. The mill was also armed with a cannon and small arms to defend itself if necessary.

Probably the most significant of these movements was the Luddite uprising in Britain between 1811 and 1813. While most other riots and frame breaking were simply the rage of a violent mob, the Luddites were organized and structured, and acted over both a wider area and a longer period of time. The Luddites acted under the leadership of their *king* or *general*, Ned Ludd, who was rumored to live in Nottingham Forest. This forest was home to another legendary outlaw 300 years earlier, Robin Hood. Like Robin Hood, the Luddites also saw themselves as supporters of the poor and, in some letters, even gave their address as *Robin Hood Cave,*

* The apprentice I talked to was able to switch professions and make a good living, but his former master was less fortunate.

Nottingham Forest. The identity of Ned Ludd is still unknown. It may be a single individual or a group of people, or even different individuals at different times.*

Due to the prosecution of the government, the Luddites acted in secrecy, and hence, we know little of their inner workings. The first action of the Luddites was on March 11, 1811, in Nottingham.† Common workers vented their frustration during a protest in Arnold near Nottingham. This protest was broken up by the cavalry. The mob returned under cover of darkness and broke 63 machines in different shops that night, followed by another 200 destroyed machines in the vicinity over the next 20 days (Sale 1995, p. 76).

Very little occurred in the next half year, but the Luddites returned for sure on the night of November 4 in Bulwell near Nottingham. In a much more organized manner than before, they assembled beforehand for a roll call before marching military style to their previously identified target, a mill owner *obnoxious to the workmen*. Upon arrival, they posted their own guards to keep bystanders away and watch out for government forces before starting *their work* of breaking frames. Shortly afterward, they disappeared again in the night, meeting again at a secret location for another roll call before disbanding. One week later, on November 10, they repeated their actions—except that the mill owner was armed and ready to defend his property. The shoot-out led to the death of one young Luddite, which enraged his fellows even more. Eventually, the mill owner had to flee, and not only his frames were smashed but also his furniture, and the whole building was set on fire (Sale 1995, p. 71).

Similar events repeated in the entire Nottingham area: Frames were smashed in a mill in Kimberley also on November 4. On November 6, a cart loaded with machines was captured and destroyed in Basford. Another group of 1000 Luddites broke numerous frames in Sutton the day after. The first written document by the Luddites appeared during the funeral of the Luddite killed in the raid of November 10. In these letters, the Luddites explained the reasons for their actions. The common hand weavers were reduced to *poverty and misery*, whereas mill owners *gain riches by the misery of [their] fellow creatures* (Sale 1995, p. 74f).

---

* The origin of the name *Ned Ludd* is also open to speculation. It may be based on a boy Ned Ludd, who broke a machine of his employer. Another possible namesake is the legendary British King Lud from the first century BCE, who founded London (Sale 1995, p. 77f).

† Although some researchers believe this was not yet a Luddite uprising but merely a regular frame breaking, it is quite believable that this revolt helped to form the Luddites.

Through their uprising, the Luddites were even able to initially achieve some of their goals. Many mill owners in the area voluntarily increased their wages and started to negotiate working conditions with their employees. However, other mill owners took a different route, calling for government actions against these *outlaws* (Sale 1995, p. 87). Overall, the initial achievements were not sustainable, negotiations ended without results, and salaries soon dropped again.

What did work out, however, was the pressure of the mill owners toward the government. The British king and the government were, for their time, rather liberal on economic regulations. This was possibly even one of the reasons for the industrial revolution happening in England rather than continental Europe. However, in the preceding years, they observed a similar revolution in France—the French Revolution, where lots of French government officials, nobility, and even the king were executed during the French Revolution. Liberal as the British government officials were, they were not *that* liberal. The government was worried about a local Luddite uprising becoming a full-blown revolution like in France, overthrowing the existing government. Hence, they undertook an enormous effort to stop the revolt in its tracks.

In February 1812, the government introduced the Frame Breaking Act, calling for the death penalty for frame breakers. This was a harsh increase compared to the previous punishment of 7 to 14 years in prison in exile to Australia. Even within the government, this bill was disputed. Lord Byron, for example, was a very outspoken supporter of the Luddites and proponent of social reforms. Nevertheless, the bill passed in March 1812, and the first Luddites were hanged in May 1812. Simultaneously, the military presence of the government in the area was increased. By May 1812, there were over 14,000 soldiers mobilized to counter the Luddites, with an additional 20,000 militia, plus numerous constables. With a total force of up to 40,000 men, the size of the mobilization was comparable more to a war than to an uprising. Clearly, the government did not want this situation to escalate into a revolution (Sale 1995, p. 148).

The Luddites were not deterred by these forces. On the contrary, their actions expanded. While before frame breaking happened only in Nottinghamshire, now it expanded to Yorkshire, Lancashire, and Cheshire, involving thousands of Luddites. As a result of the increased military presence, however, they targeted more isolated mills and took greater care to evade the military. Furthermore, while initially they aimed only at frame breaking, now in some instances they even targeted mill owners directly, and numerous mill owners were attacked. On April 28, the

Luddites murdered their first mill owner, William Horsfall. They called for a revolution and posted a death threat to Prime Minister Spencer Perceval. Although Perceval was shot only two days later, it is generally agreed that his murderer was not a Luddite but, rather, a merchant with a mental problem. In any case, this did not stop people in the streets from celebrating the death of Perceval (Sale 1995, p. 153ff).

However, the toll on the Luddites was also significant, and altogether, over 20 Luddites died in the conflict. Another 24 were executed by hanging, and another 50 were deported to Australia (Sale 1995, p. 146). Eventually, the military and legal pressure showed results, and from May 1812 onward, the frame breaking by the Luddites was reduced. The workers tried to achieve their goals through legal ways. However, every petition in Parliament fell through. Respected knitter Gravener Henson spent large sums of money to promote a bill to prevent *frauds and abuses* in order to support the common worker, including such radical ideas as a workers' union. However, his bill was watered down to nothing in the House of Commons and then thrown out entirely in the House of Lords (Sale 1995, p. 100).

The last Luddite frame breaking occurred probably around October 1814, but this is hard to say as the group always stayed undercover, and it is difficult to distinguish a regular frame breaking from a Luddite frame breaking. Overall, the Luddites were active for three years and destroyed frames worth around £100,000 (Sale 1995, p. 191), which was more than a millennium worth of wages for a common laborer (Clark, n.d.). It also cost dozens of lives. Yet, for all their effort, they achieved nothing. Working conditions and wages in the mills were still as bad as before. Artisan were unable to compete with the machines, with their financial and social structure falling apart. The Luddites were not fighting the machines, nor the millers, nor the working conditions—they were fighting the future, and this was a foe against which they could not win.

10.3 WORKING CONDITIONS IN THE FACTORIES

While the Luddites fought back—albeit unsuccessfully—many others did not have the will or the ability to do so. Rather, they allowed others to exploit their work. Overall, the effect of centralization in factories was a further rise of capital over labor. A whole new level in society appeared, the factory owner. This factory owner was no longer providing his labor but his capital to the economy. In addition, as there was no shortage of labor, the factory owner set the rules. Based on the social norms of early modern Europe, factory owners had almost the same rights over their workers as lords over their peasants. While the general idea was for the factory owner to also look out for his workers, not all of them did. Even under a benevolent factory owner, the working conditions were very harsh. A factory worker at that time was no longer treated like a human being. Rather, he was considered to be more of a machine made out of flesh and blood and was hence treated as such. While there were also, of course, higher-valued foremen or mechanically skilled machinists, the majority of the factory workers were not considered human beings but, rather, only pairs of hands.

First of all, working hours were grueling. While a craftsman was free to choose when to work and when to be idle, a worker lived by the clock. The workday was anywhere between 12 and 15 hours, often reaching up to 18 hours during summer when the days are longer, six days per week (Stearns 2007, p. 35). Their only break was usually a one hour lunch break. In addition to the working times, the worker had to commute to work, which often exceeded two miles of walking one way, or about 1.5 hours per day return. In comparison, the average modern American workday is about 7.5 hours, usually only for five days per week.* It can easily be imagined that after 12 to 20 hours of work and commute, there was simply no time left for anything else.

They did not have the time to mend things at home, do housekeeping, go shopping, or take care of the children. They did not have the time to meet friends and family, have a beer, or enjoy a neighborhood chat. They did not have time for themselves, just sitting down and putting their legs up, maybe doing some daydreaming. Speaking in modern terms, their work–life balance was nonexistent.† While the term *burnout* was not yet known at the time, or even considered to be a problem, these working conditions must have led to many cases of long-term exhaustion and depression.

---

* Japan, on the other hand, never went through the workers' movement. Subsequently, working hours for office workers are still extremely long, and they often do not have a meaningful family life. In Japanese, there is even a special word for death from overwork.

† Having worked up to 20-hour days as a consultant—albeit for much better pay and with more interesting work—I can only emphasize how important it is to have time for other chores, social contacts, and also merely yourself.

The work itself was also often mind numbing. Work was often divided into minute tasks, and a worker often had to do the very same tasks every minute of every hour of every day at work. There was often no variety or change in the daily monotonous tasks. Work often required an awkward crouched or bent posture, or heavy lifting, and many workers, especially children, developed deformities and disfigurements. There was no need to think, only to do. Worse than that, the workers did not have any say in their work and had very little opportunity to adjust their work environment. Making a suggestion for improvement to the supervisor would probably have baffled the supervisor more than anything else.

Now combine exhaustion and boredom with fast-moving steam-powered machines lacking any kind of safety equipment, and it was only a matter of time until an accident happened. If a worker had his hand in a press while the press closed, he lost his hand. Hence, he also lost the ability to work and was subsequently fired. There rarely was any social plan. Without income, he quickly dropped down to the bottom of the social hierarchy. The accident rates were high in factories, and many fingers, hands, limbs, and lives were lost to accidents, with devastating consequences for the worker and his family. For the factory owner, this was not a problem, since there was lots of labor available to replace the loss. Safety devices were not known at the time. Even if there were such devices, they may have been considered more expensive than a lost limb of a replaceable worker.

The same applies to fire safety. While some mills were built *fireproof*, meaning with metal rather than wooden beams, this was done out of concern for the expensive machines and goods rather than for the cheap labor. Often, there was only one exit from the factory to avoid workers slipping away, limiting the escape options for the workers. In the case of fire, many factories became a death trap.

Even without heavy machines, there were numerous dangers to the workers' health. The work environment was often very noisy. Some workers even learned to read lips while at work, since they could not hear their colleagues over the noise, or after a few years at work anything at all for that matter. Many factories operated at high temperatures, not only at the forges or clay ovens, but also in other industries. Cotton mills often reached temperatures of up to 80°C (Sale 1995, p. 46).

The work environment was also often toxic. Workers handled dangerous chemicals with little or no protection. Mercury, arsenic, or lead was commonly used to treat products. Dust from grindstones or cotton fibers clogged up the lungs (Alder 2010, p. 201f). To be fair, these chemicals and their long-term effects on health were not yet understood very well. Medieval artisans also exposed themselves or even the whole town to these chemicals, lacking an understanding of the health risks involved (Cipolla 1989, p. 134ff). Being *mad as a hatter* due to mercury poisoning was common for medieval headwear makers, and many artisans poisoned themselves through their work (Cipolla 1989, p. 139). The factory owners during the industrial revolution merely continued these practices, albeit on a larger scale.

In sum, working conditions significantly reduced the life expectancy of the working class. A child born in a working-class family in Manchester in 1842 could, on average, expect to become 18 years old. On the other hand, the child of a noble could expect to live more than twice as long to the age of 40 years (Sale 1995, p. 48). Even considering a high infant mortality rate, not many factory workers lived beyond 40 years of age. A contemporary account of a cotton mill workforce said that *an uglier set of men and women, of boys and girls, taking them in the mass, it would be impossible to congregate* (Gaskell 1833, pp. 161–162).

In the contemporary view, however, while the described conditions were considered horrid, they were also, to a certain degree, considered part of life. While nobody wanted to work at these factories, they needed the money. There was certainly the need to change conditions, but few had the imagination to see how it could be done. When German industrialist and cofounder of Marxism Friedrich Engels visited Manchester, he complained to a middle-class gentleman of *the bad, unwholesome method of building, the frightful condition of the working-people's quarters, and asserted that [he] had never seen so ill-built a city. The man listened quietly to the end, and said at the corner where [they] parted: 'And yet there is a great deal of money made here; good morning, sir'* (Engels 1892, p. 276).

There was, however, one aspect of factory life that created a public outrage. In order to get the workers to work, there were rules and regulations, including fines if these rules were violated. Another commonly used disciplinary measure was slapping or beating the worker, especially for child workers (Figure 10.1). In most cases, this was probably no more than a slap on the wrist or in the face of the child, which was also a socially accepted measure to discipline children at the time (Sale 1995, p. 32). Some mill owners forbade the beating of children altogether (Power 1999, pp. 22–24).

FIGURE 10.1
Addie Card, probably 10 years old, claims to be 12, working in an American textile mill in 1910. (Photo by Lewis Hine.)

However, a few sadistic foremen or mill owners took pleasure in beating children. And that was where the public disagreed. Different from other dangers of factory work, the beating of children was seen as not necessary. Contemporary reports show a battle for public opinion. Mill owners claimed that they never gave more than a slap on the wrist, if at all. Opponents claimed that the children had been beaten to an inch of their lives (Pollard 1963). True sadistic beatings were probably the exception rather than the norm, but even a single child suffering is one too many.

Overall, working in a factory was dangerous in many aspects. The work destroyed the workers' social life; their health; in many cases, their ability to work; and hence, their income. Some of them even lost their lives due to their work. Yet, in order to survive, many people had no other option than to sell their hands in a factory. Slowly, however, the government took note of the plight of the workers and eventually took action.

10.4 GOVERNMENT ACTIONS FOR THE WORKERS

Throughout most of our history, work was unregulated. Most laws and edicts regulated prices and quality. For example, the *Code of Hammurabi* around 1700 BCE regulated different prices and the obligation to teach apprentices. Similarly, the edict of Diocletian (unsuccessfully) tried to fix prices in the Roman Empire in 301 CE. If there were any rules related to working conditions, these were self-imposed, for example, medieval guild rules. The governments mostly did not concern themselves with working conditions or work-related aspects.

Besides prices, however, they also wanted to make sure that the government was in charge, and nobody else. In most countries, before the industrial revolution, trade unions were outlawed. The *Ordinance of Labourers* in the United Kingdom in 1349 was an (ineffective) attempt to fix wages to levels before the Black Death in England, but a scarcity of labor drove up wages nevertheless. Similarly, the *Le Chapelier Law* in France in 1791 banned all guilds and unions and refused workers the right to strike. In 1799, Britain again banned unions and collective bargaining through the Combination Act.

Only in 1824 did the United Kingdom officially allow trade unions. Previously underground labor organizations emerged and used their new freedom for numerous strikes. In response, only one year later, the government restricted the rights of trade unions again through the Combination Act of 1825. It was seen as the duty of the worker to work and otherwise to accept his lot in life. Not until 1871 did unions in Great Britain regain their freedom.

The first law in the United Kingdom to actually regulate working conditions was the *Factory Act* of 1802, focusing entirely on the textile industry. Most of the act regulated the use of child labor, which was widely used in cotton mills. Nobody under the age of 9 was allowed to work, and working time was limited for children. The act described somewhat vaguely the requirement for fresh air and a biannual cleaning of the workrooms. Children also had the right to a basic education, and not only were no more than two children to sleep in the same bed, but boys and girls also had to sleep in separate rooms. Unfortunately, the law was not checked or enforced, and few factory owners followed it (Cooke-Taylor 1894, p. 53ff).

However, as discussed, the working conditions for children were under public debate (Pollard 1963). As the controversy increased, the government followed up its ineffectual act of 1802 with the 1819 *Cotton Mills and Factories Act*, the 1831 *Labour in Cotton Mills Act*, and the 1833 *Factories Act*. Each act put more pressure on the mill owners. Hence, more and more of them actually followed these laws (Cooke-Taylor 1894, p. 53ff).

The 1844 *Factories Act* further reduced the time children could work. However, this act expanded its scope to also restrict the working times of women, although it was still limited to the textile industry. Another radical new idea was also to protect the worker from injuries, as the act now required machinery to be fenced in. Before, the health and safety of each worker was his/her own responsibility, and disabling or deadly accidents were considered part of the risk a worker had to take.

Around that time, aid for unemployed and disabled workers started to become part of the legislature. Different laws for vagabonds and beggars have been used since medieval times, either trying to force able-bodied people to work or to make life for them so uncomfortable that they preferred to go somewhere else. Later laws from 1600 onward even provided food for the *impotent poor* unable to work and workhouses for those who were able to but unwilling or for criminals. However, these laws were unable to cope with the urbanization during the industrial revolution, as there were simply too many poor people, and providing for them became too expensive. Primarily to reduce costs, the government established the Poor Law of 1834, trying to make conditions in workhouses as miserable as possible to reduce the number of people wanting to join a workhouse.

The first real social insurance was actually in Germany, when Otto von Bismarck, in rapid succession, established mandatory health insurance (1883), mandatory accident insurance (1884), and mandatory pension insurance (1889). While the aim in England was to keep the poor away from the rest of the population, Germany actually tried to prevent people from becoming poor in the first place. Only in the first half of the twentieth century did England follow up with similar laws (Ritter 1983, p. 77ff).

In other legal aspects, however, England was the most advanced Nation during the nineteenth century. Already around 1830, unions in England discussed an eight-hour workday, a radical change considering that 16-hour factory workdays were common during this period. Already in 1847, England passed the 10-hour bill as part of the 1847 Factories Act. However, there was significant resistance among the mill owners, who considered working times as part of their private factory regulations. Only 20 years later in 1867 was the 10-hour workday commonly accepted and also applicable not only to the textile industry but to all trades (Hobsbawm 1999, p. 103).

In 1891, the British Navy switched to an eight-hour workday (applicable to almost 30,000 employees). By 1900, the 10-hour day was standard in England, with lots of full-time workers working as little as eight hours a day (Abbe 1906). As is to be expected, working fewer hours per day improved productivity per hour. However, what totally surprised employers was that in many cases, the employees were able to produce more in eight hours than before in 14 or 16 hours, not in relative comparison but in absolute numbers! A worker, in eight hours of work with 16 hours of break in between, produced more per day than the same worker did before in 14 to 16 hours with only an eight-hour break in between (Abbe 1906; Locke 1982). Hence, the reduction to eight hours for the same wages in many cases reduced labor cost, not to mention improving the quality of life of the worker. Even nowadays, it is well known that part-time workers are often more productive than comparable full-time staff.

Overall, the industrial revolution was probably the biggest change in human history after the urban revolution (Bleiweis 1993, p. 22). A craftsman from 1700 CE worked very similarly to a craftsman from 1700 BCE and probably would have been able to make a living doing the very same thing for millennia before the industrial revolution. However, between 1750 and 1850, the fundamental rules of the game changed. Now labor was pooled together under one management. The worker contributed only to a small part of the overall process. All the skills of a craftsman from 1700 were quite likely replaced by a machine by 1900. This rapid change in technology put an enormous stress on society. The small independent craftsman vanished, and the working masses lived in cities, earning their money in factories, often doing small repetitive tasks. However, the average worker of 1900 was able to afford much more luxury, was much better fed and clothed, received better medical treatment, and lived in a better house than a craftsman 200 years earlier. The average quality of life improved significantly. There is one main reason for this improvement: mechanization made products faster, better, cheaper.

BIBLIOGRAPHY

Abbe, E., 1906. *Gesammelte Abhandlungen III: Vorträge, Reden und Schriften sozialpolitischen und verwandten Inhalts*. Verlag G. Fischer, Jena, Germany.

Alder, K., 2010. *Engineering the Revolution: Arms and Enlightenment in France, 1763–1815*. University of Chicago Press, Chicago; London.

Berg, M., 1985. *The Age of Manufactures: Industry, Innovation and Work in Britain, 1700–1820*. Fontana P, London.

Bleiweis, S., 1993. Die Europäische Gemeinschaft im Vergleich Mit Japan und den USA—Eine Strukturanalyse entwickelter Länder auf der Basis einer allgemeinen Theorie sozialer Systeme. Selbstverlag des Wirtschafts-und Sozialgeographischen Instituts der Friedrich-Alexander-Universität in Nürnberg., Nürnberg, Germany.

Burton, A., 1976. *Josiah Wedgwood*, 1st ed. Andre Deutsch Ltd, London.

Cameron, R., Neal, L., 2002. *A Concise Economic History of the World: From Paleolithic Times to the Present*, 4th ed. Oxford University Press, New York.

Cipolla, C.M., 1989. *Before the Industrial Revolution: European Society and Economy, 1000–1700*, 2nd ed. Methuen & Co. Ltd, London.

Clark, G., n.d. England Prices and Wages since 13th Century. Global Price and Income History Group.

Cooke-Taylor, R.W., 1894. *The Factory System and the Factory Acts*. Taylor & Francis, Boca Raton, FL.

Engels, F., 1892. The Condition of the Working-Class in England in 1844, with a Preface written in 1892.

Gaskell, P., 1833. *The Manufacturing Population of England: Its Moral, Social, and Physical Conditions, and the Changes which Have Arisen from the Use of Steam Machinery; with an Examination of Infant Labour*. Baldwin and Cradock, London.

Hobsbawm, E.J., 1999. *Industry and Empire*. Penguin Books, London.

Landes, D.S., 1969. *The Unbound Prometheus: Technological Change and Industrial Development in Western Europe from 1750 to the Present*, 1st ed. Cambridge University Press, New York.

Landes, D.S., 1998. *The Wealth and Poverty of Nations: Why Some Are so Rich and Some so Poor*, 1st ed. W. W. Norton & Company, New York.

Locke, E.A., 1982. The Ideas of Frederick W. Taylor: An Evaluation. *The Academy of Management Review* 7, 14–24. doi:10.2307/257244.

Pollard, S., 1963. Factory Discipline in the Industrial Revolution. *The Economic History Review, New Series* 16, 254–271. doi:10.2307/2598639.

Pollard, S., 1964. The Factory Village in the Industrial Revolution. *The English Historical Review* 79, 513–531.

Power, E.G., 1999. *Belper—First Cotton Mill Town*. Belper Historical Society, Belper, UK.

Ritter, G.A., 1983. *Sozialversicherung in Deutschland und England—Enstehung und Grundzüge im Vergleich*. C.H. Beck, München, Germany.

Rosen, W., 2010. *The Most Powerful Idea in the World: A Story of Steam, Industry, and Invention*. Random House, New York.

Sale, K., 1995. *Rebels against the Future: The Luddites and Their War on the Industrial Revolution: Lessons for the Computer Age*. Addison Wesley, Reading, MA.

Stearns, P.N., 2007. *The Industrial Revolution in World History*, 3rd ed. Westview Press, Boulder, CO.

Section III

Modern Times—Mass Production for the Masses

11

Technological Advances

Machines may be made by which the largest ships, with only one man steering them, will be moved faster than if they were filled with rowers; wagons may be built which will move with incredible speed and without the aid of beasts; flying machines can be constructed in which a man...may beat the air with wings like a bird...machines will make it possible to go to the bottom of seas and rivers.

Roger Bacon (ca. 1214–1294)
Philosopher

The Industrial Revolution between 1750 and 1850 was the biggest change to human life since the Neolithic revolution. The invention of the steam engine provided near-unlimited power anywhere, and the increase in mechanization put this power to productive use. While not everybody was better off during the early years of the revolution, the Industrial Revolution definitely increased the standard of living eventually.

This section will look in detail into the period from 1850 to 1950. The radical changes of the Industrial Revolution expanded from textiles and metalworking into almost all other areas of manufacturing and greatly increased its speed and productivity. Electricity provided a highly flexible and versatile power source, reducing the need for muscle power even more in many areas where steam power would have been highly impractical or impossible. New technologies appeared in machining and metalworking. New materials like plastic replaced wood and metal. Mechanization took two major strides through the advent of the assembly line and, later, through the invention of robotics. Above all, manufacturing management moved from trial and error toward science.

While these improvements were not as radical as the changes between 1750 and 1850, they also altered human life significantly. Hence, this period is also known as the Second Industrial Revolution. However, please be aware that the meaning of *Second Industrial Revolution* is by no means agreed on. While most apply the term to the changes in the decades around 1900, the exact range varies significantly. In addition, America usually focuses the Second Industrial Revolution on its successes with the assembly line and scientific management, while Europe defines it more through its own strengths in this period with the steel, chemical, and electrical industries. The term has also been less frequently used to describe the changes through electricity after 1850, nuclear power after 1950, robotics after 1970, or nanotechnology after 2000. Some even argue that the period of the steam engine, 1750–1850, was already the Second Industrial Revolution, and the First Industrial Revolution occurred during medieval times, in ancient China, or even during the Stone Age.*

In any case, regardless of the definition of the term, the decades around 1900 brought mass manufacturing to almost all products of everyday life and continued to increase prosperity and the standard of living.

11.1 INFINITE POWER—ELECTRICITY

The steam engine was one of the key drivers of the Industrial Revolution since it provided a source of nearly unlimited power almost anywhere. However, it did have quite a number of limitations. First, it provided primarily one type of power: mechanical movement. The power was transferred from the steam engine to the machine through a driveshaft. However, the longer the driveshaft, the higher the losses due to friction and elastic torsion. Hence, steam-powered factories were built as compact as possible with multiple floors close to the steam engine to increase efficiency.

* It becomes even more confusing with the Third Industrial Revolution, which has been applied to computer technology (most frequently), globalization, telecommunication, ethical companies, and many more. In 2007, the European Parliament even declared renewable energy to be the Third Industrial Revolution. I personally don't quite see the point, but then who am I to argue with the wisdom of the European Parliament? There are even references to the fourth, fifth, sixth, and seventh Industrial Revolutions, although without much scientific consensus.

Second, it takes both time and energy to start a steam engine. To bring an idle engine up to speed, you need to heat it up to operating temperature, which may take hours and lots of coal. Once you stop the engine, the heat slowly dissipates, and the heat energy is lost. A steam engine runs best continuously, but in manufacturing, the demand on power changes rapidly over time as you start and stop a machine.

Third, a steam engine requires a certain size to work efficiently. You cannot scale a steam engine down very much without the engine becoming inefficient, where both the cost of purchasing an engine and the cost of running it no longer justify the power generated. Finally, a steam engine is, by its very nature, a dirty, hot, and smelly operation. Both coal dust and the soot from the fire will blacken the surroundings, and the heat lost from the engine will heat up its environment.

Just imagine for a second that your blender is powered by a steam engine. The engine would probably be 10 times more expensive than the blender, and to make a smoothie, you would have to add coals to the device for at least half an hour before it would operate. While by then the heat in the kitchen would make you look forward to the smoothie, the smoothie itself might already be lukewarm, and the coal dust and soot would surely not help you in enjoying your drink. Overall, steam power by itself is not a very flexible energy source, and it is best applied to large-scale centralized steam engines.

Electricity changed all that. It is easy to transport over long distances, is clean to use, can be scaled up or down to match the demand, and can be stored in smaller quantities.* Electricity can also easily be transferred into most other forms of energy, for example heat, radiation (including light, x-rays, and radio waves), magnetic energy, sound waves, mechanical energy, and chemical energy (e.g., splitting water into hydrogen and oxygen). Less practical but also possible is its conversion into nuclear energy (via a cyclotron) or into the ultimate form of energy: mass.†

Static electricity was known already in ancient times, but the practical use of electricity requires scientific understanding thereof. Only in

---

* Storing larger quantities is still a technical problem and currently the main hindrance to electric cars. Large batteries still by far do not have the energy density of a tank of gas, and recharging the battery currently takes hours rather than the minutes for filling up your tank.

† Quick physics catch-up: Mass can be changed into energy and back, and for example, if you burn hydrogen, the mass of the resulting water is infinitesimally smaller than the mass of the hydrogen and oxygen at the beginning. The relation is described in Einstein's famous formula $E = mc^2$, and the amount of energy generated is gargantuan. For instance, the atomic bomb dropped on Hiroshima converted less than one gram of mass into energy.

the seventeenth and eighteenth centuries did modern science emerge and start to research electricity.* Numerous researchers made practical contributions in the nineteenth century and invented, among other things, electric light, generators, and electric motors.

Early use of electricity in manufacturing was hindered by the lack of a cheap source of electricity. The only practically useable source of electricity in the first half or the nineteenth century was the battery. Invented by Volta in 1800, batteries both were expensive and provided only limited power. One of the first uses in manufacturing was electroplating, where electricity was used to add a thin layer of one metal such as gold, silver, or copper onto another metal. Already in 1839, copper electroplating was used in printing bank notes (Schlesinger and Paunovic 2011, p. 34), providing manufacturing with a completely new coating technology. The first large-scale electroplating plant was established 1876 in Hamburg, Germany (Stelter and Bombach 2004).

Batteries were also used to power early electric motors in manufacturing, but due to the large expense for batteries to drive power-hungry motors, these applications were no commercial success. More successful were telegraphs since they required less power and it was possible to run them on battery. The first commercial telegraph line was installed in 1839 in London by Sir William Fothergill Cooke and Charles Wheatstone.† Even though this line communicated using only 20 letters in the alphabet (missing C, J, Q, U, X, and Z), in 1845, the description of a murderer fleeing by train was forwarded through the telegraph. The culprit was successfully apprehended at the destination. The resulting publicity greatly helped the commercial success of the telegraph (Camenzind 2007, p. 57).

* And research they did. In 1746 French researcher Jean-Antoine Nollet wanted to determine the speed of electricity. He made 180 royal guards hold hands in a chain and then connected the two guards at the end to his Leyden battery. The resulting electric shock made them all jump in the air simultaneously. Hence, he concluded that the speed of electricity was near instantaneous. King Louis XV of France was so excited about the experiment that he ordered Nollet to repeat it on a larger scale. Since for reasons lost to history the royal guard was unavailable for a second round, Nollet recruited 200 monks for his next experiment. He had them stand in a huge circle holding wires and then connected them to his battery. With 200 monks jumping in the air at the same time, he proved his theory again (Letcher 2003, p. 60). The comments from the monks or the royal guard about their valuable contribution to science are not known.

† The better-known Samuel Morse installed his first commercial line only years later in 1844, after learning about the works of Cooke and Wheatstone in 1838. Since many inventors contributed incrementally to the telegraph, often simultaneously, the true inventor is still a subject of controversy. As with so many other inventions, this is also a matter not only of history but equally of national pride (Bowler and Morus 2005, p. 404).

The first widespread uses of electricity in manufacturing were not directly connected to the manufacturing process but, rather, to signaling devices with low power consumption that could run on a battery. By 1875, the first electric fire alarms appeared, and by 1890, burglar alarms, thermometers, and clocks were common. The night watchman now also had to press switches on different stations along his route to ensure that he was not taking a nap rather than doing his duty. Any delays on these checks would lead to an alarm bell ringing (Nye 1992, p. 188).

However, more power-hungry applications had to wait for a cheaper source of electricity than a battery could provide. Only the introduction of reliable generators in the 1850s and 1860s reduced the price of electricity for wider use. Now manufacturing enterprises could solve one of their major problems with electricity: light! Until then, there was no good light source available other than daylight. Gaslight was used but had many significant drawbacks. The light was dim and flickering, making it hard to work efficiently. Turning the light on and off required lighting each individual flame, not to mention the cleaning and relighting of flames that went out.

Furthermore, the burning of gas not only removed oxygen from the air but also produced large quantities of water, acids, and soot besides heating up the air. Hence, rooms with gas lighting usually were low on oxygen, hot, sticky, smelly, and overall not a pleasant work environment. Finally, having many gas flames in the factory created a significant fire hazard. This even caused explosions in dusty environments like a textile or flour mill, not to mention leaks in the gas line creating gas explosions. Overall, gaslight was unpleasant to work with and an accident in waiting. Many factories restricted themselves to daylight hours (Nye 1992, p. 5).

Electric light resolved all these problems. First arc light and then the lightbulb created a nonflickering, bright, and reliable light. A single electric light in a shop window at night drew large crowds of customers amazed at the unbelievable spectacle of a bright, steady light. Figure 11.1 shows the back then unbelievable night illumination of the 1893 Chicago World's Columbian Exposition. The light was easy to turn on and off for the entire factory and did not smell or reduce the air quality. Also, the light was pretty safe as long as the workers kept their hands away from the uninsulated wires. But then, from the point of view of the factory owner at that time, a worker killed by electricity was easily replaceable, whereas a factory burned to the ground was financial ruin, so his priorities were clear. Besides, fire insurance rates were significantly lower for factories with electric light than with gaslight (Nye 1992, p. 5).

FIGURE 11.1
The *City of Light* during the 1893 Chicago World's Columbian Exposition was the first major use of outdoor lighting. Before, any kind of work or entertainment after sundown was difficult due to the lack of light.

Electric light allowed the factory owner to extend working hours potentially to around the clock, limited only by the endurance of the employees. While before only critical operations that needed uninterrupted attention like steel mills were staffed round the clock, electric light allowed 24-hour shifts for all industries (Nye 1992, p. 193). However, longer work hours using electric light often burned out the employees, in turn leading to strikes (Harman 2002, p. 379). In any case, from 1880 onward, electric lights were commonly installed in factories to extend working hours and to reduce the fire risk (Nye 1992, p. 186).

As the demand for electricity increased, commercial suppliers started to build power stations. While before every consumer had to build its own station, commercial power stations allowed the use of electricity without first having to invest in a whole power plant. Probably the first commercial station in the world was built in 1879 in San Francisco (Whitten and Whitten 1990, p. 315), rapidly followed in 1881 in Godalming, England, by Siemens (Landes 1969, p. 285), 1882 in Manhattan by Edison, and many more.

Electricity also has the advantage that it can be easily transferred over long distances. The first long-distance power line was built in 1891 in Germany over a distance of 225 kilometers running at 30,000 volts (Landes 1969, p. 285). While there were only 10 power stations in America in 1882, by 1889, there were 468 power stations, satisfying a rapidly growing demand for the new energy. Of course, training qualified personnel to run these plants proved more difficult, and many of the early power plant workers were ill trained and had no clue what they were doing. Electrical short circuits were common and often went unnoticed, actual measurements of voltage and current were rare, electrocution was a real risk, and overall electric efficiency was low (Nye 1992, p. 235).

The next leap forward was the electrification of movement. Until then all available power was either relatively weak human or animal muscle power, or powerful large centralized steam engines, water wheels, or windmills. The transfer of power from the source to the demand was through a driveshaft. Since the losses due to friction and torsion of the shaft increased rapidly with distance, the machines were arranged as close as possible to the source. Often, one driveshaft went vertically through multiple floors, driving one or more horizontal secondary driveshafts on each floor. Leather belts transferred the power from the shaft to the machine. Putting the leather belt on the wheels of the driveshaft turned the machine on or off. Besides loss of power over distance, these shafts also limited the options for arranging the machines; created a risk of injury for the workers; and cluttered the overhead room with shafts, wheels, and belts, making the room feel cramped. Finally, minor variations in power consumption combined with the torsion of the driveshaft increased vibration of the shaft, and the last machine on the shaft usually had the worst quality of them all. Imagine running a printing press that is jumping on the floor due to vibration, and you can easily imagine that this was a serious issue for many operations.

Electricity was much easier to transfer than mechanical movement, and the invention of the practical electric motor put the power for mechanical movement right where it was needed. The first practical electric motor used in manufacturing was invented by Thomas Davenport in Vermont in 1837* (Schiffer 2008, p. 66) and was used to power a printing press.

* To insulate the wires of his prototype motor, he used silk from the precious wedding gown of his wife Emily. However, Emily Davenport was not only a source of family heirlooms to shred to pieces but also an active inventor and contributor to science in her own right. Not only did she protocol the many experiments of her husband; she also wound the wires of the motors and had the crucial idea of using mercury as a conductor in a switch to reverse the flow of electricity after every half turn (Stanley 1995, p. 293f). Without her inventiveness, there would have been no motor by the Davenports.

Printing was a prime application for electrical motors, since the vibrations of a mechanical transmission degraded the quality of the print. However, as there were no practical generators yet, these motors ran on battery power. Due to the exorbitant price of battery power, this venture was not successful, and Davenport went bankrupt.

The introduction of generators greatly reduced the price of electricity. Following simultaneous inventions of new improved motors by Frank J. Sprague from the U.S. Navy and Galileo Ferraris in Italy, electric motors started to be used in factories again. Probably the first motor in manufacturing not driven by a battery was installed in 1884 to power yet another printing press, quickly followed by other applications like clothing factories and cotton mills (Nye 1992, p. 195). Nevertheless, electric motors remained a rarity in manufacturing, and there were only 250 motors in the United States around 1880. All motors were custom made until 1905. Afterward, however, prices dropped, and electric motors gained rapid acceptance. For example, there were 250,000 electric motors sold in the United States in 1909 (Nye 1992, p. 197ff).

Initially, these motors were installed similar to water power. A large motor powered a driveshaft, which transferred the power to numerous machines via belts. As motors became more common after 1900, machines were powered by their own dedicated motor, no longer sharing the motor with other machines. Still, the power from this single motor was transferred to different parts of the machines using mechanical links. In particular, Germany and Sweden quickly started to use individual motors for each machine, while motors powering multiple machines were still common in Great Britain until the 1940s (Nye 1992, p. 201). Eventually, a machine included multiple motors to power all the different movements individually. Turning on or off a particular action no longer required shifting gears but only the flick of a switch. Nowadays, modern machines may contain hundreds of individual electric motors, and a mechanical link to power two different actions is rare.

Initially, the preferred type of electricity was direct current, where the electricity flows in only one direction. Alternating current on the other hand has the advantage that it can easily be transformed into lower or higher voltages. Higher voltages minimize the losses during transmission, whereas lower voltages are more practical for use. However, current alternating its direction 60 times per second would be more likely to mess up your heart rhythm, leading to cardiac arrest, although direct current is only marginally safer. Until the early twentieth century, direct current was common, and the first Ford factories, for example, ran on direct current. Nowadays, alternating current is the most commonly used form of electricity, with direct current limited to battery-operated devices or inside electronics.

The transition from direct to alternating current, however, was, at its time, highly controversial. Two electricity geniuses squared off against each other in the War of Currents. Inventor Thomas Edison, who preferred direct current, had a public relation battles with George Westinghouse, favoring alternating current. Of course, besides the honor of being right, there was also much money at stake, since both sides owned numerous patents for their respective type of electricity. Hence, it was gloves off, especially for Edison, who went on a campaign warning of the dangers of alternating current. Numerous animals, including cats, dogs, cattle, and horses, were killed with alternating current by Edison to demonstrate the danger. Eventually, even a full-grown circus elephant named Topsy was executed using electricity.* Yet Edison did not even stop with elephants but also put humans to death by electricity. Of course, he was not a murderer; he even opposed capital punishment. Yet, to win this war of the currents, he put aside his beliefs and secretly supported Harold P. Brown in the invention of the electric chair. The first execution through alternating current happened in New York State in 1890.† Nevertheless, Edison lost the war of currents. By the beginning of the twentieth century, most new electricity systems installed used alternating current.

Electricity greatly changed the manufacturing workplace. It replaced the dark, damp, and humid environment including lots of moving parts with brightly lit workspaces, where the only parts that moved were the ones that actually needed to move. The quality of products increased significantly, since electric motors provided a much more stable source of movement compared to steam engines or water power. Productivity also increased, since electric motors were much cheaper than steam power or water power and were also suitable for a much wider range of applications than steam power. Due to the ease of transporting electricity, modern factories could

---

* To be on the safe side, however, Topsy was fed a liberal amount of poisoned carrots just before execution to make sure she properly demonstrated the dangers of alternating current.

† Unfortunately, the executioner lacked experience with this method of execution, and let's just say it was a very messy death.

be much larger than steam-powered factories, and the largest factories are nowadays the size of a small city.*

Electrification, however, did not only benefit large factories. Actually, it also leveled the playing field for smaller companies. Before electricity, steam and water power were the primary sources of movement besides muscle power. Both of these power sources had a high initial cost of buying and installing the power plant, and running these engines was only effective if you had lots of machines to drive. A small entrepreneur could not afford this technology and was limited to his own muscle power. In activities like sawing wood, drilling holes, etc., he had to compete with his muscle power against the power of steam engines of larger competitors. It is easy to see which wood was cut faster and cheaper, and small manufacturers were under immense competition from large enterprises.

Electricity, however, significantly reduced the entrance cost to powered machines. The generation of electricity was done by a public power plant, and the small entrepreneur had to pay only for what he consumed. Where large enterprises installed hundreds of motors to power hundreds of machines, the small craftsman could now install one motor to power his few machines at almost the same unit cost. Now small craftsmen could again compete with larger companies either by satisfying the demand of a niche product or by working as subcontractors (Landes 1969, p. 288).

11.2 NEW MANUFACTURING TECHNOLOGIES BASED ON ELECTRICITY

Besides providing light and movement, electricity also allowed completely new manufacturing technologies not possible before. Electrolysis is key in the production of aluminum. Aluminum is the third most common element on earth, occurs naturally only in oxidized from. Producing pure aluminum was almost impossible before electricity, and aluminum was more valuable than gold. Napoleon III (1801–1873) served food on aluminum plates for his most distinguished guests, whereas second-rate guests ate only from plates of gold (Hasan 2006, p. 5). This all changed rapidly

* Probably among the largest plants in the world is the Volkswagen plant in Wolfsburg, covering 6.5 square kilometers, roughly the same size as Gibraltar. The Boeing plant in Everett has the largest covered floor space with almost 3 square kilometers, larger than Monaco.

with the Hall–Héroult process, discovered in 1886. The process can produce large quantities of aluminum at low cost using electricity, and in 1888, the first major aluminum plant was opened. The price of aluminum dropped rapidly, and nowadays, aluminum rather than being worth its weight in gold, is a cheap everyday metal with over 30 million tons produced annually, second only to iron production. It is used for such mundane uses as to make car wheels, window frames, and aluminum foil to wrap sandwiches. Besides aluminum, other materials produced through electrolysis include lithium, sodium, potassium, magnesium, calcium, chlorine, sodium chlorate, and potassium chlorate.

Another completely new technology is electroplating, where electricity is used to create a thin coating of one type of metal onto another object. The process is similar to a reverse battery, where electricity removes material from the anode and places it on the cathode. Gold and silver plating are commonly used to enhance steel cutlery. A coating of tin, zinc, or nickel reduces corrosion, whereas chromium reduces abrasion. Electroplating of expensive metals like palladium, rhodium, platinum, or ruthenium greatly reduces the cost while still providing a large surface area for catalytic converters.

Another technology enormously improved through electricity is welding. This technique requires the localized melting of a workpiece while adding molten filler material. Welding was already known in antiquity, but welding iron in a coal fire was one of the most challenging and difficult processes for a master smith. Electricity greatly changed this process, as it allowed the localized application of energy to melt the product and the filler. Around the 1880s, Russian inventor Nikolai Bernardos created the arc welding process, where an electric arc created the intense heat necessary to melt the material. Initially, this welding process had the problem of oxidation, where the workpiece rusted during welding, resulting in a weaker corroded bond. This was solved by the introduction of a shield gas that locally removed the oxygen from the welding area. This shield gas is either generated from a special coating on the electrode or directly supplied from a gas tank.

A similar electric welding technology is resistance welding invented between 1886 and 1888 by Elihu Thomson (Hounshell 1985, p. 149). Rather than supplying a filler material, an electric current heats up two workpieces and melts the contact area without the need for a filler material. One of the most well-known uses is spot welding, where two sheet metals are connected through welded spots. For example, all automobile bodies rely extensively on spot welding to join the hundreds of metal sheets into one car body. Both arc welding and resistance welding are nowadays extensively used in industry, with other electric welding technologies, for example laser welding or ultrasound welding, supplementing these for certain uses. The well-known acetylene gas welding was actually invented only around 1900, after electric welding. Most manufacturing applications use a form of electric welding to permanently join metal parts.

Many other manufacturing technologies were also only possible or greatly simplified through electricity. This includes, for example, electrical discharge machining or electroerosion, where material is removed through electric discharges; cooling technologies crucial for food processing; electrochemical grinding; electropolishing; almost anything involving plasma or magnets; and any application involving lasers, including welding, cutting, heat treating, and measuring.

Finally, electricity also revolutionized the transport of goods inside a factory. Mass production also means large material flows, including raw materials, semifinished goods, finished goods, and waste by-products. Before electricity, the vast majority of this transport was through muscle power, either human or animal. In some factories, up to half of all shop floor workers could be *carriers and haulers* (Nye 1992, p. 209). Mechanical aids were limited to wheelbarrows or carts. The first overhead traveling crane in a factory was built by a Swiss, Johann Georg Bodmer, in 1833 for a factory in Manchester (Giedion 1948, p. 90ff). However, until the 1900s, most cranes were either hand operated or unwieldy and inefficient steam engine–powered cranes (Landes 1969, p. 304). The electric motor greatly simplified the installation of cranes in factories, and overhead cranes were common by 1915. Safety rules and regulations, however, were still in their infancy. At the beginning of the twentieth century, up to 5% of all accidental deaths in a factory were caused by a crane dropping its load on a worker (Nye 1992, p. 210). The electric crane provided a much faster and more efficient way to transport goods through the factory—albeit until the introduction of safety standards, not necessarily a safer one. Within a few decades, there were few carriers and haulers left in the factory.

Overall, electricity provided a clean and safe source of power that could be transferred into many other forms of energy. High-voltage networks allow great distances between the generation and the consumption of the energy while still being efficient. Numerous technologies that build the backbone of modern manufacturing would be impossible without electricity. With its versatile uses for both industry and private homes, the

National Academy of Engineering in the United States called electrification one of the greatest engineering achievements of the twentieth century.

11.3 PLASTICS AND RUBBER

Yet another new manufacturing technology appearing during the late eighteenth and early nineteenth centuries was plastics. Nowadays, plastics are one of the key materials in manufacturing. Look around you, wherever you are, and it is hard not to see products made out of plastics. Most of your pens will probably be made out of plastics, and the few that are not are probably rather expensive and prized possessions of yours. Without plastics, there would probably be no electricity in your house. Your wires are insulated with plastics rather than with silk cut from a wedding gown. You may be able to build a car without plastics, but the effort of producing everything, from your side mirror to your battery housing, from metal or wood would make any car prohibitively expensive and far too heavy. Plastics made a lot of things affordable that otherwise would be expensive, up to the point where the mind of the consumer sees plastic as the epitome of cheap.

Plastic is both cheap and very versatile. Plastics have a wide range of material properties. The chains of molecules that make up plastics—polymers*—can be interlinked to different degrees. By adjusting the structure of the molecular chains, it is possible to give these materials a wide range of properties, making them flexible, or resistant to heat, or transparent, or strong, or soft, or resistant to chemicals or ultraviolet (UV) light, etc. By adding other materials in the mix, properties can be changed even further. Initially, fillers were used to replace part of the more expensive polymers. But nowadays, fillers are also used to significantly change the properties of the material and make the polymer easier to process, or improve optical effects. Plastic computer housings, for example, often contain up to 50% fillers to make them fire retardant. Adding carbon fibers greatly increases the strength of the material and is used, for example, for high-end automotive parts.

---

* Scientifically speaking, plastics are a subgroup of polymers. For example, feathers, fingernails, hairs, and wood are polymers but not plastics, while all engineering plastics are polymers. In terms of manufacturing, *polymer* almost always refers to plastics, and vice versa, although the scientific and engineering community prefers the term *polymer*, since plastics sounds, well, cheap.

Not only are polymers very versatile; they are usually also cheap compared to their alternatives. Most often, they are created from mineral oil using technically advanced but inexpensive processes. Through these large-scale processes, raw plastic is produced. Manufacturing is also inexpensive compared to the alternatives. The most common process is injection molding, where hot plastic is injected (shot) into a steel cavity, producing the shape of the final product in only one processing step. Rather than cutting, drilling, milling, and grinding a metal part, you need only one shot, and you have the final product. For smaller parts, you may even have hundreds of products in one shot, something that would be prohibitively expensive using other techniques. The mold may also last for a million shots or more before being worn out and needing refurbishment or replacement.

Additional processes may be needed sometimes, for instance, blow molding for making bottles. An injection-molded bottle piece is heated up again and injected with air, forming the shape of a bottle much cheaper than any glass bottle. Thermoforming is the general name for changing the shape of a plastic product under heat. However, injection molding is often the only step needed before assembly. Hence, plastics are the ideal material for mass production, and it is very cheap and easy to produce millions of identical parts at low cost.

Natural polymers such as tar, wood, feathers, and cow horns have been used for millennia. The Neanderthals used birch tar as a glue. However, collecting the raw material was often cumbersome. For example, shellac—used to make records until the 1940s—was excreted by lice and time-consumingly collected by scratching it from twigs in India.

The vulcanization of natural rubber was discovered by Charles Goodyear in 1839, hence producing the first elastomer. Goodyear worked with natural rubber, trying to solve the problem of its stickiness. He thought he found the solution when he added magnesia, which not only reduced the stickiness but also gave the product a nice white shine. To cash in on his invention, he started to sell rubber shoes. Unfortunately, as his customers found out, magnesia reduced stickiness only temporarily. Adding quicklime seemed to solve the problem, but any weak acid like lemon juice made the rubber sticky again. The same for nitric acid, except that this dangerous chemical nearly killed him.

Overall, it took him a lot of money, time, and failures before he discovered around 1840 that sulfur and heat together created lasting rubber in a process called vulcanization. Yet while his method soon spread widely, he died poor. Even the world-famous Goodyear Tire and Rubber Company was opened only 40 years after his death. Frank Seiberling, an entrepreneur unconnected to Goodyear himself, established the company, naming it in honor of the inventor.

The first thermoplastic was celluloid, invented around the 1860s, involving a number of inventors and legal disputes. Celluloid revolutionized photography, but it was highly flammable, which made a risky combination with hot film projectors. It is still used in most ping-pong balls. John Wesley Hyatt was also involved in the invention of celluloid and invented the first simple injection molding machine in 1872. Other plastics invented before 1900 were, for example, rayon, linoleum, and Bakelite. During the early 1900s, the use of plastic materials slowly increased, but only during World War II did the technology really take off and make plastic the common household item it is today.

11.4 MACHINE TECHNOLOGY

During the industrial revolution, metals started to replace wood and leather. As there was a need for metal processing, different technologies emerged or improved to shape metals, including new cutting, forming, and joining techniques. The following paragraphs give a brief overview of the most significant improvements, although they can touch only the surface of the countless different techniques invented.

Different new casting techniques evolved to shape liquid iron. One of the most significant is extrusion technology, where hot or liquid metal is pulled or pushed through a die, creating a profile in the shape of the die. The first extrusion technology was patented in 1797 by Joseph Bramah. He produced lead pipes by pushing hot lead through a die using muscle power. This was mechanized in 1820 by Thomas Burr and expanded to copper and brass in 1894 by Alexander Dick (Wick et al. 1984, p. 13ff), followed by other materials such as iron and aluminum.

Another casting technology not available before the industrial revolution was die casting. In this process, molten metal is forced into a reusable cavity through pressure rather than only through gravity in order to form a part. The first die-casting process was patented in 1849 for producing metal letters for printing presses. The process was expanded in 1885 to die-cast not only single characters but an entire line of text in one piece.

Initially only designed to process tin and lead, die casting expanded into other materials in the early twentieth century.

Yet another, although more exotic, casting technology is centrifugal casting, where a reusable mold is rotated at high speed while molten metal is added. First used in 1855 to create artillery shells, the process was expanded in the early twentieth century to different metals; wax; and—a favorite with children—chocolate for chocolate eggs, Easter bunnies, and Santa Clauses.

Forging also evolved through mechanization. Drop forging shapes the entire piece of hot but solid metal with one blow, pressing the metal into a die and giving it the form of the die. This process was first used in the 1830s by Elisha Root in the Colt Armory and was patented in 1858 (Uselding 1974). A related technology is sheet metal stamping, where a sheet metal is stamped similar to a sheet of paper in a hole puncher, albeit the hole may be in different shapes. This process was used from the 1830s onward, and especially in bicycle technology, this cutting-edge technology allowed great cost reductions compared to the sturdier but heavier and more expensive drop-forged parts (Hounshell 1985, p. 8).

Another cutting technology that emerged is the planer or shaper. The process is similar to a plane on a piece of wood, but machine power was necessary before the process could be applied to stronger metals. The difference between a planer and a shaper is that in a planer, the workpiece moves, and in a shaper, the tool. The detailed history of the planer is unknown, and it may have been invented independently by different toolmakers. The earliest published sources date back to 1754 in France, and patents are known from Bentham in 1793 and Bramah in 1802 (Roe 1916, p. 50ff). The shaper was invented by James Nasmyth in 1836 and later improved by Joseph Whitworth (Roe 1916, p. 92ff).

The lathe was already known in ancient Egypt, as was its variation, the potter's wheel, but both used by muscle power. While different inventors contributed to the evolution of lathes, it was probably Henry Maudslay who first built a lathe from metal for metal around 1800 (Roe 1916, p. 17). Lathes built completely out of metal allowed for much higher precision and speeds than previous lathes where only the cutting tool was made of metal. Maudslay's lathe already had a slide rest to mount the cutting tool rather than moving the tool by hand as for turning wood. This was improved in 1845 by Stephen Fitch, who mounted a rotation tool holder on the slide rest (Hobsbawm 1999, p. 153). Hence using this turret lathe, it was possible to use different tools in rapid succession without the time-consuming need to remove the previous tool and to fit the next tool into the machine. Additional improvements include multiple spindles, copying lathes, and automatic lathes.

The usage of all metal machines and tools even allowed a completely new manufacturing process: milling. Wooden machines were simply unable to take the forces in a milling machine without bending and twisting. Metal milling machines can withstand these forces and are able to create complex geometries out of a solid block of steel. In milling, the workpiece is moved along a rotating cutter. Depending on the shape of the cutter and the movement of the workpiece, it is possible to produce complex geometric shapes with high accuracy. The idea of the milling machine evolved from filing, where rotary files reduced the time-consuming task of manual filing. Probably the earliest rotary file dates from 1760 by Jacques de Vaucanson in France (Roe 1916, p. 206). The exact inventor of the milling machine is unknown, and probably, numerous individuals contributed independently or successively to the concept. The milling machine was most likely invented around 1818* in the U.S. armories of Springfield and Harpers Ferry, involving Simeon North, John Hall, Robert Johnson, and John Blanchard (see also Section 9.4). While initially only designed to reduce hand filing, milling machines soon improved their accuracy and eventually replaced hand filing altogether. The first modern milling machine where the tool could move along three axes was designed by Joseph R. Brown in 1861, and his company dominated the booming milling machine market for the next 50 years.

The availability of cheap steel lead to many other developments in metallurgy and technology. In 1742, Benjamin Huntsman in Sheffield invented a cast or crucible steel that was harder than the previous steels used for cutlery. Since the local cutlery makers were unaccustomed to this harder steel, they refused to use it, and Huntsman sold his steel to France. While English cutlery makers were opposing the harder steel, English customers were not. Initially, English cutlery makers tried to ban the export of the harder steel while at the same time refusing to use it themselves. However, their lobbying was unsuccessful. French imports soon took over the market, and the English cutlery makers were also forced to use this harder steel (Smiles 1864). Eventually, this steel was also used for metal-cutting

* The invention is often credited to Eli Whitney, but modern historians cast doubt on this and give credit to a number of other inventors. See Section 9.5 for more.

tools. High-carbon steel tools, in 1850, were able to cut iron at speeds of 10 meters per minute.

The first true tool steel was invented by Robert Forester Mushet in 1861 by adding tungsten, resulting in a much harder tool, allowing cutting speeds of 20 meters per minute. Frederick Winslow Taylor* developed the scientific basis for cutting tools between 1880 and 1905, including the discovery of the benefits of cooling and lubricating the cutting tool with water. Together with Maunsel White, he also developed a new and faster cutting steel called high-speed steel, abbreviated as HSS† (Groover 2001, p. 560). This steel not only was able to withstand much higher temperatures but also got harder as the temperature increased, and hence allowed cutting speeds of up to 50 meters per minute. Combined with its low price, it is still the most common cutting tool on household drills.‡

By 1914, cutting speeds of up to 140 meters per minute were achieved (Landes 1969, p. 297). Even harder steel alloys with tungsten carbide were already developed around 1890, but the necessary sintering processes to make tungsten carbide tools were not developed until 1927. With the sintering technology available, numerous other metal–ceramic composite materials were developed for cutting tools in the 1950s, as were diamond tools from 1954 onward (Groover 2001, p. 560).

Similar to the advancement of cutting tools with a defined geometry, grinding materials also improved significantly. Grinding soon, developed into a full-scale manufacturing process able to remove large quantities of material with precision. This allowed the shaping of complex geometries made of even the hardest steel alloys (Landes 1969, p. 311). Probably the first modern grinding machine was designed by Joseph Brown in 1868. The grinding wheels evolved over numerous steps, improving both the abrasive materials and the bonding between them.

---

* Despite his significant advances in cutting tools, Taylor is much better known for his advances in scientific management, and hence, we will hear much more of him in Section 12.2.

† This required a significant amount of testing, and they turned literally tons of steel into chips. Initially, they used a lathe connected to the central driveshaft, and to experiment with different cutting speeds, the entire factory had to slow down or speed up, often on a hourly basis. Naturally, this did not make his experiments popular with his colleagues on the shop floor, and they soon got a lathe with an individual electric motor (Copley 1923, p. 1:237ff).

‡ If you have some drill bits at home, check them out. At least some of them will, in all likelihood, say HSS.

11.5 SCREWED—REVERSIBLE FASTENERS

Besides forming and cutting, major advances were also made in joining two parts together. Welding technologies forming a permanent bond between two parts were described in Section 11.2. Yet, the creation of a reversible connection between two parts nowadays mostly involves a screw. While the principle of the screw was known since ancient times, these were used only for moving parts, for example, in a wine press. Attaching parts to each other using screws was possible already in medieval times, but these screws were custom made and matched only their specific counterpart. Even the first mass-produced metal screws were either self-cutting, as in soft material (for example, wood), or custom-matched to their metal counterpart (Rybczynski 2000).

Critical dimensions like the diameter, the number of threads per centimeter, the depth of the thread, the shape of the thread (triangular, rounded, rectangular, or a combination thereof), and the rotational direction differed from screw to screw, making a match by chance almost impossible. However, you can easily imagine that a screw fitting only one specific bolt would have no place in mass production. This insight also occurred quickly to the screw makers of the nineteenth century. Henry Maudslay was probably the first to standardize screw sizes within his own workshop, aided by the screw-cutting lathe for making identical screws of his own design. His chief draftsman Joseph Clement improved on this standardization by clearly defining the geometric shapes of the screws to be used in Maudslay's workshop. Another protégé of Maudslay, Joseph Whitworth, proposed in 1841 to standardize the screw threads for all of England. This British Standard Whitworth was first used by railroad companies and, later, all over England (Roe 1916, p. 10).

Initially, the British Standard Whitworth was also frequently used in the United States but intermixed with many different industry or manufacturer screw standards. William Sellers proposed a new standard for the United States in 1864, and in 1898, he established the United States standard thread. This standard was based on the standard of Whitworth but with some changes in the dimensions. The British Standard Whitworth and the United States Standard Thread were merged in 1949 into the Unified Thread Standard, since different standards between the United States, Canada, and Great Britain caused much trouble during World War II (Sinclair 1969).

Also in 1898, the International Congress for the Standardization of Screw Threads in Zurich, Switzerland, defined the metric thread sizes. This metric standard was based on the original United States Standard Thread but with metric dimensions. Over time, the metric system became the dominant thread standard worldwide. While whitworth threads are still used for pipes and the Unified Thread Standard in parts of America, most globalized industrial production nowadays uses the metric screws. Even American carmakers use metric screws for their products.

Not only did the screw thread evolve, but so did the screw head. Initially, there was only the slot head in use, which had the disadvantage that the screwdriver easily slipped out and damaged the part. In 1933, Henry F. Phillips patented the Phillips head, which was self-centering and prevented overloading the screw through cam out and hence simplified assembly. Together with its improved variant, the Pozidriv, these screw heads were commonly used in manufacturing applications. Nowadays, Torx heads are more popular in industry, since screws are even less likely to slip out of a Torx head. Overall, screws are nowadays a standard product in manufacturing. Few assembled products do not contain at least a few screws. Most screws used in industry are standardized screws available from a multitude of suppliers.

The highly standardized screws are in stark contrast with the heavily individualized management. Nowadays, we still have few to no standards for managing work. Most of what we have originated with Frederick Winslow Taylor, as we will see in the next chapter.

BIBLIOGRAPHY

Bowler, P.J., Morus, I.R., 2005. *Making Modern Science: A Historical Survey*. University of Chicago Press, Chicago.

Camenzind, H., 2007. *Much Ado About Almost Nothing: Man's Encounter with the Electron*. http://www.BookLocker.com.

Copley, F.B., 1923. Frederick W. Taylor: Father of Scientific Management—Volume 1. Harper & Brothers Publisher, New York.

Giedion, S., 1948. *Mechanization Takes Command: A Contribution to Anonymous History*. Norton, New York.

Groover, M.P., 2001. *Fundamentals of Modern Manufacturing: Materials, Processes, and Systems*, 2nd ed. John Wiley & Sons, Hoboken, NJ.

Harman, C., 2002. *A People's History of the World*, 2nd ed. Bookmarks Publications Ltd, London.

Hasan, H., 2006. *Aluminum*. The Rosen Publishing Group, New York.

Hobsbawm, E.J., 1999. *Industry and Empire*. Penguin Books, London.
Hounshell, D.A., 1985. *From the American System to Mass Production, 1800–1932: The Development of Manufacturing Technology in the United States*, Reprint. ed. Johns Hopkins University Press, Baltimore.
Landes, D.S., 1969. *The Unbound Prometheus: Technological Change and Industrial Development in Western Europe from 1750 to the Present, 1st ed*. Cambridge University Press, New York.
Letcher, P., 2003. *Eccentric France*. The Globe Perot Press Inc., Guilford, CT.
Nye, D.E., 1992. *Electrifying America: Social Meanings of a New Technology, 1880–1940*, MIT Press, Cambridge, MA.
Roe, J.W., 1916. *English and American Tool Builders*. McGraw-Hill, New York.
Rybczynski, W., 2000. *One Good Turn: A Natural History of the Screwdriver and the Screw*. Scribner, New York.
Schiffer, M.B., 2008. *Power Struggles: Scientific Authority and the Creation of Practical Electricity before Edison*. MIT Press, Cambridge, MA.
Schlesinger, M., Paunovic, M. (eds.) 2011. *Modern Electroplating*. John Wiley & Sons, Hoboken, NJ.
Sinclair, B., 1969. At the Turn of a Screw: William Sellers, the Franklin Institute, and a Standard American Thread. *Technology and Culture* 10(1), 20–34.
Smiles, S., 1864. *Industrial Biography: Iron-Workers and Tool-Makers*. Ticknor and Fields, Boston.
Stanley, A., 1995. *Mothers and Daughters of Invention: Notes for a Revised History of Technology*. Rutgers University Press, New Brunswick, NJ.
Stelter, M., Bombach, H., 2004. Process Optimization in Copper Electrorefining. *Advanced Engineering Materials* 6(7), 558–562. doi:10.1002/adem.200400403.
Uselding, P., 1974. Elisha K. Root, Forging, and the "American System." *Technology and Culture* 15, 543–568.
Whitten, D.O., Whitten, B.E. (eds.) 1990. *Handbook of American Business History: Manufacturing*. Greenwood Publishing Group, Westport, CT.
Wick, C., Benedict, J.T., Veilleux, R.F. (eds.) 1984. *Tool and Manufacturing Engineers Handbook, Volume II: Forming*. Society of Manufacturing Engineers, Dearborn, MI.

12

Science Meets Shop Floor

There is perhaps no trade or profession existing in which there is so much quackery, so much ignorance of the scientific principles, and of the history of their own art, with respect to its resources and extent, as is to be met with amongst mechanical projectors.

Charles Babbage (1791–1871)
Mathematician, philosopher, inventor, and engineer,
in On the Economy of Machinery and Manufactures, 1835

Throughout history, humanity made significant advances in technology. Yet, the organization of manufacturing lags far behind. While steel bends to our will easily, we still have major problems working with other human beings. The goals of the workers, the managers, and the shareholders still diverge significantly. Of course, officially, all of these goals are aligned, but the more significant hidden objectives are not. It took until around 1900 for the problem to be at least studied, much less solved. This chapter will look in more detail at the developments of scientific management and its most significant researcher, Frederick Winslow Taylor.

12.1 THE BEGINNING OF MANUFACTURING MANAGEMENT

Until the end of the nineteenth century, coordinating work on the shop floor was based purely on experience. Wages were set by the supervisors or owners as low as they could get away with. Required production quantities depended on the mood and the speed of the worker, and at best, the only mathematical element on the shop floor was basic cost accounting. There was little understanding of the relations between work, time, and cost.

Modern science originated after the Renaissance during the sixteenth and seventeenth centuries. Men of knowledge were no longer satisfied with a religious or philosophical approach to understanding the world but, rather, tried to understand the world through observations and experiments. The view of the world changed from belief to reasoning. Fundamental new insights about the world were discovered by Nicolaus Copernicus, Galileo Galilei, Johannes Kepler, Isaac Newton, and Gottfried Leibniz, to name just a few of the many people involved in building our scientific knowledge.

However, back then, science was almost exclusively for research only, done in laboratories, or for the curiosity of spectators. Practical applications of this science were pretty much nonexistent, and products were still manufactured based on trial and error. Only with the steam engine in the eighteenth century was scientific knowledge used to improve a product by calculating pressures and temperatures rather than just trying it out. In the nineteenth century, electric products were impossible solely based on trial and error but required scientific calculations.

The introduction of science to shop-floor management followed a similar approach. Research and science of management existed long before manufacturing was organized with the help of science. For example, the thirteenth-century English writer Walter of Henley discussed agricultural estate management in great detail in his work *Le Dite de Hosebondrie*, with many possible analogies to manufacturing. Leonardo da Vinci (1452–1519), as part of an excavation project of the river Arno in Italy, studied in great detail the process of shoveling in his *Codex Atlanticus*. He found that a worker can handle 500 shovels of soil per hour, where each shovel consists of four individual steps: loading the shovel, preparing to throw the soil, throwing the soil, and returning the shovel to its original position. In this first known motion study of the world, Leonardo also noticed that some workers can do the process 30% faster but that they tire quicker (Baeck 2008).

Another medieval researcher of scientific management was Georgius Agricola (1494–1555), who discussed ratios of man power to horsepower in his fundamental work on mining and metallurgy, *De Re Metallica* (Alder 2010, p. 133).

French Abbot Gilles Filleau des Billettes (1634–1720) published a detailed study on the division of labor in pin making in 1700 in the *Journal des Sçavants* (*The Scientists' Journal*). He showed the division into individual steps and analyzed in great detail related wages, costs, productivities, and profits in pin making* (Peaucelle and Manin 2006). English mathematician and engineer Charles Babbage (1791–1871) not only invented the first mechanical computer but wrote probably the first book on manufacturing science, *On the Economy of Machinery and Manufactures*, in 1835. In this book, Babbage clearly advocates to divide tasks into smaller parts, and give skilled workers only the difficult tasks and unskilled workers only the easy tasks.

Other large organizations also became interested in scientific organization. Railroad companies were, at the time, among the largest commercial enterprises in the world. While not identical to manufacturing, many of its problems were very similar to manufacturing. From the 1850s onward, railroad companies invested resources to control time and costs by tracking a number of economic parameters in order to improve their business (Chandler 1977, p. 100ff).

Even such improbable areas as the home kitchen were subject to scientific analysis. Catharine Esther Beecher, sister of Harriet Beecher Stowe, the author of *Uncle Tom's Cabin*, was a strong advocate and pioneer of women's education.† Besides her teaching, she also studied the layout and arrangement of kitchens, publishing *A Treatise on Domestic Economy* in 1841 and *The American Woman's Home* in 1869. Analyzing movements in detail, she made fundamental changes to the kitchen layout. Back then, a kitchen had a table at the center and storage along the walls. Beecher removed the table and provided working surfaces along the wall with storage below the surface and in wall

---

* The pin-making analysis by Gilles Filleau des Billettes was enhanced and published again in 1717 by French engineer Guéroult, in 1723 by Savary, and in 1740 by Jean-Rodolphe Perronet. From there, it found its way into the encyclopedia by Denis Diderot in 1755. In 1761, it was published as part of a book, *The Art of the Pin-Maker*, by Henri Duhamel du Monceau. It also appeared again in a dictionary by Macquer in 1766. For more about French authors copying from each other, see the work of Peaucelle and Guthrie (2011). Next, the example made it into the famous pin-making story in Adam Smith's 1776 book, *The Wealth of Nations* (Peaucelle and Manin 2006). English mathematician and engineer Charles Babbage also used it for his book, *On the Economy of Machinery and Manufactures* in 1835, although he was the first who actually gave credit to previous authors.

† While she was a remarkable woman, she does not quite fit the description of a feminist. For example, she was opposed to women having political power, which she considered evil (Giedion 1948, p. 518).

cabinets—effectively inventing the modern kitchen still in use today. She also set up distinct areas in the kitchen for preservation, storage, cooking, and serving, a setup that is no longer followed nowadays (Giedion 1948, p. 518).

Even the arts were inspired by scientific management and motion studies. One of the most iconic images of this time is *Woman Walking Downstairs* (Figure 12.1) from the book *The Human Figure in Motion* by Eadweard Muybridge (1830–1904).

As mechanical products required the most manufacturing steps, mechanical engineering started to become interested in manufacturing management and, from the late nineteenth century onward, was pretty much the only science discussing manufacturing operations (Chandler 1977, p. 282). For example, James Waring See, writing under the pseudonym Chordal, discussed many shop-floor management problems in a series of columns in the *American Machinist*, which was eventually published as a book in 1883 (Waring 1883). The American Society of Mechanical Engineers (ASME) was founded in 1880 and soon was the leading institution on shop-floor management in the world (Hopp and Spearman 2001, p. 26).

Other writers were also active on this topic. Henry Metcalfe published his book, *The Cost of Manufactures and the Administration of Workshops, Public and Private*, in 1885. Henry Robinson Towne wrote about the *Engineer as Economist* in 1886 (Hopp and Spearman 2001, p. 26). Emile Garcke and John Manger Fells looked into *Factory Accounts* in 1887. Frederic Arthur Halsey published *The Premium Plan of Paying for Labor* in 1891.

Probably the first true manufacturing management book in the modern sense was *Commercial Organization of Factories* by Joseph Slater Lewis in 1896 (Landes 1969, p. 322), followed shortly by *The Commercial Management of Engineering Works* by Francis G. Burton in 1899. However, all those publications were mostly ignored by the intended audience of manufacturers. With the exception of Beecher's work on kitchen organization, there was little practical application of these studies in manufacturing. Nevertheless, the knowledge started to spread, and it was only a matter of time until scientific management entered manufacturing.

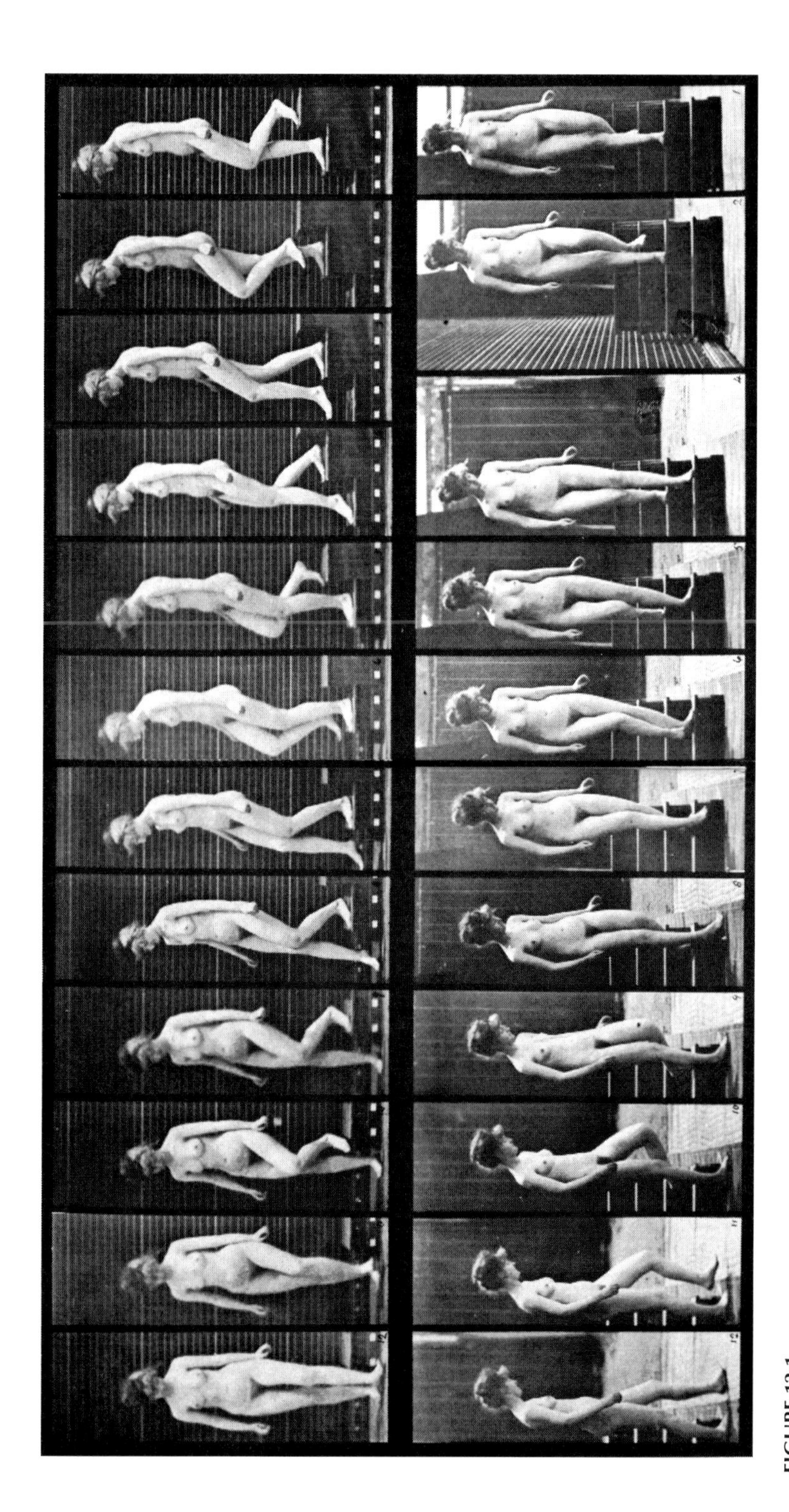

FIGURE 12.1
Woman Walking Downstairs by Eadweard Muybridge, 1887.

12.2 FREDERICK WINSLOW TAYLOR—THE FATHER OF SCIENTIFIC MANAGEMENT

The single person responsible for bringing scientific management into practical use was Frederick Winslow Taylor (Figure 12.2). He was born in 1856 in Germantown, Pennsylvania, into a well-to-do and cosmopolitan family. His maternal ancestors were among the leaders of the first settlers aboard the Mayflower. Already back then, they were wealthy, bringing their own servants along to the new world. His father was a lawyer with a master's degree from Princeton University, and his mother was politically

FIGURE 12.2
Frederick Winslow Taylor.

active to end slavery. He was schooled in Germany, France, and a private school in the United States before passing the entrance exams for Harvard Law School with honors (Copley 1923a). By all accounts, he was set up to follow in his father's footsteps and become a wealthy lawyer and member of the upper class.

However, his habit of reading books until late at night deteriorated his eyesight. Hence, while he preferred to work with his mind, his bad eyesight forced him to seek work for his hands. As such, he became an apprentice patternmaker for a small pump manufacturer. Working with metal rather than reading was apparently good for his eyes, as his vision improved.

In 1878, after his apprenticeship, he started to work as a lathe operator for Midvale Steel Works. Both due to his skills and due to his family connections, he climbed the hierarchy quickly and eventually became the chief engineer of the works in 1884.* The Midvale Steel company was presided over from 1872 onward by William Sellers,† who brought in chemists to improve the quality of steel based on scientific analysis rather than trial and error as before. Sellers also used a scientific approach to optimize angles and shapes for cutting tools, while other metallurgists worked on the steel type (Copley 1923a, p. 1:110).

It was at Midvale Steel in 1881 that Taylor started to develop the fundamentals of his scientific management approach, or what he called *task management* or the *Taylor system* (Hopp and Spearman 2001, p. 16). The basis of his approach was a detailed understanding of the work to be done, not only qualitatively but also and especially quantitatively. The first step is a time study, breaking down the task into individual steps and measuring their time. Next, unnecessary tasks are disregarded, and the technical aspect is optimized. Then, the time is adjusted for problems or deviations that may pop up and cost the worker more time than an optimal process. Finally, the times are adjusted for rest and breaks and for the level of

* While working 10-hour days, he was also socially active and a member of the cricket club. There he is best remembered for his performance in stage plays. Since the club consisted only of young men, he usually played the female parts. This he did so convincingly that most in the audience actually believed him to be a woman. His acting, however, was not only limited to the stage. Taylor also practiced swearing to impress both workers and management at Midvale Steel Works. Apparently, he achieved considerable skill in this aspect, somehow combining being classy while swearing at the same time (Copley 1923a). On top of that, he was also an expert tennis player, winning the doubles tournament of the U.S. Open championship in 1881.

† The very same William Sellers that was the father of the United States standard thread, as detailed in Section 11.4.

ability of the worker. Based on these data, Taylor set clearly defined work standards and working speeds and paid the workers well if they matched this speed.

One of his first projects was to set piece-rate wages for machinists in order to increase output. Having worked as a lathe operator before, he knew for certain that the workers were working much slower than they could. In fact, he did it himself when he was working as an operator. He estimated that the workers did only one-third of what he considered a fair day's work (Taylor 1911b). This slow work was known back then as soldiering and was common in all manufacturing enterprises. The workers did this to preserve time for their own leisure and avoid exhaustion, to keep piece wages high, to increase work and hence employment for their fellows, and to have authority and control over their work rather than give management control (Noble 1984, p. 33).

However, while Taylor did soldiering himself as a lathe operator, as a foreman, he clearly realized that he was now representing the interests of the factory. He knew that requiring a higher output would result in a huge fight with the workforce, so first he got support from company management:

> The only thing I ask of you, and I must have your firm promise, is that when I say a thing is so you will take my word against the word of any 20 men or any 50 men in the shop. If you won't do that, I won't lift my finger toward increasing the output of this shop.
>
> **Frederick Winslow Taylor**
> *Quoting himself during the congressional investigation in 1912*

Next, he determined the desired rate both by training operators and by working the lathe himself. Taylor focused first on one man; this worker refused to increase output, and Taylor fired him, replacing him with a new worker. This worker also refused to increase output and got fired. Repeating this a few times yielded no improvement whatsoever. As a next step, Taylor took untrained workers, and promised to train them as lathe operators with higher wages if they promised to produce the desired output. However, after training, all of them worked the same rate as the old workers, again with no improvement in output.

Since all newly trained workers failed to follow through on their promise of higher rates, next, he simply cut the piece-rate wages by half. Working at the desired speed, they would still make more than before, but if not, they would lose half their income. The workers continued at their old speed while complaining bitterly all the time. Taylor even received death threats, but eventually, the workers gave way and increased their speed.

This, however, was not the end of the fight but merely started the next round. Aiming to prove to management—who still sided with Taylor—that Taylor was driving them beyond their limits, they frequently and intentionally broke parts, tools, and machines, blaming it on the inhuman speed of work. As a response, Taylor had all men pay for their broken parts, tools, or machines out of their own pocket, or quit the job. He did not exclude himself from this rule. When he broke a part or tool by accident, he also fined himself accordingly. The money went into a fund for employee welfare. While he was tough to his workers, he was also fair and applied the same rules to himself as to anybody else. Of course, the workers continued to protest and find ways to show that the failure was entirely not their fault, but with the backing of management, Taylor continued to fine them. Eventually, the workers gave in and produced parts at the desired speed without any unusual rate of breakdowns. It took Taylor three years to win this conflict with his workers and to achieve a stable and higher output without soldiering (Copley 1923a, p. 1:157ff).

While Taylor was a foreman himself, he nevertheless was very critical of the role of foremen at this time. Nowadays, a foreman usually coordinates the work of a small team of 5 to 25 workers and spends most of his/her time on administrative issues and problem solving. Before the mid-twentieth century, however, a foreman was a much more powerful figure on the shop floor, more akin to a subcontractor. The foreman decided what to produce and when, hired and fired workers, set working speeds, set up the machines, maintained machines, inspected quality, tracked attendance, paid salaries, and most importantly, made his people work. He had a very high level of control of his shop. In effect, he was much more of a mini plant manager than a troubleshooter as nowadays.

Taylor was critical of this role. He believed that a foreman, no matter how smart, is easily outwitted by the collaborative intelligence of his team of workers. Additionally, since the foreman usually was a skilled worker who got promoted, he had a tendency to side with the workers and against the factory owners. Also, the foremen resisted Taylor's approach based on hard measured data, since it made them accountable and reduced their power over their shop floor (Chandler 1977, p. 271). Hence, Taylor believed that the traditional role of the foreman had to go.

Instead, Taylor envisioned splitting up his role into eight different foremen, each of them specializing in one particular aspect of production. Four of them he saw on the shop floor, and four in a planning room. On the shop floor, he envisioned

- The *gang boss*, responsible for providing work to the men and teaching them how to set up the machines efficiently
- The *speed boss*, responsible for working speeds of man and machine once the machine has been set up
- The *inspector*, responsible for quality
- The *repair boss*, responsible for fixing and maintaining machines

Regarding the four foremen in the planning department, he envisioned

- An *order of work and route clerk*, responsible for the production sequence on different machines
- An *instruction card clerk*, responsible for the standard procedures of the production process for both work and time
- A *time and cost clerk*, for tracking attendance and salaries
- A *shop disciplinarian*, responsible for proper work behavior, for example timeliness, following standards, and respectful behavior, who could also adjust salaries accordingly and furthermore had the important role of a peacemaker (Taylor 1911a)

The separation of the foremen into eight roles as described by Taylor has never been implemented in this rigidity (Chandler 1977, p. 276), but many of these roles are nowadays performed by different departments in the factory. While names of departments and organizational structure frequently differ among companies, the following is a common distribution of responsibilities:

- The role of the *gang boss* and *speed boss* are still part of the foreman's job as far as enforcing speed is concerned. However, the setting of the standards is usually part of the standards department.
- The service of the *inspector* is nowadays done by a separate quality control department.
- The function of the *repair boss* is taken over by the maintenance department for all but the simplest tasks.

- The function of the *order of work and route clerk* is nowadays usually part of logistics, in particular, production planning.
- The function of the *instruction card clerk* is nowadays part of the standards department.
- The role of the *time and cost clerk* is filled by the human resources department.
- Finally, the *shop disciplinarian* is mostly the management levels above the foreman, for example, the group leader or department leader.

Hence, while Taylor's system of eight foremen was never implemented strictly, the modern factory has a division of labor on the management level, with a similar or finer granularity than what was described by Taylor.*

Taylor stayed at Midvale Steel until 1890. Even after he left, Midvale Steel continued to use his methods and to benefit from his approach. However, without his attention to detail and drive to enforce the rules, the system slowly slipped downhill, and eventually, the piece-rate wages were again set by experience rather than measurement (Copley 1923a, p. 1:339).

Taylor's next position in 1890 was as a general manager of the Manufacturing Investment Company, an enterprise that financed and ran numerous paper mills. Employment satisfaction left much to be desired. With a distaste for financiers, Taylor left after only three years in 1893. Not having a follow-up job required him and his wife to tighten their belts. To make money and to spread his ideas into the world, Taylor became the first management consultant in the world in 1893, focusing on shop management and cost reduction as a *consulting engineer* (Copley 1923a, p. 1:391). His idea worked, and he found that there was great demand for his services. Within a few years, he made a small fortune in consulting. Additionally, he was able to further develop scientific

* At this point, it is also interesting to note that developments in Lean manufacturing actually reverse some of this granularity. In Lean manufacturing, the operators at the machine are crucial in setting and improving the standards, including parts of the roles of the *gang boss*, *speed boss*, and *instruction card clerk*. They are also responsible not only for quality control but also for preventing defects in the first place, hence taking the role of the *inspector*. Workers in Lean manufacturing even maintain and repair their machines for simple- to medium-complexity repairs, greatly reducing the work of the *repair boss*. Only the *order of work and route clerk*, *time and cost clerk*, and obviously the *shop disciplinarian* are roles not yet done by the operators themselves. We will hear much more about Lean manufacturing in Chapter 16.

management, publish different papers on this topic, and patented a number of technical ideas.

As a consultant, he also found that reducing work hours in many cases improved productivity, and the workers were able to do the same work in less time given enough rest. He reduced work hours for clerks to improve productivity, and he improved productivity at inspection of ball bearings for bicycles. The ball bearings were inspected by young women working 10.5 hours per day. The women agreed individually that they could do the same work in 10 hours at the same pay. Taylor, however, said about himself that he *had not been especially noted for his tact* (Taylor 1911b). For a change, he decided to have them vote if they wanted to work 10 hours or 10.5 hours at same pay and same daily quantities. The girls rejected this approach, distrusting the management, thinking that they would reduce pay eventually. Hence, *a few months later tact was thrown to the winds* (Taylor 1911b), and work times were successively reduced from 10.5 to 10, 9.5, 9, and finally 8.5 hours. Surprisingly, daily output not only remained constant but increased by a whopping 33%, and at the same time, the quality of inspection also improved. (See also Section 10.4 for similar experiences in other industries.)

In 1898, on the invitation of a former boss from Midvale, Taylor started to consult full time for the Bethlehem Steel Company in Bethlehem, Pennsylvania. With almost 6000 workers, the Bethlehem Steel Company was among the largest manufacturing companies of its time (Wrege and Perroni 1974), increasing to 25,000 people by World War I (Copley 1923b, p. 2:4). His reputation arrived in Bethlehem before him. Workers were worried about their jobs, while residents were worried about Taylor causing a drop in real estate values due to his bad reputation.

It was at Bethlehem where Taylor did his pig-iron experiment, probably his best known example of scientific management, albeit with some flaws, as we will see later. The plant had 80,000 tons of pig iron stored outside in a yard. This iron had to be loaded in railroad cars for sale. At this time, these jobs were still done by hand, and a group of 75 men was assigned to load iron into railroad cars. This, of course, was a lowly job, and Taylor stated that *it would be possible to train an intelligent gorilla so as to become a more efficient pig-iron handler than any man can be* (Taylor 1911b). The task was as simple as one could imagine. The worker picked up a 40-kilo piece of iron from the pile, walked to the railroad car and up a ramp, and dropped the iron before returning to the pile. This was, for Taylor, the perfect setting to demonstrate his scientific management approach.

While measuring and observing the individual steps in detail, he soon came to the problem of exhaustion and rest. Was it better to work slow and steady, or faster with breaks in between? If so, how fast to work, and how often and how long to rest? He soon discovered that it did not matter how fast the worker walked, since his arms got tired from holding the iron regardless of the speed of his legs. Eventually, Taylor calculated that a man should be able to load 48 tons per day, while they were, on average, only loading 12.5 tons per day. Taylor then calculated the piece rates that would allow a man to earn 60% more per day if he handled 47 tons per day.

Next, he selected men suitable to the task for further studies, eventually narrowing it down to a single *mentally sluggish* worker: the soon-to-be famous Mr. Schmidt.* By instructing him exactly when to work, how fast to work, and when to rest, Taylor soon managed to have Schmidt load 47 tons of iron per day without exhausting him. Taylor convinced the other workers to follow his standards and to work 47 tons per day using Schmidt as an example, effectively cutting the cost for the company in half while increasing the workers' wages by two-thirds.

At least, that is how Taylor told the story. However, his story grew with every retelling, until it was quite a bit different from what probably really happened. Already at the time of his publication of this achievement, there was a public backlash against him about the ill-treatment of the poor Mr. Schmidt, the overworking of Schmidt and his colleagues, and the unfair share of the overall benefits for the workers (Copley 1923b, p. 2:49f). Modern researchers found out that both quantities and workforce were probably different from what Taylor told. Rather than 80,000 tons, it was probably only 10,000 tons; rather than 75 men, it was probably around 15 (Wrege and Perroni 1974). The scientific analysis of rest periods sounds implausible, and the distance between pile and car was ignored, as were factors like weather and temperature. Taylor's effort saved the company only $6.05 per day or $302 altogether, a probably negative return on investment for the effort and time Taylor and others put into it (Wrege and Hodgetts 2000).

---

* Taylor was kind enough to use a fake name for the worker, but nowadays, it is known that his real name was Henry Noll (1871–1925).

These discrepancies notwithstanding, the pig-iron case probably serves as a good example to demonstrate the abilities of scientific management (Locke 1982), even though the actual implementation was probably not as successful as described. This, in fact, can be said about most of Taylor's efforts at Bethlehem Steel. He encountered much resistance and was able to implement his methods only with the lowest-paid jobs where the workers were able to put up the least resistance. Hence, his changes were limited to simple tasks like the pig-iron experiment or shoveling, but even then, some workers *forgot* how to shovel. After three years in Bethlehem, he still did not manage to achieve a change in the higher-skilled operations, for example, the machining shop, let alone in indirect areas like office work. The improvement of the machining shop was left to his protégé Henry Gantt, who did eventually implement scientific management and significantly improve productivity by using a more cooperative approach than Taylor (Nelson 1977).

In 1901, after returning from a vacation, Taylor found the following one-line letter from Bethlehem Steel on his desk (Copley 1923b, p. 2:153):

> Dear Sir:
> I beg to advise you that your services will not be required by this Company after May 1st, 1901.
> Yours truly,
> Robt. P. Linderman, President

It seems that ice-cold termination letters were already known in 1901. To Taylor, this was quite a shock. However, management and workers had difficulties with Taylor's total inability to compromise, while Taylor saw management as not supportive enough of his methods. It is not sure if this was the true reason for his termination, as there are rumors that another company wanted to take over Bethlehem Steel, explicitly without Taylor. In any case, not only Taylor but also all of the aides and assistants associated with *that man Taylor* were fired (Copley 1923b, p. 2:139ff).

However, Taylor had saved quite a bit of money and decided to reduce his consulting workload. Instead, he focused on promoting his methods. He built himself his own house, and while he was at it, he became an expert in gardening. He also developed and patented a method for moving and replanting large trees with their roots. Over a multiyear period, he scientifically analyzed the growth patterns of grass depending on the soil conditions, conducting thousands of grass-growing experiments, all in order to build his perfect green on a nearby golf course of his own design. For golf itself, he also designed a novel two-handed putter and other golf clubs (Copley 1923b, p. 2:186ff). No matter if it was botany, metallurgy, or shop-floor management, Taylor followed the principles of scientific measurement. He most certainly disliked guesswork.

In 1906, he was proposed as president of ASME. The election came as a surprise to Taylor, who, until then, had not been very active in the society, but he readily accepted the honor. As another surprise, he soon found out that the society wanted to reorganize itself and put the father of scientific management at their helm to get the most qualified man for the job. Taylor quickly sprung into action and applied his principles to the society.

However, his higher goal of organizing the entire ASME failed abysmally. Most people do not like being observed as part of a time study. However, the lowly porter shoveling or carrying pig iron cannot offer much resistance against management, and it was here that Taylor had his biggest successes. Higher-skilled labor like machinists have more power to resist such changes, and here, Taylor succeeded only sometimes. If the trouble increases as you go up in hierarchy, you can imagine how much trouble professors and doctorates can make! Hence, Taylor's proposals to conduct motion studies or high-ranking employees or distinguished committee members did not get off the ground at all, and he and his research were mostly ignored. His presidency was short lived, and a new president was elected after only one year.

The shunning of his work extended also to his research topics. While ASME was the most prominent organization promoting scientific management before his presidency, very few papers were published on this topic between 1906 and 1911 (Copley 1923b, p. 2:251). Nevertheless, Taylor continued to work in the academic field to promote his ideas, helping to set up college courses, writing papers, and speaking about his topic, including lectures at Harvard University. His most significant publication was arguably the book *Scientific Management*, published in 1911. He advised both the U.S. Navy on organizing their naval yards and the U.S. Army on organizing their manufacturing establishments, including the Springfield Armory. However, in both cases there was limited implementation, and the workers tried to resist Taylor's method as much as they could. Hence, it is not surprising that the—admittedly biased—president of the American Federation of Labor found the following (Copley 1923b, p. 2:340):

> Wherever this [Taylor] system has been tried it has resulted either in labor trouble and failure to install the system, or it has destroyed the labor organization and reduced the men to virtual slavery and low wages, and has engendered such an air of suspicion among the men that each man regards every other man as a possible traitor and spy.

Due to the controversy over Taylor's work in the U.S. Army, the U.S. House of Representatives launched an investigation in 1911 to find out if scientific management was unfair to employees with regard to compensation and bad for their emotional and physical well-being. Labor organizations tried to discredit Taylor's methods both inside and outside the hearings. They even managed, at one point, to get Taylor into an uncontrolled rage. However, the committee unanimously found that scientific management had not been around long enough to understand its impact on the labor system, and as far as they could say, there was no negative effect on the workers whatsoever. One committee member even wrote to Taylor afterward (Copley 1923b, p. 2:350):

> I feel confident that when your system is thoroughly understood, workmen themselves will rise up and demand it.

While this official investigation found no negative effects of scientific management, the method did shift power away from the common worker toward the industrialist. Hence, labor unions continued to fight scientific management. The Federal Commission on Industrial Relations started another investigation in 1914 with the goal to sort out the complaints of the labor movement against scientific management. However, relations soon turned sour, and Taylor strongly resisted the idea of joint control of both management and workers over the structure and organization of the work.

Nevertheless, Taylor continued to promote scientific management through publications, speeches, and practical implementations while traveling through America and Europe. In the winter of 1912–1913, he caught a cold, which he never got rid of. He died of pneumonia in March 1915, the day after his 59th birthday.

12.3 TAYLOR'S LEGACY

Taylor's work greatly advanced many different fields of industry, each of them securing him his place in history. He developed a new and improved cutting steel still in major use today. He improved modern landscaping. But most of all, he brought science to management. While others before him did measure work or discussed shop-floor management, he was the first to integrate the scientific approach of measuring and analyzing into management, creating the fundamental basis of modern management. Singlehandedly, he created the first modern management system ever and revolutionized management (Hopp and Spearman 2001, p. 32).

He was the first to separate the process of managing work from the actual work, hence allowing for greater specialization and, therefore, a higher level of management.* Taylor also established the completely new field of business of management consulting. He was the first management guru, and his books sold millions of copies. While there are a few older consulting companies, they focused on technical consulting rather than management consulting. For example, Massachusetts Institute of Technology (MIT) chemist Arthur D. Little founded his consultancy in 1885 but, in the early decades, did mostly chemical research and development for his customers before including management consultation.

The works of Taylor also found their way into academic teaching. The first college classes in management were probably offered by Dartmouth College in 1905 (Copley 1923b, p. 2:353), with the first graduate degrees in management offered by Harvard in 1908.

Through his fairness to everybody, he earned the respect of his workers at Midvale, where he worked among them and understood the business in detail. In his subsequent employments and as a consultant, however, he was often unable to make this connection and received much criticism and hostility from the workers.

---

* However, nowadays, this may have gone too far. There is, in most companies, an invisible boundary between white-collar and blue-collar workers, and many managers are way too far removed from the shop floor. I personally have seen numerous plant managers who avoided their own shop floor like the plague and visited the actual manufacturing only if they had a high-ranking guest to show around. Naturally, it is difficult to manage something without actually understanding the process, and I believe that having modern top executives frequently switching employment between companies in vastly different fields will be detrimental to the companies due to the lack of fundamental understanding of the processes.

Hence, already during his lifetime, his approach was criticized. His approach was—and still is—often seen as a way to cut wages and to put pressure on workers. While Taylor used his approach to change wages, he also frequently raised wages. One individual got a raise just for working well even when the boss was not around. Even workers who got fined may have gotten a raise during the same week (Copley 1923a, p. 1:179). He tried to give every worker a task at which he could excel and punished only those who worked slower than they could.

This is not to say that others used his approach with the single goal to reduce wages (Schachter 2010). Taylor disagreed with this view and wanted the workers to share the benefits, and indeed, workers often earned more than before. Nevertheless, workers and their union representatives remained wary of any moves to adjust wages, and in many cases, rightfully so. As this conflict moved into the public, the accusations grew more heated, and Taylorism was called the *tyranny of the clock* (Shapiro 2005) and *dehumanizing*. Hence, the term *Taylorism* nowadays has a negative connotation.

Aided by this negative context of Taylorism, the term *industrial engineering* became more popular than *scientific management*. The first use of the term *industrial engineering* was in 1899 (Jenks 1960), and the first college courses in industrial engineering started in 1909. Nowadays, industrial engineering is the academic field that most closely resembles scientific management.

Additionally, Taylor did not see the value of the workers' creativity. He treated the workers like machines, stifled their ability to change their workplace, and completely ignored any improvement ideas they had, stating, *You are not supposed to think. There are other people paid for thinking around here* (Copley 1923a, p. 1:189). While he saw the need for inventiveness and creative thinking, he believed that only a few gifted people are able to create something new, while the rest should just follow orders. Not only was he unable to integrate workers' creativity into manufacturing; he did not even see that these ideas had value. This is probably the most fundamental flaw of Taylorism. Modern industry has somewhat improved on this, but use of the workers' creativity is still frequently lacking.

Nevertheless, his approach was groundbreaking. For the first time, shop-floor management was given a scientific basis. Naturally, this scientific basis was still thin, and Taylor himself often deviated from this path, but as with every science, understanding has to grow over time. Applying a science not yet thoroughly understood to such a conflict-loaded topic as wages and working speeds of course resulted in significant conflicts that are still not yet fully resolved nowadays.

12.4 FURTHER PROGRESS IN SCIENTIFIC MANAGEMENT

Scientific management continued to develop, and numerous researchers contributed to this topic. First of all, there were a number of assistants and colleagues who worked with Taylor directly. Carl G. Barth significantly contributed to and enhanced Taylor's work on cutting speeds. Horace Hathaway implemented scientific management in different companies and eventually started as a consultant on Taylor's methods.

More significant was Taylor's assistant, Morris L. Cooke. While Taylor was gifted with developing the ideas of scientific management, he was not a very skilled or diligent writer. Cooke, however, had a talent for writing. Cooke started a book about scientific management, summarizing what he learned from Taylor's talks and presentations, which he often helped to prepare. This book, while mostly written by Cooke, ended up as the famous publication *Principles of Scientific Management*, with Taylor as the sole author. While the transition of authorship from Cooke to Taylor is questionable, Cooke received all profits from the book (Wrege and Stotka 1978).

Probably the most famous assistant of Taylor was Henry L. Gantt, although he preferred a much more cooperative approach to scientific management compared to the uncompromising Taylor. In later years, Gantt and Taylor lost sympathy for each other, because the uncompromising Taylor rejected Gantt's soft approach, believing that Gantt did not understand scientific management (Copley 1923b, p. 2:23). Gantt reorganized different factories, established a bonus system also for managers, and emphasized the social responsibilities of industrialists. However, he is most famous for the visualization of project steps and progress in a chart that he invented. This chart is still widely used in industry and named after him: the Gantt chart.*

Other researchers with no previous connection to Taylor also took up the ideas of scientific management. Probably the most famous of them

* Interestingly enough, Polish economist Karol Adamiecki invented a very similar tool 15 years earlier, which he called *harmonogram* (Locke 1982). However, since he published in Polish and Russian, his works were unknown in America. Even though his approach is even more intuitive, the computerized Gantt chart is nowadays the standard tool used for tracking project progress.

were a married couple, Frank and Lillian Gilbreth (1868–1924 and 1878–1972, respectively). Frank Gilbreth started with motion studies when he improved bricklaying while working as a building contractor. He reduced unnecessary motions in bricklaying using, for example, adjustable scaffoldings and improved the speed of a worker from 120 bricks per hour to almost triple the number, 350 bricks per hour (Taylor 1911b). He developed a system in which all motions in work can be split into 17 basic motions, which he humbly named after himself as *therbligs*, Gilbreth spelled backward, slightly amended.

While Frank had no education beyond high school, his wife Lillian went on to receive a PhD in psychology. Since her first dissertation was not awarded due to a formality, she wrote a second dissertation at another university and finally received her doctorate. Together, Frank and Lillian started a consulting firm on motion studies, making extensive use of modern photographic technology by using motion cameras and long-exposure shots. After studying surgeons, they also suggested the now-standard system in which a nurse hands the instruments to the surgeon, rather than the surgeon turning around to search for the correct instrument. The surgeons were very cooperative in these studies, but their egos got quite a dent, and their cooperation waned after they learned that the same method was also used to improve mundane, low-skilled labor like bricklaying (Giedion 1948, p. 100).

Lillian focused more on the human side of work, researching and consulting in both industrial psychology and human fatigue, which later evolved into the field of ergonomics. She also lectured at Purdue University on these subjects. Most amazingly, she did all that while giving birth to and raising 12 children. She even included her children in her research on home domestics and kitchen layout. Their methods also helped to organize such a large a household. They had timetables of who could use which bath for how long, including standards for efficient brushing of teeth and personal hygiene. Frank even wanted to analyze tonsillectomies by having his tonsils and those of all his kids removed, only to find out that the assistant forgot to load the film into the camera beforehand (Gilbreth and Carey 2002).*

* Two of her children wrote a family biography in 1948, *Cheaper by the Dozen*, which 20th Century Fox turned into a highly successful movie in 1950 with the same title. This was followed by a loosely related remake in 2003, which in turn had a sequel in 2005. The children also wrote a sequel to their family history, *Belles on Their Toe*, in 1950, which was also turned into a movie in 1952 as a sequel to *Cheaper by the Dozen*. That is quite a lot of movies for a single researcher couple.

Lillian was not the only researcher implementing Taylor's idea of scientific management in the kitchen. From 1910 onward, there was a movement to introduce scientific management into the household kitchen and to transform the housewife into a domestic engineer (Pursell 1995, p. 246f). A number of popular books on this topic were published, for example, *The New Housekeeping* (1912) and *Housekeeping with Efficiency* (1913). Soon, housewives all over America were counting their steps (Giedion 1948, p. 519). However, except for the layout of the kitchen without a central table—as already suggested by Catharine Esther Beecher in the 1840s—these approaches overall failed to take hold (Pursell 1995, p. 246f). The average household kitchen nowadays is still far from efficient.*

The approach of motion studies contributed significantly to Taylor's approach using time studies. However, Taylor himself did not appreciate the idea of motion studies and quarreled extensively with Gilbreth—and many other researchers—over the right way to do scientific management. This rift only deepened after Taylor's death, with his followers often clashing with other researchers (Hopp and Spearman 2001, p. 31).

Yet another researcher expanding the field of scientific management was Australian-born psychologist Elton Mayo (1880–1949), professor of industrial research at Harvard Business School. He is one of the main contributors to the human relations movement, studying the behavior of groups of people, especially in relation to work. Probably the most famous study he participated in was the Hawthorne studies, named after then-active large AT&T factory Hawthorne Works. In these studies between 1924 and 1932, researchers wanted to understand what factors affected productivity by studying workers assembling relays or other phone components.

For example, they studied the effect of illumination on productivity. They found that increasing the brightness of the workspace increased productivity. However, they found that decreasing the illumination improved productivity even more, until the women were working in near-moonlight conditions (Hopp and Spearman 2001, p. 35). This puzzled the researchers quite a bit. Then they experimented with number and duration of breaks and working times. Reducing the workday by 30 minutes increased

* I personally can confirm that finding. Shortly after getting married, my wife and I moved into a new apartment. Using all my considerable experience in work organization and layout optimization, I rearranged the contents of the kitchen according to scientific principles. Plates were close to the table, food next to the preparation area, pots and pans next to the stove, frequently used items in the top drawers, and so on. It was beautiful. Unfortunately, my wife thought otherwise, clearly stating, "This is *my* kitchen!" Without going into too much detail, just let me say that I have learned my lesson and will never try this in the kitchen again. I still love my wife, though!

productivity. Reducing it by another 30 minutes improved productivity even more. Surprisingly, productivity peaked when they reset the workday to its initial length, making it an hour longer again as in the previous measurement. Similar effects happened when they changed other factors. In sum, no matter what they changed, in which direction they changed, or if they changed it back, output increased. Only when they stopped the experiments did productivity drop back to the initial level.

Researchers were confused. Taylor treated humans like machines, and they expected a system like a machine. If you turned the knob one way, the machine went faster. If you turned the knob the other way, the machine slowed down. They only wanted to know which knob to turn and in what direction in order to improve productivity. Instead, they found out that no matter which knob they turned, productivity increased as long as they turned knobs.

The results are still under discussion nowadays, making the Hawthorne Studies one of the most re-researched studies in the world. However, most researchers believe that the attention the researchers and supervisors paid to the workers increased productivity much more than any other factor like illumination or work times. Of course, there are still a host of possible alternative explanations under discussion like seasonality, learning curve, room temperature, etc. The lack of a direct relation between influences on a worker and his/her reactions makes it difficult to influence work in a desired way. Yet, the most likely conclusion is that we have to treat workers not like machines but like human beings.

This lack of understanding of human behavior in the workplace gives rise to lots of people who pretend to be knowledgeable, often combined with buzzwords or fancy terminology. Often, these *improvements* will lead to actual but short lived improvements in cost, quality, or delivery performance. Hence, the person conducting the improvement will feel confirmed that his/her actions were right, and the sponsor is likely to spend more money or effort on similar actions. However, in most cases, this is only the result of the Hawthorne effect, named after the Hawthorne studies. No matter which knob you turn, the system will improve as long as you keep turning. Once you stop turning, the system most likely will revert to its old self, although by then, the initiator of the improvements will likely be somewhere else.

By the 1930s, there were too many researchers in Taylorism or related fields to name them all. For example, Harrington Emerson introduced scientific management to railroads and established the distinction between line and staff employees. Hugo Münsterberg established the field of industrial psychology. Walter A. Shewhart established statistical quality control. Agner Krarup Erlang invented a queuing theory, providing a mathematical model to calculate waiting times and queue lengths in a system. His approach is mathematically beautiful but practically rarely useful, being more theoretical than applied science.

Based on Taylor's initial work on scientific management, many different fields of research emerged, including industrial psychology, operations research, statistical quality control, industrial engineering, management science, time study, etc. Some fields became more academic, drawing inspiration from the shop floor but rarely making a lasting impact in shop-floor operations. Others increased in detail, but covering these details would exceed the scope of this book. Some of these fields, however, are still directly used for the day-to-day business on shop floors and will be discussed in more detail in the following section. In any case, while the field of scientific management is still developing, it all started with Frederick Winslow Taylor.

12.5 THE DEVELOPMENT OF TIME STUDIES—REFA AND MTM

Time studies are used to determine the ideal speed for workers doing manual work. Since they are often the basis for determining piece-rate wages, they also have large potential to escalate or calm down labor conflicts. Taylor was the first to measure the duration of work, enhanced by motion studies by the Gilbreths.

These methods not only measured how fast the worker was actually working but also provided a framework on how fast he/she *should* work. As such, they aimed to solve the crucial problem of determining a fair working speed. Naturally, the employer wanted to get the maximum out of his/her paid time and preferred a faster speed, whereas the worker preferred a slower pace.

Frederick Taylor's system of scientific management was based on measurements, and Frank and Lillian Gilbreth refined these measurements in their time studies. However, the efforts of these and many other researchers were still on an individual basis, measuring times either themselves or with the aid of their assistants. There was no overarching, generally accepted system for measuring work times, and hence, opinions differed between labor and management.

The German engineering association Verein Deutscher Ingenieure (VDI) was influenced by Taylorism and established a committee in 1921 to define and refine the time-study approach. This committee eventually established itself as its own organization in 1923, *Reichsausschuß für Arbeitszeitermittlung* (*Imperial Committee for Working Time Studies*), abbreviated as REFA. They provided a framework for time studies in the workplace and trained thousands of people each year in their method. Therefore, they successfully established themselves as a brand on its own, something neither Taylor nor the Gilbreths did.

As in most organizations, the early years were difficult, and the influence of REFA spread only slowly. For example, in 1923, they trained only 1650 people in their methods. However, the idea of a structured, thorough method to measure work was very compatible with Nazi ideology, as were organized systems like assembly lines. Hence, from 1933 onward, the influence of REFA increased as Hitler gained power, and in 1943, REFA trained a total of 12,000 REFA timekeepers (Hachtmann 2008). Workers objecting to either working speeds revised by REFA or the use of stopwatches to measure their speed risked imprisonment and death as slowpokes (*Bummelanten*) in concentration camps. In the concentration camps themselves, their working speed was also often determined using REFA methods, whether they liked it to or not. Unsurprisingly, from 1939 onward, there were very few workers objecting to REFA methods (Lüdtke, Marßolek, and von Saldern 1996, p. 61).

After the allied forces ended the madness of the Third Reich, REFA quickly dropped the *Reich* part from their name while keeping the abbreviation REFA. Nowadays, a few name changes later, REFA is no longer an abbreviation but the full name of the organization. After the war, REFA also expanded into other related fields like ergonomics and shop-floor organization.

One of the weaknesses of REFA is that they timed the actual work, and hence, measurements were subject to the individual speed of the worker. They eased this problem through measuring multiple workers at different times. However, this is time-consuming, and can be done only after the workplace is established. They also adjusted the times based on an expert estimate of the work speed, which is another way of saying that the REFA timekeeper used his/her gut feeling to determine if the worker was slow or not.

Finally, usually only a few measurements are taken over a short period of time. In reality, however, the worker, the tools, and the material can change over time. In painting, for example, humidity and temperature can change the process. New tools behave differently than older tools. Materials may differ slightly between batches. Technical changes to machines and tools require a new measurement, etc. Overall, the long-term actual work times can differ from the short-term measurements.

Many of these flaws were resolved by a slightly different approach based on predetermined motion times. Rather than measuring the actual working time by observing a worker, you break down the work into its most basic elements. For these basic motion elements, you determine the times needed to do them. With this knowledge, you are now able to estimate the time of other work without actually having to measure the work. The first person to establish such a predetermined motion time system was A. B. Segur in the 1920s (Schmid 1957, p. 10). He assigned times to the 18 basic therbligs motions established by Gilbreth. However, his motion time analysis (MTA) system is nowadays largely forgotten.

Herold Bright Maynard, John Schwab, and Gustave Stegemerten reinvented a similar system in the 1940s while working at a railway equipment manufacturer. They summarized their complete method in 1948 in a book named after their method, *Methods—Time Measurement*, better known by the abbreviation MTM (Maynard, Stegemerten, and Schwab 1948). Shortly thereafter, they established the MTM Association for Standards and Research, which is nowadays active in over 20 countries and is the largest time-study organization in the world. One curiosity of their system is that they do not measure time in seconds, minutes, or hours, but rather, in a unit created especially for MTM. This time measurement unit is called—surprise—*time measurement unit* and abbreviated as TMU. One TMU represents 1/100,000 of an hour, or 36 milliseconds.

Their approach does not rely on therbligs but also breaks down work into small elements like reach, grasp, move, position, and release. For these, the time either is known or can be calculated easily by adjusting for distance, weight, size of the object, etc. For example, the task of taking a screw out of a box and putting it into a hole could be broken down in the following steps:

- Reaching 20 centimeters to an object jumbled in a group of objects will take 0.414 seconds.
- Grasping an object less than 2 centimeters jumbled in a group of objects will take 0.327 seconds.

- Moving an object 20 centimeters to an exact location will take 0.425 seconds.
- Positioning the symmetric object exactly will take 1.674 seconds.
- Releasing the object will take 0.072 seconds.
- Returning the empty hand 20 centimeters to the original position will take 0.284 seconds.

Hence, the total process will take 3.197 seconds or 89 TMU. Using these calculations, it is possible to estimate the time needed for a task without ever watching the actual work. This has a couple of advantages. Mostly, the system is independent of worker performance. With REFA, a worker can intentionally work slowly to artificially increase the time allotted for the task. While the REFA timekeeper is trained to estimate the effort of the worker, it is nevertheless a source of ambiguity. MTM is independent of the individual worker's performance. This fact alone makes it a very useful tool to establish a neutral target speed for the worker, eliminating the constant conflict between labor and management. MTM can also be used before the workplace is established and can even be used to optimize the workplace by positioning parts so that they are closer or easier to reach. And finally, in many plants worldwide, unions have prohibited the use of stopwatches, which makes it difficult for REFA to measure times but not for MTM to calculate them.

On the other hand, MTM also has disadvantages. Some work is very difficult to estimate theoretically, for example welding, painting, or filing. For these, MTM also has to rely on actual measurements. MTM also does not incorporate the width of the learning curve but rather assumes an experienced worker. Most problematic is, however, that MTM calculates the individual steps without much consideration for the transition between steps. In real life, however, workers can—quite literally—cut corners and work faster than MTM. In my experience, workers often can consistently do 30% more than what MTM considers normal speed.

Both MTM and REFA have to be recalculated whenever the workplace changes, which may be rather frequently. While both MTM and REFA go to great lengths to determine workers' performance, they are still subject to differences in human performance.

While REFA started out by measuring work, they have now also included a predetermined motion time system similar to MTM called *work-time* to calculate work. With the rising popularity of Lean manufacturing (see Chapter 16), both MTM and REFA also tried to include the Lean aspect in their tool set, although with mixed success. Lean manufacturing is nowadays in the spotlight of shop-floor management, whereas REFA and MTM appear—while still useful—a bit old-fashioned. Some REFA and MTM researchers take it rather hard that they no longer enjoy top management attention, resulting sometimes in almost comical bickering about which method is better than the other.

Overall, both MTM and REFA provide a valuable service to daily shop-floor operations by giving an unbiased answer to the critical question of how fast the worker is supposed to work. Thus, many performance-related conflicts between labor and management can be diffused using a structured time study using REFA or MTM.

In addition, they provide excellent methods for optimizing working speed by understanding the time components of work and by improved ergonomics. For example, rather than putting a box of screws on the table, it is better to have a small chute where a few screws are lying on foam rubber. This makes it much easier to grasp the screw, which, in the long run, will more than pay for the effort of installing a chute with foam rubber. Using the detailed calculations, it is also possible to determine the best location for the screws in the workplace to shave a few seconds off the work time. Hence, both REFA and MTM can greatly reduce the time needed to handle this screw, increasing worker performance. Unfortunately, they never ever ask the question of whether the screw is really necessary.

BIBLIOGRAPHY

Alder, K., 2010. *Engineering the Revolution: Arms and Enlightenment in France, 1763–1815.* University of Chicago Press, Chicago and London.

Baeck, A., 2008. How Leonardo da Vinci Invented (and Can Reinvent) Our Design Profession, in: Proceedings of CHI 2008. Presented at the CHI 2008, Florence, Italy.

Chandler, A., 1977. *The Visible Hand: Managerial Revolution in American Business*, New edition. Harvard University Press, Cambridge, MA.

Copley, F.B., 1923a. *Frederick W. Taylor: Father of Scientific Management*—Volume 1. Harper & Brothers Publishers, New York.

Copley, F.B., 1923b. *Frederick W. Taylor: Father of Scientific Management*—Volume 2. Harper & Brothers Publishers, New York.

Giedion, S., 1948. *Mechanization Takes Command: A Contribution to Anonymous History.* Norton, New York.

Gilbreth, F.B., Carey, E.G., 2002. *Cheaper by the Dozen, Perennial Class.* Harper Perennial Modern Classics, New York.

Hachtmann, R., 2008. Fordismus und Sklavenarbeit: Thesen zur betrieblichen Rationalisierungsbewegung 1941 bis 1944. *Potsdamer Bulletin für Zeithistorische Studien* 43–44.

Hopp, W., Spearman, M.L., 2001. *Factory Physics*, 2nd ed. McGraw Hill Higher Education, New York.

Jenks, L.H., 1960. Early Phases of the Management Movement. *Administrative Science Quarterly* 5, 421. doi:10.2307/2390664.

Landes, D.S., 1969. *The Unbound Prometheus: Technological Change and Industrial Development in Western Europe from 1750 to the Present*, 1st ed. Cambridge University Press, Cambridge and New York.

Locke, E.A., 1982. The Ideas of Frederick W. Taylor: An Evaluation. *The Academy of Management Review* 7, 14–24. doi:10.2307/257244.

Lüdtke, A., Marßolek, I., von Saldern, A., 1996. *Amerikanisierung: Traum und Alptraum im Deutschland des 20. Jahrhunderts*. Franz Steiner Verlag, Stuttgart, Germany.

Maynard, H.B., Stegemerten, G.J., Schwab, J.L., 1948. *Methods-Time Measurement.* McGraw-Hill Book Co., New York.

Nelson, D., 1977. Taylorism and the Workers at Bethlehem Steel, 1898–1901. *The Pennsylvania Magazine of History and Biography* 101, 487–505.

Noble, D.F., 1984. *Forces of Production: A Social History of Industrial Automation.* Transaction Publ, New York.

Peaucelle, J.-L., Guthrie, C., 2011. How Adam Smith Found Inspiration in French Texts on Pin Making in the Eighteenth Century. *History of Economic Ideas* 19, 41–67.

Peaucelle, J.-L., Manin, S., 2006. Billettes and the Economic Viability of Pin-Making in 1700, in: Proceedings of the Eleven World Congress of Accounting Historians. Presented at the Eleven World Congress of Accounting Historians, Nantes, France.

Pursell, C., 1995. *The Machine in America: A Social History of Technology*, Illustrated edition. Johns Hopkins University Press, Baltimore.

Schachter, H.L., 2010. The Role Played by Frederick Taylor in the Rise of the Academic Management Fields. *Journal of Management History* 16, 437–448. doi:10.1108/17511341011073924.

Schmid, R.O., 1957. *An Analysis of Predetermined Time Systems (Master of Science in Management Engineering)*. Newark College of Engineering, Newark, New Jersey.

Shapiro, E.D., 2005. J. D. Chalfant's Clock Maker: The Image of the Artisan in a Mechanized Age. *American Art* 19, 40–59.

Taylor, F.W., 1911a. *Shop Management.* Harper & Brothers Publishers, New York and London.

Taylor, F.W., 1911b. *The Principles of Scientific Management.* Harper & Brothers Publishers, New York and London.

Waring, J., 1883. *Extracts from Chordal's Letters. Comprising the Choicest Selections from the Series of Articles Entitled "Extracts from Chordal's letters," Which Have Been Appearing for the Past Two Years in the Columns of the American Machinist.* John Wiley & Sons, New York.

Wrege, C.D., Hodgetts, R.M., 2000. Frederick W. Taylor's 1899 Pig Iron Observations: Examining Fact, Fiction, and Lessons for the New Millennium. *The Academy of Management Journal* 43, 1283–1291. doi:10.2307/1556350.

Wrege, C.D., Perroni, A.G., 1974. Taylor's Pig-Tale: A Historical Analysis of Frederick W. Taylor's Pig-Iron Experiments. *The Academy of Management Journal* 17, 6–27. doi:10.2307/254767.

Wrege, C.D., Stotka, A.M., 1978. Cooke Creates a Classic: The Story behind F. W. Taylor's Principles of Scientific Management. *The Academy of Management Review* 3, 736–749. doi:10.2307/257929.

13

The Assembly Line and the Era of the Industrial Empires

I will build a motor car for the great multitude. It will be large enough for the family but small enough for the individual to run and care for. It will be constructed of the best materials, by the best men to be hired, after the simplest designs that modern engineering can devise. But it will be so low in price that no man making a good salary will be unable to own one—and enjoy with his family the blessing of hours of pleasure in God's great open spaces.

Henry Ford (1863–1947)
Automotive industrialist, in My Life and Work

Henry Ford, with his famous Model T, is generally known as the father of the assembly line. However, while he implemented the most advanced and technically complex assembly line the world has ever seen, there were many other examples of assembly lines before Ford. Like many other manufacturing methods, the assembly line evolved over time, including numerous different engineers and entrepreneurs. In that respect, the definition of the *first* assembly line depends heavily on how you define assembly line.

First of all, while the term *assembly line* bears *assembly* in its name, it should not be limited only to assembly operations, since many modern assembly lines also includes other processes, including cutting and forming operations. An assembly line in its most basic definition would be a number of manufacturing steps performed in close proximity by different workers, where the workplaces are arranged in the sequence of the manufacturing steps. Using this most basic definition, the first assembly lines would probably date back to ancient times, for example, with the production of Roman oil lamps or Egyptian papyrus. If we further refine this definition to require the passing of the product to the next station immediately after completion, then the first assembly lines we know of were in medieval Europe, with the Arsenal of Venice being a prominent example* (see also Section 7.1).

We can then further narrow down the definition of assembly line by requiring mechanization. In this case, the earliest assembly lines would appear during the industrial revolution, with the block production in Portsmouth being the first prominent example. If we require, additionally, mechanization not only for production but also for transport of the parts between the workstations, we would end up with the production of cigarettes or canned soup at the end of the nineteenth century. Finally, if we additionally require an assembly line to have interchangeable parts, the first assembly line would be part of the production of the Model T with Henry Ford.

Hence, depending on your definition of an assembly line, different sites and times would be the first assembly line ever. In the following, we will look closer at the earliest examples of assembly lines.

13.1 THE FIRST ASSEMBLY LINES—CONSUMER PRODUCTS

For the first assembly lines using mechanized transport, we have to go back to the beginnings of the industrial revolution to Oliver Evans (1755–1819) (Figure 13.1) and his automated mill. Evans was a highly creative inventor. He invented, among other things, an improved machine for wool processing, built his own steam engine, designed an amphibious steam-powered vehicle,† and was the first to put the fire for his steam engines inside of the boiler rather than underneath to improve efficiency (Rosen 2010, p. 285).

However, probably his most influential invention was the automatic mill (Figure 13.2). Before Evans, milling was a labor-intensive process.

* It is quite possible that this happened already in ancient times, but our knowledge of ancient manufacturing technology is not detailed enough. We simply don't know (yet).

† Some claim he built a working amphibious vehicle, but there is little to no evidence of what should have been quite a sight for the local population.

FIGURE 13.1
Oliver Evans, the *Watt of America*. (Engraving by W. G. Jackman.)

While the millstone was usually water powered, the transport of the grain and flour was mostly on the backs of the apprentices. Running a mill required lots of hauling of grain and flour while at the same time losing flour and grain in the process and contaminating the product with dirt and sweat. The mills worked mostly seasonally, selling their service to the farmer rather than buying the grain from the farmer and selling the flour.

As part of his many endeavors, Evans bought an old water-powered mill in 1782 near Wilmington in Delaware. Over the course of the next few years, he improved all aspects of material handling. A total of five types of devices eliminated all manual labor in material handling. He used cups on a leather belt loop to lift grain and flour upward (which he called *elevator*), Archimedes screws to move goods horizontally (which he called *conveyor*), and a conveyor belt to move goods downward (the *descender*). Wooden plates along an endless loop pushed goods through a wooden channel horizontally (the *drill*). Finally, he used a device with rotating arms to dry and cool the ground flour (the *hopper boy*) (Evans 1795).

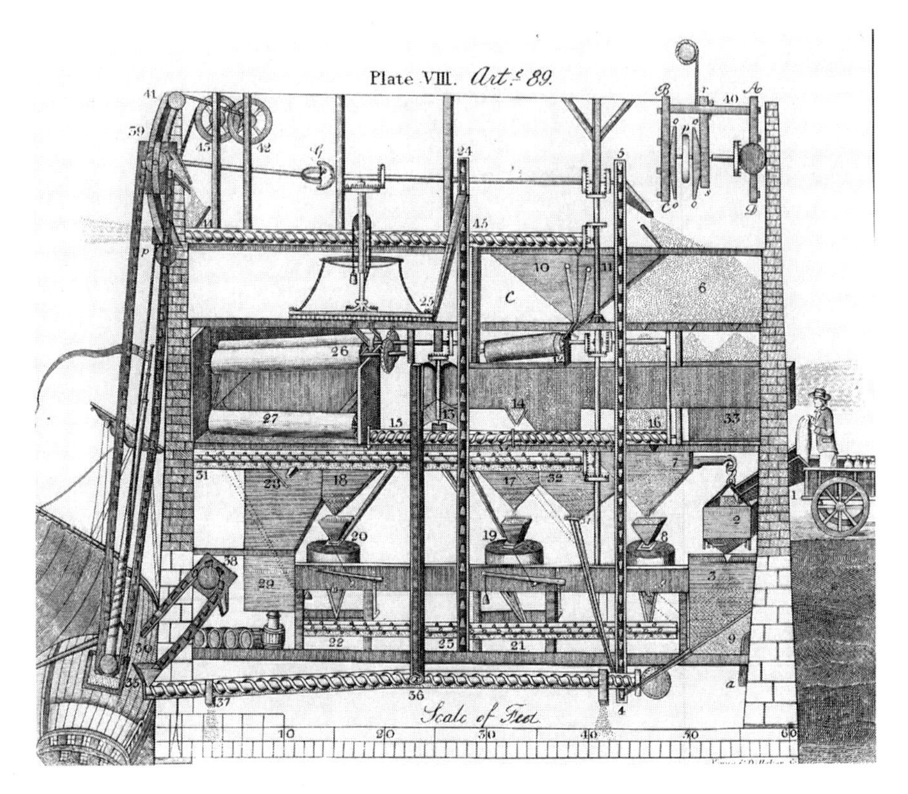

FIGURE 13.2
Overview of the first automatic flour mill by Oliver Evans in his book, *The Young Mill-Wright and Miller's Guide*, 1834.

He completed his project in 1785 (Pursell 1995, p. 27). In his fully automated mill, no hand touched the grain except for loading and unloading from the delivery carts. Through his inventions, he reduced labor to a fraction, got more flour out of the same grain (Gage 2006), and produced better and more uniform grain compared with manual mills. Most of his workers were busy simply closing barrels of processed flour (Chandler 1977, p. 55).

Evans patented his ideas in different states and eventually with the U.S. Patent office in 1790, one of the earliest U.S. patents. He also published his works in a highly successful book, *The Young Mill-Wright and Miller's Guide*, to help popularize his invention. He was able to license his patent to many different millers, including President George Washington. However, as with most inventors at that time, there were many more who copied his invention without proper license. In any case, the much more profitable automatic mills became widespread from 1810 onward (Gage

2006), buying large quantities of grain from farmers and selling the flour directly rather than merely selling the milling service to the farmer.

Evans died in 1819, after having a stroke upon receiving the news that his workshop had burned down. His mill was the first assembly line in the world with mechanized transport of goods—except for the small fact that nothing was assembled, but rather, the grain was *disassembled* into flour and bran.

The next major example of an assembly line was the biscuit bakery at the victualling yard of the British Navy in Deptford, England. If *biscuit* makes you envision a delicious scone (if you are American) or a tasty cookie (if you are British), think again. For the British Navy, biscuits were also known as hard tack and consisted of water and flour, sometimes a bit of salt, and—depending on the storage conditions—quite a number of worms. It was baked two to four times to make it as hard as possible, to extend shelf life. This also made it nearly impossible to eat unless softened with some liquid. Nevertheless, many sailors lost teeth to hard tack, especially if they suffered from scurvy.

However, the British Navy did not care much for the palate of their sailors but valued other qualities in their provisions, namely, price and shelf life. Hard tacks were very cheap and lasted extremely long even in the hot and humid conditions found on ships. Therefore, these biscuits were one of the key provisions on board. The navy needed large quantities of biscuits to keep their ships running. The classical biscuit production was a highly manual process, the only machine besides the oven being the kneader for the dough.

In the early nineteenth century, the British Navy decided to build a new depot near Plymouth, England, named the Royal William Victualling Yard after King William IV. Responsible for the design was Sir John Rennie. Construction started in 1826. In order to provide a full range of provisions, the yard was designed with a brewery, a slaughterhouse, and a bakery. Rennie reviewed different existing bakeries before designing the bakery at the yard.

Rennie combined both milling and baking in the same building, a novel concept at the time. He also designed a fully automatic baking process with 12 ovens in parallel (Miele 2006). By 1833, the bakery was fully operational. The dough was prepared in an automatic kneader, and the only hand touching the unfinished product was that of the worker removing the dough from the kneader. Afterward, the process was fully automatic, with a machine slicing the dough, molding the biscuit, stamping, splitting, and eventually baking the biscuit in an endless-chain oven* (Giedion 1948, p. 176).

However, for some reason, the bakery was running at full speed only once in 1833 before a long break and a refurbishing in 1843, after which it continued to produce until 1925. In any case, by the 1860s, most ships were equipped with an onboard bakery, surely to the great pleasure of the sailors (Miele 2006). Again, while the biscuit bakery did merge different components, it depends on your definition of assembly line whether you consider the bakery in the victualling yard the first assembly line or not.

Another example of an early assembly line are the meat packing *disassembly lines* that started to appear in the nineteenth century in Chicago. In traditional butchery, the butcher takes apart a single animal. These new Chicago slaughterhouses, however, divided the labor among different workers. The first worker looped a chain around the hind leg of the pig to lift it in the air. The second opened the artery. After the animal died, the third cleaned the carcass. The fourth opened its belly, the fifth took out the entrails, and so on, until the finished meat products emerged at the other end of this disassembly line. By 1837, 20 men slaughtered and processed a pig every 1.3 minutes, increasing in 1850 to a pig every 35 seconds (Giedion 1948, p. 90).

However, there was little mechanization in this line. Besides the division of labor, the biggest improvement was the mechanical railing that transported the animal through the process. Minor improvements were made by cleaning off bristles using hot steam rather than fire. Beyond that, there was pretty much no mechanization until the late twentieth century. It is not that people have not tried. There are numerous patents and designs for mechanically killing the animal, scraping bristles off, sawing animals in half, skinning, and other steps. However, few of them proceeded beyond the design stage. The few that did get produced, unintentionally, mostly minced meat (Giedion 1948, p. 230ff). With respect to assembly lines, both Oliver Evans and the British Navy were more technologically advanced than the Chicago slaughterhouses, although these slaughterhouses did inspire Henry Ford for his famous automotive assembly lines.

* The endless-chain oven was invented before around 1810 but was never used in an automatic line as in the victualling yard.

It is hard to say where the first assembly line of the world was installed.* However, the time was ripe for automation. Within a short period after 1880, numerous fully automated processes appeared where no human hand touched the product except in the adding of the raw material and the removal of the complete, packaged product.

For example, James Bonsack patented a working cigarette-rolling machine in 1881. Before Bonsack's invention, tobacco was mostly consumed as cigars or chewing tobacco. Cigarettes were a handmade luxury item. A skilled worker could roll 3000 cigarettes per day. Bonsack's machine, however, was able to roll 120,000 cigarettes a day, or the equivalent of 40 skilled workers. Despite the cost of the machine, automation reduced the production cost to one-sixth of the manual process, and fifteen machines were able to supply the entire U.S. market in 1884 (Chandler 1977, p. 249). Initially, the packaging process was still manual, but Bonsack continued to work on this process and eventually created a fully automatic assembly line from raw materials to packaged cigarettes.

Another consumer product transformed by automation was matches. The modern friction match was invented in 1827 and enabled consumers to conveniently start a fire anywhere anytime, rather than using sparks from a flint or a magnifying glass, or by carrying around smoldering tinder in a box. Hence, matches were in high demand to light a fireplace, a cigar, a candle, or anything else needed for cooking, lighting, or heating. Workers producing matches, however, ran the risk of getting *phossy jaw*, a disease caused by exposure to the poisonous white phosphorus used in matches. The disease caused the jawbone to rot away painfully, while at the same time giving off a strong foul odor that was hard to conceal. It did not help either that the bones started to glow in the dark. The only known treatment was to completely remove the jawbone, which left the victim disfigured for life. The problem was only solved when white phosphorous was replaced by the nonpoisonous red phosphorus (Moss 2011). While the first matches with safe red phosphorus were available from 1891, industry preferred the cheaper but poisonous white phosphorus. Only at the Berne Convention of 1906 was the use of white phosphorus for matches finally prohibited.

---

* There are even more claims for the first assembly lines, for example, the arranging of workers in sequence at the Colt Armory of Samuel Colt around 1855, the conveyor system in the Westinghouse air brake foundries as early as 1890, Oldsmobile in 1897, the Keim bicycle plant around 1900, and many more (Brinkley 2003, p. 152).

FIGURE 13.3
Norton's automatic can-making line, 1883. (Picture from American Machinist, July 14 1883 or 1885.)

Health risks notwithstanding, a worker before 1880 was able to produce 4000 matchboxes per day by hand. Different inventors worked on creating an automated machine to produce matches, and by 1881, a machine was able to automate the entire production process from blocks of wood to finished, packaged matches. Productivity went up sevenfold, and by 1890, one worker was able to produce almost 30,000 matchboxes per day with the help of a machine (Chandler 1977, p. 293).

Another example of assembly lines was the canning lines of brothers Edwin and Oliver Norton in 1890 as shown in Figure 13.3. The tin can* as a method to preserve food has been used since 1813, but since each can was made by hand, it was a slow and expensive process. A tinsmith was able to produce five to six cans per hour in the early nineteenth century, with production increasing to 60 cans per hour in the 1870s (Busch 1981). Edwin Norton (1845–1914) opened a tin plate company in Toledo, Ohio, in 1868, soon expanding into tin cans and being joined by his brother Oliver. While Oliver took care of the business side, the more mechanically gifted Edwin worked on the mechanical problems of producing tin cans. In 1883, he developed an automatic can-making machine, soldering side seams, tops, and bottoms at an incredible 3000 cans per hour (Dunlavy 1980, p. 19f).

* Or, more precisely, tin plate can. Until the 1950s, almost all cans were made from tin-plated steel. Nowadays, plastic coatings and aluminum cans are also frequently used.

However, this machine still required some manual work, for example soldering the cap, an additional lid common in early tin cans. Edwin Norton continued working on improving these and other processes. By 1890, he developed a fully automated canning line where tinned metal sheets were cut, formed, assembled, soldered, tested, dried, and shipped automatically. The entire line was able to manufacture 30,000 cans per day without anybody touching the parts (Twede 2009).

Edwin Norton continued to improve canning technology, patenting a vacuum-packed tin can. In 1898, he merged numerous small canning companies into the market-dominating American Can Company, only for him to resign as president and start its biggest competitor, the Continental Can Company. He had over 300 patents and was a proponent of shorter working hours (Encyclopedia Britannica 2012). The cheap and fast production of canned goods made the tin can a modern icon in the American household.*

The list of consumer products whose production was automated in the late 1900s can be easily expanded. Besides cigarettes, matches, and canned soup, there would be soap, photography films, sugar, beer, toilet paper,† chewing gum, candy, and many more. The production of pretty much most consumer goods of modern life was automated in the late nineteenth century. One of the few exceptions was the automobile as one of the most complex products of modern life. However, before we look in more detail at Henry Ford and his automotive empire, it is worthwhile to look at the consumer side. Production works only in combination with selling, and as mass production evolved, so did mass selling.

13.2 BUILDING AN INDUSTRIAL EMPIRE—MASS PRODUCTION NEEDS MASS SELLING

Until the 1840s, all manufacturing in the United States was done in small workshops, mostly family owned, with few employees. There were very few industrial sites, most of them in New England, with access to ample water power, for example Springfield Armory, Harpers Ferry Armory,

* And also in art, most famously in Andy Warhol's *32 Campbell's Soup Cans.*

† It was sold as *therapeutic paper*, contained aloe to cure sores, and had the name of the inventor, Joseph Gayetty, printed on every sheet. What a homage!

the Boston Manufacturing Company in Lowell and Holyoke, textile mills in Lawrence, and both textiles and Samuel Colt's Patent Arms Manufacturing Company in Paterson, New Jersey (Chandler 1977, p. 60). Except for these sites, all other manufacturing was owned by an entrepreneur or a small partnership, with usually less than 100 employees. Due to the small size of these companies, the ownership needed no more than one level of management in the form of foremen to run their businesses, and there were no middle managers.

The first companies whose size outgrew this traditional system in America were railway companies. Their geographic size and the number of employees were no longer manageable with only one level of management. Starting in the 1840s, railroad companies started to use middle management. Being a manager became a new profession no longer directly related to the product (Chandler 1977, p. 87).

In manufacturing, however, companies took longer to increase in size. Until the 1880s, most manufacturing was local. Most materials were purchased from nearby, and the goods also were sold in nearby markets, sold on order for customers in the region, or sold to wandering peddlers, who distributed the goods. However, as mechanization and automation increased productivity, it became harder and harder to sell goods to nearby customers due to a lack of demand. For example, 15 of Bonsack's cigarette making machines were able to supply the complete U.S. market in 1880 (Chandler 1977, p. 249), and four match manufacturing plants produced enough matches for the nation in 1900 (Chandler 1977, p. 293).

As mechanization reduced the cost of goods and railroads provided cheap transport, it became more and more viable to sell goods long distance. As Alfred Chandler beautifully describes in his Pulitzer Prize–winning book *The Visible Hand* (Chandler 1977), many companies that started to mass-produce consumer goods at the same time started to establish sales organizations and dealers across the United States. They no longer relied on other merchants and peddlers for the distribution of their goods but established their own dealerships, provided mail-order catalogues, supplied department stores, and overall took control of distribution.

To their surprise, they found that this was highly profitable. Most companies that sold cheap mass-produced goods through their own supply chains found that this was very lucrative. Of course, this was only lucrative for the first company in its field. Once a company established dominance for a product range, it was very difficult for a latecomer to rise to profitability. The latecomer first had to establish both a mass production and a mass selling system, while the established competition could use their enormous profits to squeeze the newcomer.

Many of these companies and brand names that started to appear during the late nineteenth century are still famous nowadays, for example, Diamond Match, Campbell Soups, Procter & Gamble and their Ivory Soap, Quaker Oats, H. J. Heinz Company and their Heinz Tomato Ketchup, Swift & Company meat packing with their refrigerated railroad cars, Budweiser, and many more. Starting with the U.S. railroad companies, these were the first industrial empires of the world.

13.3 SEWING MACHINES, TYPEWRITERS, AND THE BICYCLE CRAZE

To design a soup can, a match, or a cigarette is rather straightforward, and the competitive advantage of these companies was to produce them cheaper than the competition while at the same time selling these products nationwide. However, during the late nineteenth century, another group of products also became widely available and affordable to the general public—complex mechanical goods like sewing machines, typewriters, and bicycles. For these products, the problem was not so much the selling, and only partially the manufacturing, but mostly the design of the product.

Except for clocks, the world had never produced such complex mechanical devices in larger quantities, let alone mass-produced them for the consumer market. However, due to the efforts of the Springfield and Harper Ferry Armories, a body of knowledge in metalworking was building up. Many workers and foremen from these armories, over time, took different jobs in other industries and took their knowledge with them. However, while interchangeable parts were the norm in the armories by 1850, private industry, for a long time, still required qualified fitters to file parts to match during assembly.

One of the first complex consumer products was the sewing machine. Like many other products, the technology evolved over decades, with numerous inventors contributing. By 1850, many different companies were selling functional sewing machines. Unfortunately, none of them had all the patents necessary to produce a modern machine, and most companies put in a lot of effort in either suing or being sued by other companies, or both. Hence, this period in industry is known as the Sewing Machine War (Mossoff 2009). Eventually, the companies realized that this legal wrangling benefitted mostly the lawyers, and in 1856, four companies pooled their patents in the newly founded Sewing Machine Combination. Not only could these four companies now produce sewing machines without constant legal threats; they also charged a hefty $15 license from other companies making sewing machines.

Probably the most successful company of the combination was the Singer Manufacturing Company. In 1874, Singer produced 250,000 machines per year; by 1880, half a million; and by 1886, 1 million machines annually (Hounshell 1985, p. 109). They pioneered a *hired purchase*, where less wealthy customers could make a down payment and pay the remainder in installments (Hobsbawm 1999, p. 141). In 1873, Singer built the largest factory in the United States for a single product in Elizabeth, New Jersey, producing a quarter million machines in 1874. They also established a sales and marketing system that outperformed their competition.

It was their sales system that made Singer successful, since the manufacturing side was less well organized. Some competitors built better machines than Singer did and used interchangeable parts earlier. Up to the 1880s, Singer sewing machines did not have interchangeable parts but relied on fitters for final assembly. This was due to a number of reasons. First, Singer used inside contractors, who were more independent and less likely to follow standards. Second, this independence also applied to different manufacturing plants, and for example, the manager of the Montreal factory preferred filing by hand over the use of gauges, despite a Singer blue book describing standard practices (Hounshell 1985, p. 118). In any case, the gauges Singer used were unsuitable for the task. Rather than the go/no-go gauges checking for an upper and lower limit of the tolerance of the dimension, Singer used pass/fail gauges that checked only the target dimension, not the tolerance (Hounshell 1985, p. 121). Hence, hand finishing was used at Singer until the 1880s, much later than other sewing machine companies. Only at the end of the century were interchangeable parts consistently used (Pursell 1995, p. 91).

Another major product was the typewriter, also benefitting from numerous different contributors. Commercial typewriters started to be sold in 1867, with mass production beginning in 1881. With its thousands of parts, a typewriter was probably the most complex commercial product of its time, including also new materials like rubber and steel springs, delicate parts, and high-precision parts like ball bearings (Hoke 1990, p. 133).

To make such a complex machine work, even tight tolerances were not enough, and typewriters contained internal adjustment screws and points to fine-tune the quality of the product during and after assembly (Hoke 1990, p. 152).

Both sewing machines and typewriters were completely new products that were used both in companies and also in small private businesses. A woman who could afford a sewing machine was able to sew clothes faster and better than by hand and, by offering her services, was able to contribute to the family income. Hence, these products sold very well. However, the market consisted mostly of companies and individual entrepreneurs that sold their goods and services rather than individuals purchasing these machines for private use.

The bicycle, on the other hand, was of interest to almost everybody. Before bicycles, the main way to get around was on your own two feet. A horse, more comfortable and faster, was very expensive both to buy and to keep, and was hence affordable only to wealthy people. Public transport was sometimes an option, although this required you to conform to the timetable while still walking from and to the stations. Imagine for a second not having a car or a bike, and being limited to walking and public transport. That may work if you live in Manhattan, where most necessities are within a two-block radius and public transport is plenty. However, if you are in one of the suburbs, it will already become tricky, more so living in the countryside. Having a bicycle would make quite a difference in the reach of your activities and, therefore, in your life. For late-eighteenth-century Americans, too, a bicycle was a great time saver, expanding both the possible location for work and the range of social activities. In fact, for many Americans, it was the first time ever that they were able to leave their town and travel to the next city.

The bicycle went through different stages of evolution, from the pedalless dandy horse to three- and four-wheelers (including multiseaters) to the adrenaline-pumping high wheel. However, only the invention of the modern safety bicycle shown in Figure 13.4 changed the product from an eccentric toy to a useful tool.

One of the first large bicycle companies in the United States was the Pope Manufacturing Company, producing high-quality bicycles including many forged parts. Pope also pioneered in the field of testing and quality control, using both destructive and nondestructive tests for its parts. Initially, Pope tested the strength of all its bicycle chains up to the breaking point, ending up with lots of broken chains, before they figured out

FIGURE 13.4
The high wheel and the safety bicycle. (From Lexikon der gesamten Technik [*dictionary of technology*], 1904.)

that it would be better to test just slightly below the breaking point. Some parts were inspected up to 12 times during production. Vibration testing also led to the development of swaged spokes, where the middle is thinner than the ends, reducing weight without sacrificing strength (Hounshell 1985, p. 208).

The main competitor of Pope, Western Wheel Works, used a different, lower-cost approach. Rather than high-quality forged parts, Western Wheel Works pioneered the use of sheet metal stamped parts. A forged part is much stronger but requires expensive tools, time-consuming manufacturing, and is much heavier. Stamped sheet metal is much easier to make, lighter, and cheaper, with its strength being still adequate for regular use (Hounshell 1985, p. 209ff). Hence, Western Wheel Works soon outsold Pope, since their bikes were both cheaper and lighter. Eventually, Pope also started to use stamped sheet metal parts. Western Wheel Works also improved its manufacturing system by providing materials to the workers, rather than workers leaving their post to get more materials. Material handlers provided press operators with parts, and the operators remained stationary in front of their press, improving the overall efficiency of the process (Hounshell 1985, p. 211).

As the bicycle filled a large need of the general population, it sold well despite being not quite cheap. The bicycle boom started slowly in 1880, where two companies had about $2 million in revenue. A bicycle in 1880 cost about $100, or the equivalent of four months' salary. Both demand and production increased greatly during the 1890s bicycle craze, with a whopping 312 firms making, in sum, over $30 million in revenue

(Rosenberg 1963). New laws and regulations for bicycles were established, and bicycle paths were built, while regular roads were improved for bicycles. Bicycles helped the women's movement by giving women greater mobility; the freedom to move around on their own; and the relief of not having to wear stiff corsets, wire underskirts, and long skirts. Social patterns changed, and social outings became popular.

Of course, with every social change comes a downside. Streetcars lost customers to the bicycle. Laundries lost business since starched and stiff shirts fell out of fashion with cyclists. Bars and amusement centers had fewer customers, since the customers enjoyed their bicycle trips. Tobacco dealers lost revenue due to more health-conscious cyclists, and the bicycle replaced the watch as the coming-of-age gift. Naturally, some religious authorities denounced bicycles as the *diabolical devices of the demon of darkness*, and doctors warned about *jarred internal organs*, while other priests and doctors became cyclists themselves. Most shockingly, women started to wear pants for easier riding of bicycles. And, let's not forget, politicians started to tax bicycles (Spreng 1995).

The bicycle craze peaked in 1896–1897. By 1905, there were only 101 bicycle firms left, one-third of the number five years earlier, with each firm, on average making, half the sales compared to six years earlier (Rosenberg 1963). The reason for the downturn is not quite clear, but it seems that the movement just lost its momentum. Besides, everybody who wanted a bicycle had one. Sales of new bicycles were now mostly to replace worn-out bicycles. Hence, sales slowed down significantly, and many of the produced bikes were exported (Harmond 1971). However, there was something even better around the corner. It had four wheels, went faster and further than any bike, and started a craze that is still going strong 100 years later. And, best of all, it did all the pedaling for you.

13.4 THE BIRTH OF THE AUTOMOBILE

The next step in the quest for personal mobility was the automobile. During the end of the nineteenth century, tinkerers experimented with a large number of different power sources to power a vehicle. Numerous steam cars were developed, as were land sailing, electric vehicles, and different gas-powered vehicles using hydrogen or coal gas. Yet another path followed was the gasoline-powered internal combustion engine, which turned out to dominate the twentieth century. However, during the end of the nineteenth century, this was only one of many different methods tried.

During the 1860s and 1870s, German Nicolaus Otto developed the four-cycle internal combustion engine, on which most car engines nowadays are based. Shortly thereafter, Dugald Clerk from Scotland invented the two-cycle engine, now common for scooters, most motorcycles, and other small vehicles. Finally, Rudolf Diesel from Germany invented the diesel engine in 1893, which is nowadays the primary propulsion system for trucks and heavy-duty vehicles.

The first combustion engines were stationary engines, too heavy, cumbersome, and unreliable to power a vehicle. The first true automobile was the Benz Patent-Motorwagen by Carl Benz, a tricycle powered with a two-stroke gasoline combustion engine presented in Mannheim, Germany, in 1886. While Carl Benz designed the car, he and his company were powered by his wife, Bertha Benz. While still engaged, Bertha got an advance on her dowry, hence financing the company of her fiancé. However, Carl did not market his invention. Hence, Bertha decided to take public relations into her own hands.

In 1888, Bertha Benz (Figure 13.5) drove the car from Mannheim to Pforzheim and back to visit her mother, taking along her two sons aged 13 and 15. She did not ask her husband for permission, which he would surely have refused. During the trip, she had to overcome numerous difficulties. She cleaned a blocked pipe using her hairpin, replaced the brakes herself, had a chain fixed at a blacksmith, insulated a wire using her garter, and refueled at a pharmacy, the only place where gasoline* was sold as a cleaning agent.

While a 200-kilometer round trip sounds uneventful nowadays, back then, it was groundbreaking. All previous cars were only driven on trial runs around the block, or at best across town, before retreating to the safety of the garage for repairs and refueling. No car had ever made such a long-distance trip, and her stunt made considerable headlines, boosting her husband's confidence, and—most importantly—generating sales.

Within a few years, there were hundreds of car companies around the world, making customized cars for wealthy gentlemen—and despite Bertha Benz, it remained mostly a men's domain. These cars were designed and built individually, based on customer specifications. Each

* Technically speaking, she purchased ligroin, also known as benzine, which is very similar to gasoline.

FIGURE 13.5
Bertha Benz, first long-distance driver. (Image by Bühler from Mannheim, around 1870.)

car was handcrafted, with few or no interchangeable parts, and every car was different. Assembly included lots of filing down of parts until they fit. Naturally, this was a very expensive process. The closest thing nowadays to a handmade car is a Rolls Royce, which costs easily 10 times as much as a regular car, even with a mass-produced engine, gear train, and car body. Only the wealthiest could afford to own a car, which usually also meant hiring a mechanic, building a garage, and other maintenance costs.

However, due to the success of the U.S. armories, sewing machines, typewriters, and bicycles, there were large numbers of workers, foremen, and technicians in the United States skilled in metalworking, mass production, and interchangeable parts. Hence, the automotive industry moved rather quickly from a craft-based system to mass production. Division of labor was already common in U.S. carriage makers, where the largest firm around 1890 made up to 50,000 carriages per year (Chandler 1977, p. 248). Many of these carriage makers expanded into the automotive business.

One of the first companies in the United States to produce automobiles in large quantities was Olds Motor Works, founded by Ransom E. Olds in 1897. Olds developed numerous different prototypes, including steam engines, electric cars, and of course combustion engine vehicles. Bringing all of them to the market simultaneously would have been suicidal to the young company, but Olds had difficulty deciding which model to pursue. A factory fire in 1901 made this decision for him, as it destroyed all prototypes but one: the small and inexpensive curved-dash motor car.

In the same year, Olds applied the principle of assembly lines to the production of his curved-dash motor car (Nye 1992, p. 221). In the motor assembly, motors were moved from station to station along an assembly bench, with workers adding parts at every step. In the final assembly, vehicles moved on wooden platforms through the different assembly steps, again with workers always performing a certain set of tasks at their stations (May 1977, p. 192). Hence, Ransom E. Olds is often credited as the inventor of the first assembly line.

However, as described earlier in this chapter, it was definitely not the first assembly line, as Owens, Norton, Bonsack, and many others used assembly lines much earlier. Furthermore, while Olds moved the product from station to station, his system did not use interchangeable parts.* Instead, he was rather proud of every Oldsmobile being different (May 1977, p. 191ff). Hence, there still was lots of fitting in the final assembly, far from modern mass production. Nevertheless, Olds was a highly successful car manufacturer, producing about 7000 vehicles in 1903. This made him the biggest carmaker with a market share of 20–30% of all U.S. automobiles (May 1977, p. 190).

Yet, interchangeability was not far down the road, and the first motors with interchangeable parts were produced in the Henry Leland Cadillac plant in Detroit in 1906 (Womack 1990, p. 36). To test this claim, three Cadillac Model Ks built in 1907 were tested on a racetrack in Weybridge, England. After the tests, the cars were completely disassembled, and

* There are quite a number of references mentioning interchangeable parts together with the Olds assembly line, but these are mostly modern, and it is not certain if there were truly interchangeable parts or if the authors interpreted that an assembly line can only function with interchangeable parts. Due to the technical challenges and complexity of interchangeable parts and the lack of hard evidence, combined with credible sources rejecting interchangeability, I believe that the assembly line by Olds had few or no interchangeable parts.

the parts mixed. Furthermore, 89 high-precision parts were completely removed and replaced with new parts from the dealer before the three cars were rebuilt. The rebuilt cars not only worked but also completed a 500-mile test run successfully. Moreover, one of the rebuilt cars competed in a 2000-mile reliability test, where it won the first prize. For this outstanding proof of interchangeability, Cadillac won the prestigious 1908 Dewar Trophy promoting advancement in the automotive industry.

Hence, the automotive industry was leapfrogging technical development on the shoulders of men trained in rifles, typewriters, sewing machines, and bicycles. From the first prototype car in 1886 to mass production in 1901 to interchangeable parts in 1907, the advancement in the automotive industry could not have been as quick without the pioneers in other industries. Nevertheless, the car remained a rich man's toy, affordable only to the upper class, with only 43,000 cars produced in 1907 and 140,000 cars registered, or less than 1 car for every 600 citizens. However, this was about to change with Henry Ford and his Model T, the *Tin Lizzie*.

13.5 THE FATHER OF MASS PRODUCTION—HENRY FORD AND HIS MODEL T

Henry Ford, the father of mass production, was born in 1863 near Detroit. He went on to become the founder of not one, not two, but three automobile companies. From an early age, he was skilled in mechanics. Also, since he was thrown off a horse at age 9, he disliked horses (Bak 2012). In 1879, he started as an apprentice machinist, where he quickly became better than his foreman. Hence, the foreman fired him (Bak 2003, p. 9). This did not slow down his career very much, and he soon repaired steam engines for Westinghouse. In 1893, Ford became a chief engineer for the Edison Illuminating Company.

It was in this position that he built his first prototype automobile in 1896, an ethanol-powered vehicle with no brakes. This brought him to the attention of Thomas Edison, who supported Ford's experiments with vehicles. One year later, Ford quit and started his own automobile company, the Detroit Automobile Company, which went belly up only two years later after building a total of 20 vehicles. Not discouraged, Ford started his next automobile company less than 10 months later, called the Henry Ford Company, based on the reorganization of his first company

(Brinkley 2003, p. 43). Already back then, Ford focused on efficiency and utilization rather than luxury, advertising his car as *made of few parts and every part does something* (Bak 2003, p. 48). However, due to quarrels with his financiers, he had to leave his own company after only six months, taking only $600 and the brand name *Ford* with him. The company, however, was renamed as the Cadillac Automobile Company, and is still going strong as a luxury brand of General Motors (GM).

Finally, in 1903, Ford started his third automobile company, which would become the most successful automobile company of the twentieth century: the Ford Motor Company. In the first years, cars were produced in a craft-type fashion, where all necessary parts were brought to one spot and the car was assembled on the same spot. About 15 such teams produced a car each in parallel, surrounded by piles of material that usually had to be filed down by hand to match the car (Hounshell 1985, p. 220).

Ford soon started to experiment with different manufacturing organizations. Rather than a team of 15 workers at one station, he had only one worker assembling a complete car. Another experiment had the fitter remain stationary, with material being supplied by other workers rather than the fitter walking away from his vehicle to get parts (Womack 1990, p. 27f). Ford supposedly also was impressed by the Sears department store, which had implemented a 15-minute window for parts deliveries (Hopp and Spearman 2001, p. 22). One of his key machinists, Walter E. Flanders, also rearranged the manufacturing process from departments to a flow-oriented layout in 1906. Rather than having all spinning machines in one department and all drills in another, the machines were arranged in the sequence in which the parts were processed. Flanders also installed the first single-purpose machines at Ford, where a machine could do only one task but did that task faster than a multipurpose machine (Hounshell 1985, p. 221). Finally, Flanders also insisted on the plant not having more than 10 days' worth of raw materials on hand. Overall, there was a continuous process to improve the manufacturing operations from the very beginnings of the Ford Motor Company.

In 1906, Ford built a new multistory plant for the Ford Model N, the Piquette Avenue Plant. The Model N was designed for interchangeable parts, although it took some time from the idea to true interchangeability (Hounshell 1985, p. 221). In 1908, Ford introduced a new model that would revolutionize automobile production, the famous Ford Model T (Figure 13.6). The Model T was designed for easy manufacturing and low

FIGURE 13.6
Henry Ford and his Model T in 1921.

cost. The entire car had less than 100 different part numbers,* significantly simplifying the supply chain and the production process (Pearson 1992, p. 108). In terms of complexity, it was about as complex as a modern-day washing machine, excluding electronics. As a side effect, this also meant that the car could easily be repaired. Furthermore, the Ford Model T achieved interchangeability of parts, greatly improving the efficiency of the workers, since they no longer had to fit a part using a file and the interchangeable part fit by default.

Production, however, was initially still very inefficient. The cars were assembled in one spot, with materials delivered to the cars for assembly. Workers were assigned different tasks, and after finishing with one vehicle, they walked to the next vehicle to repeat their tasks. It is easy to see that coordinating these actions was difficult, and the overall process was not very efficient. Workers often had to wait if the worker on the previous task was not yet done or if the required material had not yet arrived. Or worse, the worker finished his tasks not knowing or ignoring that the previous tasks had not yet been completed due to a delay with the worker

* A modern car consists of around 30,000 parts.

for the previous task (Womack 1990, p. 28). For example, if you installed the hood before the motor, you could no longer install the motor, since the hood was in the way. In the best case, the worker lost time waiting; in the worse case, the worker lost time undoing previous work; and in the worst case, the customer got a defective product. On top of that, lots of time was lost with the workers walking between stations.

However, Ford continued to improve. Since the Piquette Avenue Plant was a multistory building, with the raw materials entering on the top floor and the finished product leaving on the ground floor, gravity chutes were used extensively. Like a slide, they connected upstairs and downstairs areas, and gravity did the transport of materials. The plant even experimented with an assembly line in 1908 where the vehicle moved on a skid and was pulled by ropes, following a suggestion by Ford engineer Charles E. Sorensen (Bak 2003, pp. 73–75).

As sales increased, the Piquette Avenue Plant became too small. In 1911, Ford opened a new plant in Highland Park, Michigan. It was nicknamed the *crystal palace*, due to a revolutionary design by industrial architect Albert Kahn,* who used reinforced concrete for the structure, which allowed for large and spacious windows. While the traditional brick-built New England mill used only 20–25% of the walls as windows, Kahn's reinforced concrete buildings used 70–80% of the wall area for windows, giving a much brighter work environment (Hyde 1996). This style of industrial architecture was highly influential, until Kahn came up with an even better idea for the next Ford plant. Nevertheless, reinforced concrete was still used, for example, in 1927, with the Fiat assembly plant in Italy. However, for some unknown reasons, Fiat decided to have the materials come in at the bottom and the completed car emerge at the top (Hildebrand 1975). Hence, instead of using simple gravity chutes, they had to carry work-in-progress material upward all the time. But perhaps it was more important to Fiat that there be a widely visible test track on top of the roof (Figure 13.7). In any case, as we will see many times in manufacturing history, merely copying an idea does not help if you do not understand the concept behind it.

Another novelty uncommon for the time was that the plant was powered by a 3000-horsepower gas engine generator through centralized electric motors and driveshafts (Hounshell 1985, p. 228). The system used a

* No relation to U.S. architect Louis Isadore Kahn.

FIGURE 13.7
The test track on the roof of the Fiat Lingotto plant. (Image by Dgtmedia–Simone and licensed under the Creative Commons Attribution-ShareAlike 3.0 Unported license.)

direct current system, which Ford continued to use for decades in all of his factories (Nye 1992, p. 198).

Around this time, Ford also started to use assembly lines in manufacturing. Unfortunately, exactly when and where the first assembly line was installed at Ford is shrouded by history and relies mostly on interviews conducted 40 years later (Nye 1992, p. 418). Besides the skids at the Piquette plant in 1908, it may have been the motor assembly in Highland Park that installed an assembly line in 1912. Or it may have been a 1913 assembly of the generator needed to generate electricity for the spark plugs—back then, at Ford, called *magneto* (Figure 13.8). Or, it may have been with the final assembly also in 1913, where cars stood on two strips of metal that moved them from station to station. Henry Ford himself once remarked that he got the idea from a slaughterhouse, where the pig carcass was moved along a disassembly line while hanging from a chain (see Section 13.1). Other sources also indicate that he may have gotten the idea from the Waltham Watch Company (Hoke 1990, p. 255).

In any case, the idea proved to be so good that Ford installed dozens of assembly lines in the following years. The productivity improvements after installing an assembly line were enormous. For the magneto line, the total labor time was reduced from 20 to 5 minutes. For the front axle,

FIGURE 13.8
Ford magneto assembly line in Highland Park, 1913.

the time was reduced from 150 to 26.5 minutes, or less than one-fifth. The transmission assembly was cut from 18 to 9 minutes, and the engine from 594 to 226 minutes. Most impressive was the final assembly line, where the time needed for one car was reduced from 12.5 to 6 hours, to three hours, and eventually to 93 minutes, or an increase in productivity of 800% (Hounshell 1985, p. 246ff). Overall, through the installation of the assembly lines, productivity increased between 50% and an unbelievable 1000% (Pursell 1995, p. 92). Eventually, a Model T left the plant every 40 seconds.

In effect, this meant that you could produce up to 10 times as many products with the same workforce, or use only one-tenth of the workforce for the same quantity. In any case, this greatly reduced labor time. Additionally, since the worker no longer needed to know how to assemble a complete car or how to make a part fit, the required qualifications went down, and with them, the labor cost per hour. Most of the workers at Ford were unskilled immigrants, usually with an agricultural background, often with little or no English skills, and hence the cheapest labor on the market. Finally, by using an assembly line, there was less space to place parts, but also, fewer parts were needed. As parts tie up capital but do nothing except take up space, Ford was able to produce a car with less material on hand and hence less investment. Therefore, the assembly line generated enormous profits for Ford.

Curiously, while the approach by Ford had some similarities with Taylor's scientific management, Ford claimed to never have heard of Taylor (Hopp and Spearman 2001, p. 25). At first glance, Fordism and Taylorism seem to have much in common. However, there are significant differences. While Taylor focused on improving the efficiency of the worker, Ford mechanized the process and removed the need for workers. For example, in the famous pig-iron experiment, Taylor optimized the movement of the workers carrying the iron. Ford would have simply installed a conveyor belt.

Other automakers soon started to copy Ford, although not all of them understood the benefit of an assembly line. Some automobile manufacturers with an annual production of less than 2000 cars started to install assembly lines, completely missing the point of mass production. A company producing less than 2000 cars per year needs an assembly line as much as you need a two-ton crane in your kitchen.*

The working conditions of an assembly line, however, were tough. The worker was subject to the speed of the assembly line, and to frequently irrational foremen who treated workers badly and set the speed of the line as fast as the worker was able to cope with.† The work was mentally boring, with the tasks repeating every 40 seconds for nine hours per day, while at the same time physically exhausting. Health and safety of the workers was of no concern to the management. For example, in 1916, a total of 192 fingers were lost in accidents at Ford (Bak 2003, p. 75). Both housing and the work environment were far from clean. Ford did not treat his workers as human beings, but—as was common at the time—as machines. In his autobiography, he sighed, *How come when I want a pair of hands I get a human being as well?* (Ford 1922).

---

* Sure, you can use it as intended, but is it worth the investment? Merely copying what worked for someone else does not mean that it works for you, but it can make things worse if the concept behind is not understood.

† Charlie Chaplin parodied this wonderfully in his film *Modern Times* from 1936. The president orders the foreman to speed up the line, and the little tramp played by Chaplin cannot keep up and eventually loses his mind in the process.

The workers were unable and unwilling to do this for long, but without representation, their only way to react was to leave Ford, and leave they did. For example, in 1913, the annual turnover at Ford was 370%. Hence, Ford had to hire and train over 50,000 men every year to have a constant workforce of 13,000 men. Moreover, while 20% of the workers leaving were fired and 10% quit officially, 70% of the workers leaving just did not show up at work anymore. In this case, Ford assumed after five days that they quit (Raff and Summers 1987). In 1913, only one in 20 employees was with Ford for more than three years (Hounshell 1985, p. 258). While Ford was able to hire new workers rapidly, they also had to be registered by accounting, trained, and familiarized with the work. This required both time and money. Furthermore, it is difficult to staff an assembly line—or, for that matter, any kind of workplace—if you never know who will show up the next day and how well they are trained. Overall, productivity suffered significantly due to the constant flux of workers.

To change this constant outflow of workers, Ford started in 1914, in his own words, *the greatest revolution in the matter of rewards for its workers ever known to the industrial world* (Raff and Summers 1987). And, what he did was truly bold. Just for a second, think about how much you earn annually. Got a number?

Now double it!

That's exactly what Ford did.* He set a minimum wage of $5 per day, roughly doubling the average wage of the workers. Additionally, he reduced the work time from nine to eight hours daily. Results were stunning. This $5 day made national headlines, and large numbers of workers from all over the continent moved to Detroit. Police even needed to break up riots at the factory gates (Brinkley 2003, p. 170). The $5 day applied only to males over the age of 22, with at least six months with Ford, and of good moral standing, but up to 90% of the eligible workforce received the $5-a-day wage.

The turnover rate dropped dramatically from 370% to 20–50% (Raff and Summers 1987). Morale increased, and workers proudly wore the company badge in public (Bak 2003, p. 71). Ford clubs started to pop up. Despite insourcing of parts, one hour less per day, and 14% fewer workers, productivity increased 15%. It is estimated that including all these factors, overall productivity improved by 40–70% (Raff and Summers 1987).

* Actually, it was one of the few instances where Ford decided to accept an idea from one of his employees, James Couzens, rather than deciding everything himself or just leave everything to his managers (Brinkley 2003, p. 166).

However, this still means that overall labor costs increased per vehicle produced.

So why did Ford do it? There are a number of possible reasons, and probably all of them were important to Ford in one way or another. First, Ford wanted to reduce employee turnover and to generate a more stable and motivated workforce. Moreover, also not unlikely, Ford wanted to be famous by being generous, and he enjoyed public attention (Raff and Summers 1987). For example, already in 1905, he gave workers a Christmas bonus of a $1000, more than their annual salary. In addition, by increasing his workers' salaries, they could now afford to buy a car, hence generating higher sales and returning at least part of the money to his company. Finally, as the Ford Motor Company was highly profitable, he simply could afford to be generous.

It was with this background of economic success that Ford started to build his next plant in 1917, the largest and most thoroughly organized factory the world has ever seen: the Ford River Rouge Complex. Ford again hired architect Albert Kahn, who already revolutionized industrial architecture using reinforced concrete for Highland Park. Amazingly, Kahn managed to revolutionize industrial architecture again, by switching from multistory concrete factories to single-story steel structures. Rather than using multiple floors, this approach had only one factory floor. While this required more land, the structure did not have to support additional floors but only the roof. Using steel frames, this permitted very wide spans with few supporting pillars. In addition, as anyone who ever was in charge of the shop floor layout can tell you, pillars are always in the way of machine placement, and the fewer pillars you have, the better. In fact, some large products like aircraft cannot even be produced in an assembly hall with pillars.

An additional benefit of single-floor factories is easier material flow. With multistory buildings, materials always have to be moved between floors. Unless there is a dedicated conveyor belt or something similar, elevators are a constant bottleneck in material flow, besides using up valuable floor space. Finally, natural light can come in through the roof, making the single floor factory brighter than ever before. Kahn was also the first to use the now-ubiquitous sawtooth roof of industrial buildings. Overall, Albert Kahn was probably the most influential industrial architect of the twentieth century. He designed over 1000 large buildings for Ford and hundreds of other structures for GM, Chrysler, and others, in addition to hundreds of nonindustrial buildings (Hyde 1996). However, since his industrial designs were highly utilitarian rather than aesthetic, he is not well known even among architects.

In any case, the first thing built in the Rouge complex was not cars but ships. President Woodrow Wilson asked Ford for support in the World War I shipbuilding efforts. Ford insisted on simple parts and started building submarine chasers in the Rouge complex. While he insisted on simple ship designs and simple parts, his effort to install an assembly line for ships failed. It took over 10 years to complete the construction of the River Rouge plant. For most of the time, the plant produced only parts for the Model T, whereas the final assembly line remained at Highland Park. None of the 60 ships built saw action during World War I.

With his River Rouge plant under construction, and his Highland Park plant assembling the Model T, Ford and his team continued to optimize the assembly line. They employed the principle of the assembly line not only for assembly but also for machining. They installed single-purpose machines, arranged along the manufacturing line. These machines were highly customized for their tasks. For example, the machine drilling holes in the engine block did not only drill one hole but had dozens of drill bits in different sizes and at the right places. Hence, the machine drilled all the parallel holes in one pass. Overall, every machine was highly customized and able to do only one task, but did this task with unparalleled efficiency. Similarly, the foundry also used assembly-line principles, with the molds being moved to the molten metal for pouring, rather than the other way around as it was done traditionally.

The tightly packed assembly lines also had another advantage. Due to the limited space between machines, it was not possible to build up piles of material. Before the assembly line, manufacturing usually had large quantities of materials on hand to cover disruptions of preceding or succeeding processes. Yet, materials are basically tied up cash. The firm has paid for it already but cannot sell it, since it is not yet completed. Having more materials than necessary is a significant economic disadvantage.

At the Rouge, Ford demonstrated his ability for fast material flow by having a load of ore delivered on Monday and selling the completed car made from that ore on Thursday night, a lead time of less than four days (Hopp and Spearman 2001, p. 25). However, it seems that this was a staged demonstration, with the prioritized material cutting in line in front of other materials (Bak 2003, p. 110). As it happens, while the assembly line had few materials in progress, there were huge batches of materials between the different assembly lines (Fujimoto 1999, p. 59). Nevertheless, it is a highly impressive feat, which even modern car companies would have difficulty repeating.

At the beginning, Ford also purchased many components from suppliers. Over time, however, he insourced more and more parts for the Model T, until all parts were made by Ford. Rather than purchasing components from suppliers, Ford wanted to make every part in his own factory. He owned the mines for the metal ore and the rubber plantation for the rubber to make tires. To use a modern management buzzword, he had an almost complete vertical integration of his firm. Bids from outside suppliers were mostly used to benchmark prices of his own internal suppliers (Hounshell 1985, p. 272).

However, the synergies from the vertical integration were much smaller than expected. Ford tried to extend his highly successful assembly-line principles almost everywhere. Among other places, he built a huge plantation, *Fordlandia*, in the Amazon rainforest. He made his employees wear ID badges, established strict working times, optimized lunch breaks by building a cafeteria, and prohibited alcohol and prostitution. While Ford had the best intentions, and, in fact, took exceptionally good care of his workers compared to other plantations, he was not able to bridge the cultural gap with the Brazilian workers. The workers disliked the unfamiliar food, the ID badges, and especially the prohibition on alcohol and prostitution. Hence, the workers rioted, destroyed the cafeteria, and rampaged through the plantation (Galey 1979). Additionally, neither Ford nor his managers had any knowledge about rubber plantations, and they planted the trees too densely packed on unsuitable soil. The rubber plantations never worked out for Ford. The only beneficiaries of his efforts were the South American leaf disease (*Dothidella nisi*), the lace bug, different fungi, ants, caterpillars, and lice, who never before had seen so many delicious rubber trees in one spot (Galey 1979).

Besides expanding into unknown businesses like rubber plantations, the vertical integration was hampered by Ford's need to be personally in control of everything—the supply chain was simply too complex for a single man to handle (Womack 1990, p. 39).

Despite these setbacks in vertical integration, the manufacturing system developed by Henry Ford and his team was exceptional. Fordism, as his approach was called, was able to create products cheaper and in larger quantities than ever before. Already when it came out in 1908, the Model T was cheaper than any other available car. The sales price of $850 in 1908 would be the equivalent of about $23,000 in 2016. While this was still more than the annual income for an average American, competing cars cost at least three times as much (Bak 2003, pp. 55–56). Overall, the Model T was just barely affordable for the upper middle class.

However, remember, this was in 1908, before the efficiency improvements of the assembly line. In 1912, when the first assembly lines started, a Model T was only $590. In 1916, the price dropped to $345, and in 1925, you could get it for $260. Hence, the price was reduced from $850 in 1908 to only $260 in 1925. If you take inflation into account, the price of the Model T dropped to one-sixth of its original price, a stunning 85% price reduction. The average American had to work for less than three months to afford a Model T. Now, almost everybody could afford a car.

Sales went through the roof, making Henry Ford one of the wealthiest people on earth, and the Ford Motor Company, the greatest industrial empire by 1926. While in 1909 Ford sold a respectable 17,000 vehicles, in 1924, he sold 1.7 million vehicles, a hundredfold increase in 15 years and a two-thirds market share in the United States (Hopp and Spearman 2001, p. 25). Before 1908, a car was a rare sight, and most commercial transport was by horse or by railway. By 1910, traffic jams were common, and horses started to disappear. The automobile industry was not even listed in the 1900 U.S. census but was one of the largest industries by 1929 (Bak 2003, p. 190), with half of all automobiles being made by Ford. It is estimated that the efficiency improvements by the assembly line generated the equivalent of a 2% increase in the gross domestic product (GDP) (Economist 2011).

While the car was very basic by today's standards—lacking a fuel gauge, shock absorbers, doors, and a heater among other things—it was able to get you around. Before the Model T, farmers often needed a whole day to bring one load of produce to the market by horse cart. However, using a Model T, they could easily make multiple trips and sell more in the same time. The ease of transport by using a car increased income for merchants, farmers, and retailers (Bak 2003, p. 57).

Of course, other businesses suffered. In 1900, marrying a horse breeder was considered a very good catch for any young lady, since, you know, people always need horses. Unfortunately, Henry Ford eliminated the principle underlying this economic approach to marriage. Urban legend claims that in the 1920s, meat prices dropped due to the large quantities of horsemeat becoming available. However, while food prices did indeed drop in relation to the GDP (Rayner, Laing, and Hall 2011), it is unlikely that horsemeat is the primary cause behind this drop. In any case, nobody wanted a horse anymore, since a car was cheaper both in purchase and in maintenance, more reliable, and also way more fashionable.

By the 1920s, even some high school kids were able to afford a used Model T to drive their girl around. Similar to modern college students, these cars were often personalized with text like *Four wheels, no brakes*, *Chicken, here's your roost*, or—my personal favorite—*Welcome to the Mayflower—Many a little Puritan has come across it* (Gilbreth and Carey 2002, p. 173ff). In all aspects, the automobile, and especially the affordability of the Model T, greatly expanded the reach of Americans and catapulted the United States and most of the rest of the world into the twentieth century.

In the 1920s, Henry Ford enjoyed his success from turning the industry upside down, but the next turn was already around the corner, and this time, he and his company would be caught on the flipside.

13.6 THE FLAWS OF FORDISM AND THE RISE OF GM

The Model T was designed for simplicity. Ease of manufacturing was part of the strategy of Henry Ford. He believed that one universal automobile would satisfy the needs of everybody. There was no need for an update, and there was no plan for a change. While the Model T was eventually available in different body styles and even as a truck, the underlying technology of the last Model T in 1927 was the same as in 1908. The spare parts of the first Model T from 1908 fit also in the last Model T built in 1927. In comparison, the Volkswagen Beetles in 1960 shared only three common parts with the first Beetles built in 1938 (Hengstenberg 2010).

Once, his engineers tried to surprise Henry Ford with an updated prototype of the Model T, having some technological changes. In response, Ford methodically started to smash the car into bits (Micklethwait 1998, p. 118). The message was clear: the Model T was not to be touched. While there were some changes, they were minor, and mostly for cost reduction, except an upgrade from acetylene lamps to electrical lamps around 1910 (Hounshell 1985, p. 273ff). If Henry Ford could have had his way, you would still be able to go to your local Ford dealership today and buy your very own brand-new Model T, with all the best technology from 1908. However, in this case, it would be more likely that there would be no Ford dealership anymore anywhere.

Even in the 1920s, despite the advantages of the design in 1908, it was essentially an outdated car. The competitors now offered cars with a much more advanced technology like safety glass, electric starters rather than hand-cranked starters that could break your arm, and other features that were considered luxuries in 1908 but necessities in the 1920s. Sales of the Model T peaked in 1923 and then declined. Eventually, Ford could no longer resist the forces of the market and also offered the Model T in colors other than black.* Even in advertising, Ford resisted changes until 1925 (Hounshell 1985, p. 273ff).

But by then, it was too late. The Model T was outdated in the view of the customer, and market share was far below its peak. The River Rouge plant was oversized for the number of cars produced. The customer clearly wanted something more modern. In 1927, Ford gave in, and after 19 years and 15 million produced, he announced the end of the Model T. However, due to the large inventories between the assembly lines, Ford had to make another 50,000 Model Ts, simply because the parts were already produced (Hounshell 1985, p. 279). The successor of the Model T was to be named Model A. Ford strongly felt that after the unforgettable Model T, there should not simply be a Model U.

The problem was that they did not yet have a Model A. Henry Ford was a gifted engineer, but he aimed for perfection. He wanted to repeat the success of the Model T, and the Model A had to be perfect. And since he knew the car in all detail, he easily found all the flaws, however minor.† Construction was repeatedly delayed.

Worse than the delays were the design changes. Both the Highland Park plant and the River Rouge plant were designed exclusively for the Model T. Every machine was custom-made to do one task and one task only for the Model T. Out of 32,000 machine tools in the inventory, a quarter had to be thrown out, a quarter could be reused as is, and half needed extensive reconstruction. Four thousand five hundred new machines had to be

---

* From 1914 until 1926, all Ford Model Ts were painted black. Ford famously said, *Any customer can have a car painted any color that he wants so long as it is black* (H. Ford 1922, p. 72). The main reason for the black color was that it needed fewer coats and dried faster than any other color.

† This is still a problem nowadays. Top executives of major carmakers are usually gearheads aiming for perfection. It does not matter that 99.9% of all customers will never hear the sound of a particular generator gear, but it has to be plastic coated since the CEO hears it. All parts in the engine have to have an anticorrosion coating, lest the customer see a speck of rust when he/she crawls underneath of his/her car. Even the managers responsible for low-end car designs have to be almost forced to pick a cheap carpet for the floor rather than the luxury version. And don't even get me started on the application of chrome decors.

purchased. There were not enough toolmakers in America to do all the work (Hounshell 1985, p. 286ff). Often, after the tool was completed, Ford ordered a design change, and the tool had to be rebuilt or scrapped. And, just for the record, when I say *tool*, I do not mean a screwdriver or a wrench but, for example, a 20-ton block of steel in the shape of the front hood.

Changeover costs mounted and, worse than that, delayed the start of the Model A. The last Model T was built on May 26, 1927, and the first model A was sold only in December 1927. Not a single car was produced and sold for over six months. Even then, the plant was running far below capacity. If you are working for a manufacturing company, imagine for a second not producing any product for half a year within the entire company. In all likelihood, the company would be bust, and you would be looking for a new job.

Fortunately, Ford had plenty of cash stashed away to finance all of this. Nevertheless, he fired large numbers of workers, including key managers and foremen in manufacturing. Therefore, he also fired the knowledge on how to design and run assembly lines and terminated a crucial body of knowledge in his firm. His manager, Charles E. Sorensen, phrased it to the point when he said, *We are going to get rid of all the Model T sons-of-bitches* (Hounshell 1985, p. 263). This, of course, increased the problems during the model change and the following ramp-up of production.

Eventually, the production ramped up during 1928, and the efficient manufacturing system of Ford showed its effect. The car sold very well. Ford was pleased to have a new model with equal success and longevity as his beloved Tin Lizzie. At least, he expected the car to run as long as the Model T. Unfortunately for Ford, the customer disagreed, and after five years the car was outdated, sales declined, and the production of the Model A stopped. The Model B succeeded the Model A.

What followed, however, was a repetition of the chaos of the previous model change. Plants closed in August 1931, with the first car being sold only in March 1932. It took five months and $50 million to reconstruct the tools and change production. This is the equivalent of almost $1 billion in 2016. While this cost and delay was better than during the change from Model T to Model A, it was much more expensive than at the competitors. It clearly took the company some time to establish at least a basic flexibility in its operations, as changeovers in other plants in the United States and abroad took similar efforts.

The competitors of Ford, however, took a different and more successful path, most notably GM. GM was founded in 1908 by William C.

Durant. He purchased a number of smaller car companies, including Ford's old company, Cadillac, but also Oldsmobile, Pontiac, and half a dozen others. Yet he never integrated them into one coherent company. Eventually, he overstretched himself financially and lost control of the company in 1920.*

After Durant, Alfred P. Sloan took over as the president of GM. His successful restructuring of GM made him into one of the first management gurus, who is still revered nowadays. Sloan took a distinctively different path than Ford. While manufacturing was also based on the assembly line, Sloan not only realized the desire of the customer for new products but actively promoted it. While Ford wanted to have one car that fits everybody, Sloan promoted a *car for every purse and purpose* (Finkelstein 2004). His company offered a wide array of different models, with frequent technical and cosmetic changes, plus regular larger model changes. Rather than making the same car forever, GM planned the obsolescence of their cars. Where Ford insisted that every Ford vehicle is identical, Sloan celebrated the arrival of new models. Where Ford sold vehicles in black, GM sold cars in fancily named colors like Ardsley Green, Lindbergh Blue, Saint James Gray, and many more.

Naturally, with frequent changes in design, it is not possible to construct highly specialized machines for every design change. Former Ford manager William S. Knudsen, who left due to Ford constantly overruling his decisions, joined GM and was responsible for manufacturing, eventually becoming president himself. It was Knudsen who insisted on not using specialized machines but, rather, general-purpose machines. While this was slightly less efficient during production when compared to Ford, changing to a new model was much quicker. Additionally, at GM, the model changes were planned strategically beforehand, whereas Ford only reacted much later, when the customers stopped buying.

For example, at GM, for a 1929 major changeover, planning started in 1927. A pilot line was constructed to perfect tools and techniques. The changeover at the main plant happened between October 25 and November 15, 1929. While Ford required six months for a changeover, GM did it in 20 days (Hounshell 1985, p. 266f).

Another major change of Sloan was to decentralize the company. Ford tried to manage everything himself and by far exceeded his abilities. Durant also overstretched himself by trying to manage the conglomerate

* Even nowadays, GM is a conglomerate of brand names, comparable only to Volkswagen.

of companies. Sloan and his team on the other hand structured the brands into separate cost centers, each responsible for their own profit and loss. At the same time, each brand received directions from central management and was responsible for a certain market segment to avoid cannibalization. For example, Cadillac was aimed at the luxury market and Chevrolet at the low-end market.

The decentralization was not only with respect to brands but also components. Since 1910, GM started to heavily rely on external suppliers, thereby simplifying their own management tasks, resulting in cheaper and better components. Even critical components manufactured in house like the engine and the transmission were divided into different plants. Where Ford's vision was a centralized industrial Moloch at River Rouge, GM distributed its components to different plants in the Detroit region. This not only simplified the changeover process but also made the entire system more robust.

Most of the ideas of Sloan are now standard management practice in industry. Most products have a planned obsolescence, regardless if it is your car, computer, furniture, or other possessions, with few exceptions. Yet, most of them were probably mass-produced on an assembly line, albeit with at least some machine flexibility. Outsourcing is also common nowadays, often with between 25% and 75% of the added value outsourced to suppliers. Finally, for larger companies, a decentralized structure with plants all over the world is the norm. All of this started in America and was quickly applied in overseas plants of American companies. European companies and, later, Asian companies copied the system, albeit often with a significant delay. European carmakers achieved mass production only around 1960, about 50 years after Ford. Even Wolfsburg, the main plant of Volkswagen, was modeled after the River Rouge plant, with the cornerstone being set by Adolf Hitler himself in 1938.

GM continued to prosper, and Ford eventually learned the advantages of flexibility. For Ford, however, it was a more difficult path. He also frequently had labor trouble. As he tried to suppress unions through intimidation and violence, many people got hurt in clashes with his *security force* of ex-criminals and boxers. This, of course, only increased resistance, and eventually, he had to give in to the demands of the unions. His popularity also declined, partially due to his clashes with the unions but also due to his general attitude toward workers and due to his heavy anti-Semitism and support for Hitler. Already in 1928, the New York Times called Ford an *industrial fascist* and the *Mussolini of Detroit** (Bak 2003, p. 112). Ford resigned in 1945, confused and senile after multiple strokes, and died in 1947 (Brinkley 2003, p. 496).

However, Ford was primarily and foremost an industrialist. In Europe, however, real fascism and other extreme political tendencies were gaining power, with millions of deaths to follow.

BIBLIOGRAPHY

Bak, R., 2003. *Henry and Edsel: The Creation of the Ford Empire.* John Wiley & Sons, Hoboken, NJ.

Bak, R., 2012. Detroit's Ride from Horse to Horsepower. *Hour Detroit.*

Brinkley, D., 2003. *Wheels for the World: Henry Ford, His Company, and a Century of Progress*, 1st ed. Penguin, New York.

Busch, J., 1981. An Introduction to the Tin Can. *Historical Archaeology* 15, 95–104.

Chandler, A., 1977. *The Visible Hand: Managerial Revolution in American Business*, New edition. Harvard University Press, Cambridge, MA.

Dunlavy, C.A., 1980. *Food Machinery Corporation's Central Research Department: A Case study of Research and Development (Bachelor of Arts).* University of California, Berkeley.

Economist, 2011. Economics Focus: Marathon Machine. *The Economist.*

Encyclopedia Britannica, 2012. Edwin Norton. *Encyclopedia Britannica Online.*

Evans, O., 1795. *The Young Mill-Wright and Miller's Guide: Illustrated by Twenty-eight Descriptive Plates.* Printed for and published by the author, Philadelphia, PA.

Finkelstein, S., 2004. Case Study: GM and the Great Automation Solution, in: *Why Smart Executives Fail: And What You Can Learn from Their Mistakes.* Portfolio Trade, New York.

Ford, H., 1922. *My Life and Work.* Doubleday, Page & Company, Garden City, New York.

Fujimoto, T., 1999. *The Evolution of Manufacturing Systems at Toyota*, 1st ed. Oxford University Press, Oxford; New York.

Gage, F., 2006. Wheat into Flour: A Story of Milling. *Gastronomica* 6, 84–92. doi:10.1525/gfc.2006.6.1.84.

Galey, J., 1979. Industrialist in the Wilderness: Henry Ford's Amazon Venture. *Journal of Interamerican Studies and World Affairs* 21, 261. doi:10.2307/165528.

Giedion, S., 1948. *Mechanization Takes Command: A Contribution to Anonymous History.* Norton, New York.

Gilbreth, F.B., Carey, E.G., 2002. *Cheaper by the Dozen, Perennial Class.* Harper Perennial Modern Classics, New York.

Harmond, R., 1971. Progress and Flight: An Interpretation of the American Cycle Craze of the 1890s. *Journal of Social History* 5, 235–257.

Hengstenberg, M., 2010. Zeitreise ins Wirtschaftswunderland. *Spiegel Online.*

* He probably would have been called the *Hitler of Detroit*, but back then, Adolf Hitler was still a rather unknown fringe radical, getting less than 3% of the votes in the 1928 general election.

Hildebrand, G., 1975. Albert Kahn: The Second Industrial Revolution. *Perspecta* 15, 31. doi:10.2307/1567012.
Hobsbawm, E.J., 1999. *Industry and Empire*. Penguin Books, London.
Hoke, D.R., 1990. *Ingenious Yankees: The Rise of the American System of Manufactures in the Private Sector*. Columbia University Press, New York.
Hopp, W., Spearman, M.L., 2001. *Factory Physics*, 2nd ed. McGraw Hill Higher Education, New York.
Hounshell, D.A., 1985. *From the American System to Mass Production, 1800–1932: The Development of Manufacturing Technology in the United States*, Reprint. ed. Johns Hopkins University Press, Baltimore.
Hyde, C.K., 1996. Assembly-Line Architecture: Albert Kahn and the Evolution of the U.S. Auto Factory, 1905–1940. *The Journal of the Society for Industrial Archeology* 22, 5–24.
May, G.S., 1977. *R. E. Olds, Auto Industry Pioneer*. William B. Eerdmans Publishing Company, Grand Rapids, MI.
Micklethwait, J., 1998. *Witch Doctors: What Management Gurus Are Saying, Why It Matters and How to Make Sense of It*, New edition. Random House, UK.
Miele, C., 2006. Bold, Well-Defined Masses: Sir John Rennie and the Royal William Yard. *Architectural History* 49, 149–178.
Moss, D.A., 2011. Kindling a Flame under Federalism: Progressive Reformers, Corporate Elites, and the Phosphorus Match Campaign of 1909–1912. *Business History Review* 68, 244–275. doi:10.2307/3117443.
Mossoff, A., 2009. The Rise and Fall of the First American Patent Thicket: The Sewing Machine War of the 1850s. *SSRN eLibrary*.
Nye, D.E., 1992. *Electrifying America: Social Meanings of a New Technology, 1880–1940*. Mit Press Paper, Cambridge, MA.
Pearson, S.A., 1992. *Using Product Archeology to Identify the Dimensions of Design Decision Making [MS]*. Massachusetts Institute of Technology, Cambridge, MA.
Pursell, C., 1995. *The Machine in America: A Social History of Technology*, Illustrated edition. Johns Hopkins University Press, Baltimore.
Raff, D.M.G., Summers, L.H., 1987. Did Henry Ford Pay Efficiency Wages? *Journal of Labor Economics* 5, 57–86.
Rayner, V., Laing, E., Hall, J., 2011. *Developments in Global Food Prices, Bulletin*. Reserve Bank of Australia, Sydney, Australia.
Rosen, W., 2010. *The Most Powerful Idea in the World: A Story of Steam, Industry, and Invention*. Random House, New York.
Rosenberg, N., 1963. Technological Change in the Machine Tool Industry, 1840–1910. *The Journal of Economic History* 23, 414–443.
Spreng, R., 1995. The 1890s Bicycling Craze in the Red River Valley. *Minnesota History* 54, 268–282.
Twede, D., 2009. Cereal Cartons, Tin Cans and Pop Bottles: Package-Converting Technologies that Revolutionized Food and Beverage Marketing, 1879–1903, in: 14th Biennial Conference on Historical Analysis and Research in Marketing. Presented at the Conference on Historical Analysis and Research in Marketing, Leicester, UK.
Womack, J.P., 1990. *The Machine That Changed the World: Based on the Massachusetts Institute of Technology 5-Million-Dollar 5-Year Study on the Future of the Automobile*, Later Printing. Rawson Associates, New York.

14

Centrally Planned Economies—War, Communism, and Other Catastrophes

Sustained economic growth is not compatible with predatory cleptocratic government.

Karl Gunnar Persson
Professor of economic history, in An Economic History of Europe, p. 97

Manufacturing works within the society and economy. As such, the society and economy do have a strong influence on manufacturing. In this chapter, we will look at different periods where the situation of the economy influenced manufacturing. The biggest disruptions in the twentieth century were the two World Wars, with many of the participating countries switching at least partially to a centrally planned wartime economy. We will also look at what rose from the ashes after the destruction of World War II (WW II), in particular, the economic miracle in West Germany.

A larger and longer-lasting planned economy was attempted by communism. Choosing a different path from capitalism, socialist and communist countries started to declare all resources as shared public goods. However, these countries had to find out the hard way that communism does not work. Many communist countries nowadays have a communist political system but allow an at least partially capitalist economy. At the time of writing, only two countries enforce a strict communist economy, North Korea and Cuba, neither of which has any significant economic strength.*

Overall, in centrally planned economies, political leadership takes over the economy. Their aim is either to generate a better system than possible through capitalism or to strengthen part of the economy needed for

* But instead sizeable black markets. Hence, they also had an underground market economy.

political reasons. The boundary between capitalism and planned economies, however, is fuzzy, as most countries exert at least some control over their economies.

14.1 WORLD WAR I

Throughout history, military expenditure often drove technological change. However, until the industrial revolution, war was severely constrained by the availability of food and materials as well as the speed of communication. The spread of industry and new technologies after 1850 allowed a completely new type of war. Probably the first major industrial war was the American Civil War, 1861–1865. Weapons were mass-produced, railroads and steamships transported troops and equipment at a never-before-seen speed, and telegraphs allowed commanders to stay in control of large and geographically distributed armies. With this new technology, it was now possible to wage war across the entire world. Shortly thereafter, the possibility turned into reality through World War I (WW I). Many nations in Europe were eager for war, and the assassination of Archduke Franz Ferdinand of Austria in 1914 was a welcome excuse to start hostilities.

Both information and materials flowed at an unparalleled speed, allowing combat in many different locations simultaneously. Rather than a local conflict with a few short battles, a completely new type of war emerged. Battles took not days but months. Casualties would be not a few thousand men but hundreds of thousands of men. For example, the Battle of Verdun lasted for about 10 months, with almost 1 million casualties, yet this was only one of dozens of major battles in WW I. Modern technology made killing each other easier, and winning the war depended in large part on having better technology. Research expenditures by the government tripled (Cameron and Neal 2002, p. 341). New technologies like airplanes, tanks, and radios were developed for the war. Shortages of raw materials led to the development of new technologies, for example, the Haber-Bosch process for obtaining nitrogen from the atmosphere for gunpowder production.

Even more important than research was the strength of the economy. To win the war, it was necessary to convert industry toward military production. The generals had to learn the hard way that war was no longer only war of men but also war of materials. Most major wars, including the American Civil War, WW I, and WW II, were won by the stronger military industry.

In a free market, the change from a peacetime industry to a war industry would be through a large customer—the military or the government. They had significantly increased demand for military-related products. They also had the ability to pay for these through taxation or simply by printing more money. Hence, in theory, industry would adjust its resources to increase military production in order to increase profit. In practice, however, this process was often too slow. Factories were set up for civilian goods, and converting factories or building new factories required significant investments. Yet, it was foreseeable that as soon as the war ended, demand for weapons, ammunition, and military gear would drop. Making large investments in machines and production facilities for a demand that may drop quickly and significantly was risky.

In some cases, there was not even enough knowledge available to establish new production facilities. Not only unskilled labor but also skilled engineers, designers, and machinists are needed to design products and machines, build them, and control production. Training new engineers or retraining existing engineers toward military production takes time. Overall, the change from peacetime industry to war industry was normally slow and sluggish, usually not satisfying the needs of the government.

Fortunately for the government, they can change the rules of the game by changing laws and, in effect, force industry to convert. However, civilians are equally skilled in finding loopholes to circumvent laws. In peacetime, free trade of goods was the norm in most countries. During the war, on the other hand, the government took at least partial control of raw materials, labor, and production facilities in most countries (Cameron and Neal 2002, p. 340).

In the United States, the War Industries Board was created among others during WW I to take control of the economy. They influenced prices, wages, and labor relations to improve military output. To allocate the scare resources according to the highest need, production was prioritized. As the officers and bureaucrats in charge were able to set their own priorities, soon, all production orders by the government were classified as top priority A, since, of course, everything was the most important product to the official ordering them. Subsequently, a finer grading was used with priorities A1, A2, and A3 to rectify this problem. This resulted again—unsurprisingly—in all orders being prioritized A1 (Lozada 2005).

The U.S. Food and Fuel Control Act in 1917 created the Food Administration and the Fuel Administration. They also influenced prices but, even more so, appealed to the patriotism of the citizens. They promoted meatless Tuesdays, porkless Saturdays, and the consumption of victory bread using inferior flour. When eating an apple, citizens were reminded to be *patriotic to the core*. At the end of the war, a quarter of all food production was used for the war effort.

Military output multiplied, and many factories were converted to produce military goods and supplies. President Wilson asked Henry Ford for support in military production. Hence, the first products produced at the Ford River Rouge plants were not cars but eagle boats, designed to chase German submarines. However, it takes about five years for a ship to go from design to combat readiness. Ramping up production was slower than the war, and only seven boats were produced during the last year of the war, with another 53 following the year after. None of them saw combat during WW I, although some were used for WW II. Often, the industry had difficulty ramping up production quick enough for the war effort. Similarly, the demand for military goods vanished at the end of the war. Many private entrepreneurs had invested large sums in machines for military production and lost money as these machines were no longer needed after the war. Shipyards were hit especially hard financially. Hence, when the next world war started, private industry was reluctant to put their full weight behind the war effort in fear of losing money again (Carew 2010, p. 169).

14.2 WORLD WAR II

Only 25 years after the horrors of WW I, Germany, under the leadership of lunatic Adolf Hitler, started the next world war. While 16 million deaths due to WW I seemed unbelievable back then, WW II quadrupled this number with over 60 million deaths.

To switch the American industry to a war industry for WW II, the War Production Board was established in 1942. As the War Industries Board 30 years earlier, they also allocated resources according to priority. The results were also the same as in WW I; everything was top priority. Starting out with a basic A, B, C...grading system, the grading went ever finer until it made a difference if your task was priority AAA-1 or only AA-1a. For manufacturing, this created lots of additional paperwork to get what was needed, and hence, the bill of materials was often inflated to get at least some material (Carew 2010, p. 237ff).

The workplace itself also changed. With many of the able-bodied men abroad fighting the war, there was a serious shortage of labor. To solve this problem, the government campaigned to housewives to join the war industry. The percentage of women in industry increased significantly to 37% by 1943 (Zeitlin 1995). *Rosy the Riveter* was, at that time, a cultural icon (Figure 14.1). Production also relentlessly focused on output, even at the cost of other factors. During the war, 88,000 workers were killed and 11 million injured through their work, about 11 times the number of U.S. combat casualties, with around 1 million dead or wounded (Noble 1984, p. 23). Work safety clearly had to take a step back for the war.

FIGURE 14.1
We Can Do It!—famous 1942 wartime propaganda poster by Howard Miller, nowadays often erroneously referred to as Rosie the Riveter.

During WW II, shipbuilding efforts in the United States were more successful than in WW I. While in 1941, 92% of all U.S. ships were older than 20 years, the shipbuilding efforts nearly tripled the number of navy ships (Carew 2010, p. 41ff). In particular, one standard ship design for cargo ships was produced rapidly. A total of 2710 Liberty ships were built between August 1941 and September 1945, using interchangeable components. This is by far the largest number of ships built of the same design.

Another mass-produced product was aircraft. Negligible in WW I, they shaped the form of WW II. Due to the lack of an aircraft industry, many automotive firms converted to aircraft production. One prominent example is the Ford Willow Run plant, built by Henry Ford to produce B-24 bombers. Using Ford's experience with mass production, B-24 bombers were produced in an assembly-line fashion, with up to 650 planes per month.

However, despite the impressive output, automotive mass production technologies proved inadequate for the still rapidly changing aircraft technology. Inexperience with aluminum technology in combination with frequent design changes led to many scrapped tools. Overall, as much a half of all tools produced for Willow Run were never used for production (Zeitlin 1995). Additionally, the planes produced were outdated by the time they rolled off the assembly line. Since it was easier to modify planes after completion than to constantly change the production line, modification centers were used to update existing planes to the latest technology. This was, of course, also very time consuming, and between 25% and 50% of all military aircraft workers were not building planes but updating existing planes (Zeitlin 1995).

Nevertheless, in the end, the allied forces had an industrial output more than three times larger than the axis powers (Broadberry and Harrison 2005). For WW I in 1917, the United States could build boats faster than Germany could sink them (Wren 1980). During WW II, by 1944, the United States produced five times as many planes by weight than Germany. Willow Run alone represented the equivalent of half of the German aircraft production capacity. The United States was the primary supplier of war materials to Great Britain, France, and—reluctantly—the Soviet Union.

Germany often had the more advanced technology, including Tiger and Panther tanks with superior armor and weaponry, the world first supersonic V2 rockets, the first and then-fastest ME 262 combat aircraft with a jet engine, and unparalleled U-boats with homing torpedoes. Nevertheless, their manufacturing system was still mostly craft based, far inferior to American mass production (Zeitlin 1995). Additionally, they were hampered by material shortages. Overall, the allied forces were able to field a constant onslaught of men and materials against Germany and, by 1945, were able to stop the madness of Nazi Germany.

Providing the lion's share of the equipment besides fielding up to 12 million soldiers in two theatres of war put a great stress on the manufacturing capabilities at home. Three major efforts were undertaken to increase output while the means of production were limited: *Training within Industry* (TWI) and statistical process control (SPC) in the United States and operations research (OR) in Britain.

TWI reduced the shortage of skilled workers by increasing and standardizing training methods. The basic idea goes back to WW I, when Charles Allen developed a four-step system to train shipyard workers. Recognizing a lack of trained workers, the United States Department of War established the TWI service in 1940. The service was headed by the *four horsemen of TWI*, Channing Rice Doley, Walter Dietz, Mike Kane, and William Conover, three of whom learned the methods of Charles Allen during WW I. Expanding Allen's approach they developed four training programs:

- Job Instructions (JI), teaching supervisors how to train workers effectively
- Job Methods (JM), teaching workers how to improve their work (Figure 14.2)
- Job Relations (JR), teaching supervisors how to interact with workers on a fair and rational basis
- Program Development (PD), as an overarching training for the trainers

The TWI service was a smashing success. Training times were often significantly reduced, providing much-needed skilled workers to the war industry. One of the first uses was to ease a shortage of skilled lens grinders. Traditionally, it took many years for a worker to become a skilled lens grinder. TWI was able to reduce the training time from years to months, increasing the output of skilled workers and, subsequently, lens production (Huntzinger 2007). The service trained over 1.7 million people throughout the war in 16,500 plants, successfully reducing shortages of skilled labor. It would have been very worthwhile to continue the program after the war, but the service ended in September 1945. The United States experienced an

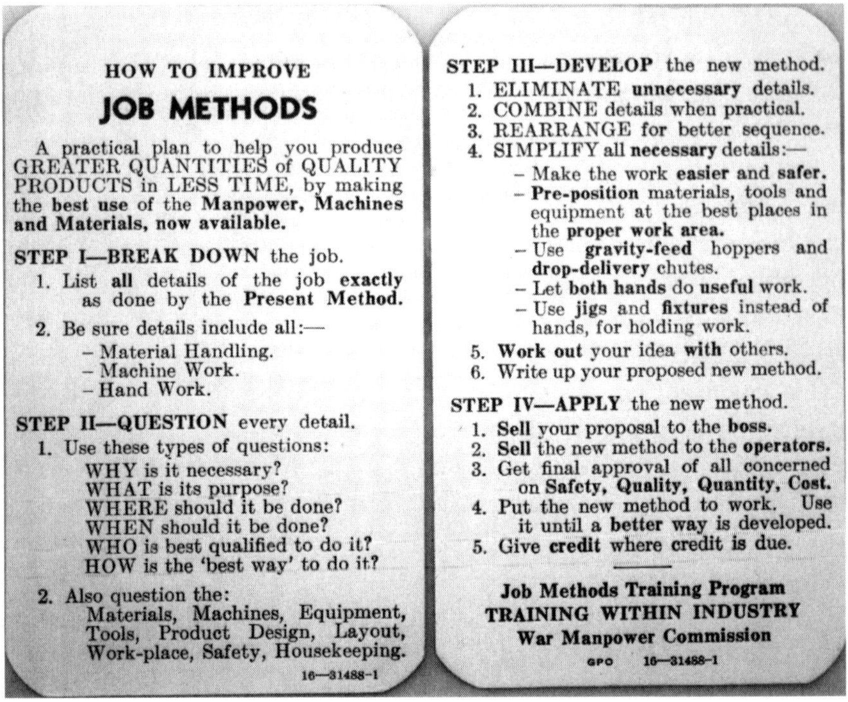

HOW TO IMPROVE

JOB METHODS

A practical plan to help you produce GREATER QUANTITIES of QUALITY PRODUCTS in LESS TIME, by making the **best use** of the **Manpower, Machines and Materials, now available.**

STEP I—BREAK DOWN the job.

1. List **all** details of the job **exactly** as done by the **Present Method.**
2. Be sure details include all:—
   - Material Handling.
   - Machine Work.
   - Hand Work.

STEP II—QUESTION every detail.

1. Use these types of questions:
   WHY is it necessary?
   WHAT is its purpose?
   WHERE should it be done?
   WHEN should it be done?
   WHO is best qualified to do it?
   HOW is the 'best way' to do it?
2. Also question the:
   Materials, Machines, Equipment, Tools, Product Design, Layout, Work-place, Safety, Housekeeping.

16—31488-1

STEP III—DEVELOP the new method.

1. ELIMINATE **unnecessary** details.
2. COMBINE details when practical.
3. REARRANGE for better sequence.
4. SIMPLIFY all **necessary** details:—
   - Make the work **easier** and **safer.**
   - **Pre-position** materials, tools and equipment at the best places in the **proper work area.**
   - Use **gravity-feed** hoppers and **drop-delivery** chutes.
   - Let **both hands** do **useful** work.
   - Use **jigs** and **fixtures** instead of hands, for holding work.
5. **Work out** your idea **with** others.
6. Write up your proposed new method.

STEP IV—APPLY the new method.

1. **Sell** your proposal to the **boss.**
2. **Sell** the new method to the **operators.**
3. Get final approval of all concerned on **Safety, Quality, Quantity, Cost.**
4. Put the new method to work. Use it until a **better** way is developed.
5. Give **credit** where credit is due.

Job Methods Training Program
TRAINING WITHIN INDUSTRY
War Manpower Commission
GPO 16—31488-1

FIGURE 14.2
Training within Industry Job Methods card handed out after the completion of a training. (From the U.S. Civil Service Commission, 1945.)

industrial boom after WW II; millions of soldiers returned to the United States, bringing their civilian skills back into industry and eliminating the need for training. The U.S. government also withdrew its funding from TWI. While some experts continued to pursue TWI, overall, TWI lost a lot of its steam. A few TWI experts moved to Japan to offer their services there, among them W. Edwards Deming and Joseph Juran. From there, TWI would come back with a vengeance as part of Lean manufacturing.

SPC solved another problem of the U.S. war industry by improving quality. The method originated from Walter A. Shewhart during the 1920s at Bell Laboratory. SPC monitored a process using statistical tools to ensure compliance with specifications and to catch quality problems early on. Like TWI, the U.S. government taught SPC to improve product quality, with a strong focus in the aircraft industry. The program was also a huge success, significantly reducing defects in U.S. production. Yet, as the economy boomed after the war, efforts in quality diminished. The war effort of SPC was, in the view of the U.S., industry no longer needed after the war. Consultants specializing in SPC found it difficult to make a living after the war and offered their services abroad. Like TWI, the approach of SPC fell on fertile grounds in Japan and also came back with a vengeance as part of Total Quality Management (TQM).

Scientists in Great Britain and the United States also extended the methods of Taylor and used analytical methods to support their decision making, naming their approach operations research (OR). Initially, this approach was most active in military decision making. In Britain, they optimized the placement and use of antiaircraft artillery, reducing the number of rounds per downed enemy aircraft from 20,000 to 4000. Similarly, they optimized antisubmarine warfare by optimizing convoy size, depth charge patterns, or aircraft design (Kirby 2003, p. 94ff).

After the war, OR expanded into industry, although with less spectacular results. The methods are best suited to problems of limited complexity but have difficulty dealing with the complexity of modern production. Nevertheless, many companies nowadays use OR as part of their toolbox to solve problems of limited complexity. There is a lot of research for more complex problems but with few practical applications. Most methods had been developed before the 1950s (Waring 1994, p. 27), and modern OR has a reputation of involving highly complex mathematics with a lack of practical use. Using lots of sexual metaphors, Russell Ackoff, one of the coauthors of the first OR textbook in 1957 stated in 1979 that OR was stuck in *mathematical masturbation* and *impotent* to modern complexity, understanding management as a *eunuch understood sex*, and only *intercourse with management* could solve this problem (Waring 1994, p. 38).

During the war, the allies developed numerous methods to improve production, but none of them managed a successful transition to the postwar period. OR got stuck in mathematics. Flexibility developed for aircraft manufacturing was no longer needed. TWI and SPC were abandoned—except for a spark that went to Japan. There, however, this spark fell on fertile grounds and helped to start the success of Lean manufacturing (see Chapter 16).

14.3 POSTWAR ECONOMIC MIRACLES

After the war, the industry faced an abrupt change in demand. While many feared another great depression after WW II, the U.S. industry was able to convert back to a peacetime economy successfully. Furthermore, due to rising consumer demand, a housing boom, and the automotive industry, the economy actually increased after the war. Italy, France, and Russia also recovered. The economy in Britain, however, was already weak before the war and did not recover.*

Most surprisingly, Germany was able to create an economic miracle (*Wirtschaftswunder*). Directly after the war, the industry was in shambles. Large parts of the male population died during the war or were imprisoned in Russia. The United States and Russia captured key engineers and scientists, forcing them to work for the United States or Russia†—although in many cases, very little *force* was necessary to bring them to a better life in America. Money was in short supply. Factories were bombed to the ground. Russians carried the still-useable equipment in East Germany away. Overall, Germany was in shambles, and the economy was ruined.

Yet, Germany did not lose the most important ingredients for economic success: a highly trained labor force and a solid work ethic. The German mindset was focused not on the horrors of the past but on rebuilding the future. Aided by a monetary reform and the U.S. Marshall Plan in 1948, the economy soon started a race to catch up with the rest of the world. This rapid increase in wealth is known as *Wirtschaftswunder.* Key to the quick recovery were medium-sized firms, known as *Mittelstand*, and comparatively good labor relations. By 1960, the West German standard of living was comparable to the rest of the industrialized world.‡ While Britain introduced in 1891 the country-of-origin label on their products to shy people away from German goods, *Made in Germany* soon recaptured its reputation for high quality at an affordable price.

Japan was equally defeated, with two nuclear bombs on Hiroshima and Nagasaki ending the war. Yet, similar to Germany, Japan also had the two most important ingredients for economic success. The Japanese population also contained a large pool of trained labor, and its work ethic is—unbelievably—even higher than in Germany. Hence, the Japanese

* Food rationing in Great Britain ended only in 1954, almost 10 years after the war. Possibly because of this long-term rationing, British cuisine has a low reputation internationally, although in my view, this is slowly changing. You can find some really good British food if you know where to look for.

† These scientists, besides medicine and electronics, were most active in rocket science. The most prominent example is Wernher von Braun, inventor of the V2 rocket and key developer of the Saturn V as part of the National Aeronautic and Space Administration (NASA) space program. Others, like Albert Einstein, fled to the United States already before the war.

‡ This, however, applies only to West Germany. Russian-occupied East Germany fared less well. In the communist system, shortages of products were the norm. Despite much propaganda to the contrary, the standard of living was far below West Germany.

economy also recovered by the 1960s and, like Germany, is now one of the strongest economies in the world.

14.4 THE FAILED EXPERIMENT—COMMUNISM

The industrial revolution deepened conflicts between the workers and the factory owners (see Chapter 10). This conflict troubled not only the have-nots but also some of the haves, and many intellectuals were looking for better ways to organize society. One of the most successful early social reformers was Robert Owen (1771–1885). Already at age 18, he was the successful manager of Manchester's largest cotton mill. In 1799, he and his partners bought the cotton mills in New Lanark southeast of Glasgow. While the workers in the mill were treated comparatively well, Robert Owen wanted to form the mill according to his vision.

From early on, Owen distanced himself from religion. Instead, he believed that a man does not form his own character; it is formed for him by the circumstances that surround him. The logical conclusion was that praise or blame has no effect, but rather, it is necessary to provide a good environment that forms the human character. New Lanark was to be that environment. He had the workers' cottages renovated, provided health and retirement benefits, shortened work time to a (back then) radical 10.5 hours per day, established a shop with inexpensive goods for daily use, stopped employing children below the age of 10, and limited the consumption of alcohol. Pretty much the only reason for being fired was drunkenness (Pollard 1964). To motivate his workers, he installed colored cubes above the machines where the color visualized the amount and quality of the work.

The experiment was a success. Productivity went up, thefts decreased, and the mill made a healthy profit. His partners, however, preferred to make even more and wanted to reduce the expenses toward the workers' well-being. Owen, however, was able to buy his partners out with the help of new investors more sympathetic to his visions. While working in New Lanark was still demanding, conditions were much better than in any other mill. His mill gained fame throughout Europe, and many visitors and dignitaries came to see his mill.

In 1825, Owen left for the United States, planning to establish an even more radical collective community in New Harmony, Indiana, where fields and tools were shared among its members. Unfortunately, the vision of being provided for regardless of the amount of work attracted workers who were more interested in being provided for than in working. Combined with a mix of radical and religious members, the society started to quarrel with itself. New Harmony disbanded completely only two years later in 1827. Other similar attempts by Owen in Mexico and the United States faltered even before they started. Owen returned to England, where he had numerous disagreements with his New Lanark partners and eventually cut his connection to New Lanark. Having spent most of his wealth on New Harmony, he never again managed a factory or community. Instead, he focused on promoting his ideas and visions, but visions they remained.

Two other visionaries heavily influenced by the social conflicts during the industrial revolution were the German philosophers Karl Marx (1818–1883) and Friedrich Engels (1820–1895). At least partially influenced by Robert Owen (Harman 2002, p. 327) they developed the theory that humanity only advances through class struggle. In the past, primitive societies evolved into slave-owning societies ruled by kings, which then evolved into feudalism ruled by aristocracy, and finally into capitalism ruled by the middle class. The next step they envisioned was socialism, where the lower social classes ruled. Finally, the highest form would be communism, where, due to public ownership of the means of production, there would be no class distinction and, hence, no class struggle.*

The idea of all being equal, sharing resources, and being provided for was very appealing to many. While it took half a century for the first state to convert to socialism, many other states followed, and during the twentieth century, at least 40 nations turned to socialism or one of its variants. Hence, the experiment of Robert Owen was repeated on a much grander scale, as was its epic failure. Nowadays, there are only two countries with a socialist economy left: poorhouse Cuba and the even poorer hermit kingdom North Korea.

14.5 COMMUNIST RUSSIA—FIVE-YEAR PLANS FOR CHAOS

The first nation to embark on the path to communism was Russia during the Russian Revolution of 1917 under the leadership of Vladimir Lenin

* For the remainder of the document, we will refer to nations following the vision of Karl Marx as having a communist system, following common western phrasing. In reality, these nations consider themselves often only a socialist system, having not yet reached the full communist ideal.

and Leon Trotsky. The Russian Revolution was followed by a bloody civil war where the Bolsheviks took control of the country—and while they were at it, also of some neighboring countries. During WW I, Russia implemented a system called *war communism*, which basically gave them a legitimization to steal anything needed to win the war. Unsurprisingly, economic outputs dropped, exchange of goods reverted to barter and black markets, the value of the ruble fell sharply, and about 6 million people died in the famine of 1921 (Davies, Harrison, and Wheatcroft 1994, p. 6ff).

While the revolutionaries won the war by 1922, the economy had collapsed. To rebuild the economy, the government started a *New Economic Policy* in 1921, giving limited freedom for private manufacturing and trade, lowering grain taxes, and overall setting the path for an economic recovery. By 1926, the economy was back to the same level as in 1913 (Wren 1980).

During the New Economic Policy, the wave of scientific management and Taylorism also reached the Soviet Union. Lenin promoted Taylorism already in 1914 and saw it as a tool to convert the Soviet Union from an agricultural society to an industrial society (Sochor 1981). Lenin even planned to write a book on Taylorism (Bailes 1977). One of the key researchers supported by Lenin was Aleksei Gastev, who founded the Central Institute of Labour in 1920. As the head of numerous research groups, he wanted to promote scientific management as the path to the industrial success of the Soviet Union.

Yet there were few successes. Pretty much the only tools of scientific management used in the Soviet Union were Gantt charts and functional foremen. The Soviet Union invited a U.S. expert on scientific management, Royal Keely, in 1919 to observe the industrial progress. Apparently, his findings did not impress him. Even more so, his findings did not please the big wigs of the Soviet Union. Hence, shortly before he was about to publish his report, he was arrested without charges and sentenced to two years of labor camp. Lucky for Keely, the United States was able to negotiate his release. Therefore, we know the state of the Russian economy in 1919: lack of food, lack of fuel, lack of materials, bad housing, low technical skills, ignorance of management, opposition to Taylorism by workers and management alike, and—as he experienced himself—censorship and oppression (Nelson 1992, p. 25). Another researcher, Russian-born but U.S.-trained Taylorist Walter Polakov, supported the Soviet Union between 1929 and 1931, although also with no results to speak of. However, since he grew up in Russia, he probably was culturally sensitive enough to know when to keep his mouth shut and was able to avoid prison.

Aleksei Gastev, the key driver of Taylorism, however, was not so lucky. After his supporter Lenin retired in 1922 and died in 1924, opponents of the capitalist system of Taylorism grew stronger. In 1938, Joseph Stalin emerged as the winner of the power struggle after Lenin's death. Stalin was a clear opponent of Taylorism. His modus operandi was also to execute all opponents. Aleksei Gastev was arrested in 1938 on the all-popular charge of being a *counterrevolutionary terrorist* and executed in 1939, along with most of his followers (Nelson 1992, p. 25).

Already in the year after Lenin's death, Stalin stopped the New Economic Policy. Instead, he collectivized private industry and agriculture and established a system of five-year plans. A five-year plan is a master plan of the economy, detailing which goods to produce and where to produce, setting prices and wages, and defining where to invest the resources. It replaced the invisible hand of Adam Smith with a centralized planned economy. However, planning the economy of a small town is already a daunting task, but planning the economy of a nation as large as the Soviet Union is impossible. Hence, most five-year plans were a mixture of wishful thinking, lack of understanding of economic relations, forecasts based on wrong assumptions, and communist propaganda. Throughout its existence, the Soviet Union—and for that matter, any communist or socialist economy—was plagued by constant shortages of consumer goods in demand and wasteful oversupply of other products no one wanted.

Nevertheless, it did have some success with large-scale projects, for example, the Soviet space program. Furthermore, it managed to transform the Soviet Union from an agricultural toward an industrial society, and industry output rose. However, the needs of the common people were constantly different from the five-year plan. Except for large-scale public projects, waste and mismanagement were widespread.

The goal of the economy changed from making a profit to looking good to the leaders of the communist party. Since politically, only quantity was important, quality was shoddy overall. Numbers were fudged everywhere and on every level of the hierarchy. The lowest level lied about its production to the next level, which fudged the numbers even more, and so on, until at the end, the data had no connection anymore to the economic reality. Any machine breakdown was often attributed to sabotage (Wren 1980). Corruption was widespread. Qualified scientists and managers often were imprisoned or sent to the Gulag and replaced by politically opportune *apparatchiks*. The mind of the economy was not on profit but on politics.

Production was also far from constant but fluctuated widely. At the beginning of the month, factories lay idle due to a lack of raw materials and absent workers. In the middle of the month, normal production started. However, to reach the monthly quota, it was necessary to rush during the end of the month, and workers often worked double shifts and more, including weekends. After the end of the month, workers rested again to recover from the rush. Even if they wanted to work, there often would have been no materials to work with. Hence, the industry constantly shifted from idle to 200% and back on a monthly cycle. This also overlapped with a yearly cycle, and often, all of December was an enormous rush, followed by a complete standstill in January. In some years, the police even forced workers to go to their workplace, although they were unable to actually force them to work (Don Van Atta 1986).

This mad cycle of extreme pressure followed by idleness was ripe with waste. To achieve the goals during the end-of-month rush, managers hired more people than they actually needed. Wages were inflated to get even more workers; necessary qualifications were ignored. Firing for lack of work was rare; the biggest risk for losing your job was having an opinion different from the official party guidelines. Similarly, additional machine capacity was needed to allow the rush, which then lay idle most of the rest of the month. Often, workers received bonus payments to participate in the rush. Factories lay idle 70% of the time due to a lack of materials, labor, and equipment (Wren 1980). It was a terrible waste of labor and resources.

In 1935, coal miner Alexey Stakhanov (Figure 14.3) made national headlines by mining 102 tons of coal in less than six hours, more than 14 times his quota. He then broke his own record a few weeks later with over 200 tons of coal. For this achievement, he was awarded the *Order of Lenin* and the *Order of the Red Banner*, and became a *Hero of Socialist Labour*. He also gave his name to the *Stakhanovite movement*, in which unbelievable feats of productivity of individual workers were achieved throughout the industries of the Soviet Union. And, unbelievable they were indeed. Most if not all of these achievements were staged. One popular method was for all the other workers of the plant or mine to prepare work and set everything up so the *hero* could then do the finishing touch to achieve his quota. Alternatively, the production of a longer period was recorded

FIGURE 14.3
Propaganda photo of Alexei Stakhanov (center) explaining his system to a fellow miner.

in one day. Or, probably simplest of all, the output of many workers was attributed to the hero to achieve publicity. Yet since much effort went into this dog-and-pony show, overall productivity suffered.

While mismanagement of the industry led to constant shortages of consumer goods, similar mismanagement of agriculture lead to serious shortages of food. During the Soviet famine of 1932–1933, approximately 6 million people died from starvation. Some people dropped dead in the street. The hungry ate dogs, cats, frogs, mice, and other vermin, and even the leaves of trees. People resorted to cannibalism, and parents even ate their children. The government suppressed news of the famine. At the end, 25% of the population in the affected region were dead. Of course, cadres of the party ate well throughout the entire time.

The communist mismanagement in the Soviet Union throughout the most of the twentieth century seriously slowed down economic growth. While in 1950 the Soviet Union had 70% of the gross domestic product (GDP) of Western Europe, this shrank to 50% by 1990. The economy was in shambles, and all attempts to fix the communist system failed. Finally, during the 13th five-year plan in 1991, the morbid construction broke apart, and the Soviet Union split into 15 different states. The communist system of the Soviet Union had failed. While that transition was mostly peaceful, the transition from a communist to a capitalist system showed different results. Some nations took the road to economic success, for example Estonia, Latvia, or Lithuania. Others are struggling. Russia itself turned into a *third-world kleptocracy*, with most important positions and properties taken by fat cats of the former regime (Cameron and Neal 2002, p. 396).

14.6 COMMUNIST CHINA—THE GREAT LEAP FORWARD INTO THE GREAT FAMINE

Events in China took a very similar path. After a long war, Mainland China, under the leadership of Mao Zedong (1893–1976), became the communist People's Republic of China in 1949. With the help of the Soviet Union, the economy was collectivized, and the first five-year plan started in 1953. The second five-year plan, starting in 1958, was also known as the Great Leap Forward. The goal was to catch up with industrialized nations, but the results were a total disaster. Throughout the long history of China, there were many disasters and catastrophes, but the Great Leap Forward outdid them all.

As with Stalin, one of the first actions was to prosecute intellectuals. Many scientists and engineers were killed or imprisoned. Private farmland was turned into public property, machines and tools were collectivized, animals were seized, and even household kitchens were destroyed in favor of communal kitchens. The idea of the communal kitchens was that everybody could eat as much as they wanted and would work on the public farms in return to restock supplies. Food was plenty. Villagers ate better than ever before in their lives, consuming large amounts of food, with pigs eating the leftovers. Unfortunately, they ate eight months' worth of food in five months, and then supplies ran out (Jisheng 2012, p. 189).

As for the public fields, why work if you would get food anyway? Work ethic was clearly lacking. Similarly, animals were not cared for. Since they belonged to everybody, it had the same effect as if they belonged to no one. Additionally, pseudoscientific regulations for farming were handed down through the political system, often based only on a small remark by Mao. One new *technique* was *deep plowing*, burying the seed grains a meter underneath the ground. Besides burying fertile topsoil and putting less fertile soil on top, the shoots never made it to the surface.

Another *technique* was *close planting*, setting the plants much closer to each other than usual. The planting positions had to be measured precisely using a ruler, so all plants lined up perfectly with exactly the same distance from each other. If the results did not please the cadres, the plants were pulled out and planted again. However, since they were planted much too tightly, these plants grew much slower due to a lack of light and nutrition. Similarly, the government decided what to plant where and when. Pumpkins were planned on wet rice fields. Rice was assigned to dry pumpkin patches, with little regard for the best time to plant the crops. Some fields received excessive fertilizer, burning the crop, while others were not fertilized at all. The harvest was very meager and often returned less grain than what was planted. In one village, following official orders, villagers planted 600 kilograms of peanuts, but the total harvest afterward was only 2 kilograms (Jisheng 2012, p. 274).

During the *Four Pests* campaign, millions of sparrows were killed as a grain-stealing pest. However, sparrows ate lots of insects. Due to the near elimination of sparrows, insects multiplied. Especially, locust swarms devoured entire fields and also deforested the countryside.

As in the Soviet Union, the five-year plan set targets in agricultural production, but numbers were fudged everywhere. Local cadres exaggerated the harvest reports, which were further exaggerated by midlevel politicians, and the top leaders had no clue what was really going on. Since it was easier to exaggerate individual results, reports of peak harvests achieved a staggering level, similar to the Stakhanovite movement in the Soviet Union. At the end, some claimed that they harvested so much grain that it would have covered the field ankle deep with rice, or a meter high with potatoes, or almost 2 meters high with sugarcane (Jisheng 2012). At the end, even Mao started to question the validity of the results. Until then, however, officials believed food to be plenty and set delivery quotas accordingly, with midlevel and low-level cadres pressing all the food out of the villages to achieve their quotas.

Similarly, cadres presented labor-saving devices as major successes. These tools were, for example, the Great Leap stoves, Great Leap vegetable washing and cutting machines, Great Leap millstones, and Great Leap noodle cutters. Unfortunately, none of them worked, but the publicity was great.

The government also undertook large-scale projects, especially in building irrigation canals or dams. The government forced tens of thousands of people to work on these projects, often in the middle of the planting or harvest season. The food rotted on the fields since nobody was available to harvest it. The dams and canals, on the other hand, were often ill planned. Dams broke, flooding villages downstream. Canals were destroyed by landslides due to faulty construction. Many canals were never finished.

Mismanagement was also ripe in industry. Mao was a fan of small, localized production. Probably best known are his backyard furnaces. The idea* was that small blast furnaces in many households would produce steel. In reality, due to a lack of education and training, most backyard furnaces merely converted useable metal tools into scrap iron. Often, even the pig iron was no longer useable, due to high sulfur content. Yet again, the government ordered tens of thousands of people off the fields to contribute in the steelmaking process. Whole forests were cut down to fuel the furnaces, resulting in erosion and landslides. In one case, they even filled a steep ravine with a mixture of coal and iron ore and set it on fire, hoping to obtain steel, when all they got were hot rocks with ash (Jisheng 2012, p. 201). Similar events happened with the production of electricity, cement, and other industry goods, while tens of thousands died in industrial accidents.

The resulting famine—now known as the Great Famine—was on an unparalleled scale. Cannibalism was widespread in most villages. Dead people were dug up from the grave for food. Parents ate their children. In one village, for example, a father killed his two children and ate them. After the man died from starvation, he was eaten by a neighbor. After the neighbor also died from starvation, he was eaten again by yet another neighbor. Parents instructed their children to eat them after they died. Other families deliberately starved one child to save the other, with children begging their parents not to eat them (Jisheng 2012, p. 143). Exact numbers are disputed, but anywhere between 20 and 45 million people died from starvation between 1958 and 1961,† with another 40 million not being born due to the women becoming infertile from malnutrition (Jisheng 2012). Naturally, cadres ate well throughout that period, and pretty much the only children born in the countryside were from the well-nourished wives or courtesans of government officials.

The Great Famine ended only in 1961, when individual households were allowed to farm for their own profit, allowed to have their own livestock,

* The idea was developed in one of the many meetings at the pool of the Mao residence. A lot of decisions were made at such pool parties.

† Naturally, the Chinese government states that *only* 15 million people died due to the famine, but in reality, the number is probably much larger.

and generally working according to their own plans. The laws of supply and demand worked quickly, and food was soon available again at affordable prices. Yet overall, the Great Leap ended with a crash landing. Besides the millions of deaths, the economy was reduced to two-thirds of its former size.

Currently, China is on a shift away from communism toward a market economy, while still keeping the one-party political communist government. Besides the improvements that ended the Great Famine, most changes occurred after Mao's death in 1978 (Beattie 2010, p. 267; Stearns 2007, p. 234ff). The government permitted free markets in special economic zones, starting with the Shenzhen Special Economic Zone in 1980. Many partially government-owned companies operate with little or no political influence, for example, different carmakers and electronic companies (Gan, Guo, and Xu 2011). Some cities—using a plethora of flowery words—completely privatized some of their companies from 1992 onward (Economist 2011). However, the official legal status of private firms is still unclear.

Nevertheless, the economy in China boomed, and many urban Chinese are now living a comfortable middle-class life. Some people even wonder if a state-managed market system would be more competitive than a pure capitalist system, but the GDP per capita is still only about $6000, whereas in the United States, it is almost $50,000. China still has a lot of catching up to do, while at the same time trying not to disintegrate into many smaller nations like the Soviet Union or Yugoslavia did.

While some die-hards still believe in communism, it can be safely said that the great experiment of communism failed. Most people are willing to share, but only with people they know. When sharing with an entire nation, individual greed is the stronger drive than goodwill to others, especially when following the bad examples set at the top. Most importantly, replacing Adam Smith's invisible hand with centralized planning leads to mismanagement, chaos, mayhem, and death.

Currently, there are only two nations on earth with a communist or socialist economy: poor Cuba, where a teacher earns the equivalent of $20 per month, and the nutcase-of-a-state North Korea. The latter also had a major famine due to mismanagement between 1994 and 1998, where probably 330,000 people died (Goodkind, West, and Johnson 2011). In any case, both existing communist nations are far from an economic success. Communism is dead. The few people left in the west promoting communism live a very comfortable capitalist life they probably would not want to give up.

BIBLIOGRAPHY

Bailes, K.E., 1977. Alexei Gastev and the Soviet Controversy over Taylorism, 1918–24. *Soviet Studies* 29, 373–394.

Beattie, A., 2010. *False Economy: A Surprising Economic History of the World*, 1st ed. Riverhead Trade, New York.

Broadberry, S., Harrison, M., 2005. The economics of World War I: A comparative quantitative analysis, in: Proceedings of the Annual Meeting of the Economic History Association. Presented at the Annual Meeting of the Economic History Association, Toronto, Canada. doi:10.1017/S0022050706000210.

Cameron, R., Neal, L., 2002. *A Concise Economic History of the World: From Paleolithic Times to the Present*, 4th ed. Oxford University Press, USA.

Carew, M.G., 2010. *Becoming the Arsenal: The American Industrial Mobilization for World War II, 1938–1942*. University Press of America, Lanham, MD.

Davies, R.W., Harrison, M., Wheatcroft, S.G., 1994. *The Economic Transformation of the Soviet Union, 1913–1945*. Cambridge University Press, Cambridge, UK.

Don Van Atta, 1986. Why Is There No Taylorism in the Soviet Union? *Comparative Politics* 18, 327–337. doi:10.2307/421614.

Economist, 2011. Privatisation in China: Capitalism Confined. *The Economist*.

Gan, J., Guo, Y., Xu, C., 2011. What Makes Privatization Work? Evidence from a Large-Scale Nationwide Survey of Chinese Firms1 (Working Paper). Hong Kong University of Science and Technology (HKUST).

Goodkind, D.M., West, L.A., Johnson, P., 2011. A Reassessment of Mortality in North Korea, 1993–2008. Population Division U.S. Census Bureau, Washington, D.C.

Harman, C., 2002. *A People's History of the World*, 2nd ed. Bookmarks Publications Ltd, London.

Huntzinger, J., 2007. The Roots of Lean: Training Within Industry: The Origin of Japanese Management and Kaizen, in: Proceedings of the TWI Summit. Presented at the TWI Summit.

Jisheng, Y., 2012. *Tombstone: The Great Chinese Famine, 1958–1962*. Farrar, Straus and Giroux, New York.

Kirby, M.W., 2003. *Operational Research in War and Peace: The British Experience from the 1930s to 1970*. Imperial College Press, Covent Garden, London.

Lozada, C., 2005. The Economics of World War I. *NBER Digest*.

Nelson, D., 1992. *Scientific Management in Retrospect, in: A Mental Revolution: Scientific Management since Taylor*. Ohio State University Press, Columbus, OH.

Noble, D.F., 1984. *Forces of Production: A Social History of Industrial Automation*. Alfred A. Knopf, New York.

Pollard, S., 1964. The Factory Village in the Industrial Revolution. *The English Historical Review* 79, 513–531.

Sochor, Z.A., 1981. Soviet Taylorism Revisited. *Soviet Studies* 33, 246–264.

Stearns, P.N., 2007. *The Industrial Revolution in World History*, 3rd ed. Westview Press, Boulder, CO.

Waring, S.P., 1994. *Taylorism Transformed: Scientific Management Theory Since 1945*. The University of North Carolina Press, Chapel Hill, NC.

Wren, D.A., 1980. Scientific Management in the U.S.S.R., with Particular Reference to the Contribution of Walter N. Polakov. *The Academy of Management Review* 5, 1–11.

Zeitlin, J., 1995. Flexibility and Mass Production at War: Aircraft Manufacture in Britain, the United States, and Germany, 1939-1945. *Technology and Culture* 36, 46–79. doi:10.2307/3106341.

15

**Click* Let-Me-Do-This-for-You *Clack*—Computers in Manufacturing*

If [...] the shuttle would weave and the plectrum touch the lyre without a hand to guide them, chief workmen would not want servants, nor masters slaves.

Aristotle (384–322 BCE)
Philosopher, in Politics, Book I

Mechanization improved manufacturing by replacing muscle power with water power, steam engines, or electricity. Yet, in almost all cases, manufacturing still needed a human brain to control the process. There were very few exceptions, for example, the Jacquard loom from 1801, where cards with holes designed the weaving pattern, or the Thomas Blanchard copying lathe from the 1820s, able to copy the shape of existing objects. Yet, the vast majority of products were shaped and assembled using human intelligence to guide the tools.

This changed with the invention and subsequent rise of electronic computers. In the first half of the twentieth century, science continued to build the body of knowledge on electricity. In the 1940s, the time was ripe for electrical computers. Like many other inventions for which the time was ripe, electric computers were developed independently in three different countries, although due to World War II, there was little sharing of this classified military technology.

The first electric computer was the Z3 built by Konrad Zuse in Berlin, Germany. The Z3 was based on electric relays and started calculating in 1941. Since Zuse was German, his country was at war with pretty much the rest of the world, and Zuse could neither share his invention nor gain from the knowledge of other researchers in the United States or Britain.

Britain also developed computers for military use, and their first Colossus computer was switched on in 1943. Since the Colossus helped to decipher encrypted German messages, the British government naturally kept this technology secret. To be on the safe side, they kept the computer secret until well in the 1970s, 30 years after the war. At that time, the first Apple computers easily outperformed all the Colossuses combined.

In the United States, at least three different computers were developed. The Atanasoff–Berry Computer started operating in 1942, followed by the Mark 1 in 1944 and the Eniac in 1946. Soon, many other computers followed all over the world. Computers in the 1940s used either relays or vacuum tubes as basic elements. These switches were rather large. Even the most basic computer would easily fill a large room with tubes or switches. This improved in 1947 when the transistor was invented, reducing size, weight, energy consumption, and cost while increasing speed and reliability. During the early 1970s, the first microprocessors appeared, integrating thousands of transistors in the space of a fingernail, roughly the same as the first Zuse had in a whole room, albeit the microprocessor was significantly faster. Modern microprocessors have 2 billion transistors or more.

Initially, computers were used to crunch numbers and return information about the calculations. However, it was not long before computers were used not only to give information but also to control other hardware. By the end of the 1940s, computers were used for military flight simulators or to control ship-mounted antiaircraft guns (Noble 1984, p. 106ff).

15.1 CONTINUOUS PROCESSING INDUSTRY

As we have seen in Section 13.1, continuous processes are much easier to automate than discontinuous processes. It is much easier to pump oil, wheat, or water through a factory than it is to transport engine blocks, yarn, or clocks. Hence, mechanization of continuous processes was much easier than of discontinuous processes. This also applies to computer control.

The first examples of computer-controlled chemical processes date back to the 1950s. The Texaco Fort Arthur Refinery was completely controlled by computers by 1959 (Noble 1984, p. 60ff). Computers controlled temperatures, pressures, speeds, and other processing variables. At first, a so-called open loop was common, where the operator adjusted the setting according to the computer, but soon, the operator was dropped out of the picture, and the computer controlled the system directly.

As with any new technology, there were still many problems to sort out. Programming errors led to processing faults and to lost products. To rectify these problems, additional alarms were added to inform the operators when a process went wrong. However, due to bugs, the system often had false alarms, and the overall number of alarms overwhelmed the operators, who then chose to ignore alarms (Noble 1984, p. 60ff).

Eventually, these problems were brought under control, and managers started to reduce the number of workers. Unions decided to fight back. During the 1960s, numerous strikes in the chemical industry aimed for job security. The longest of these strikes lasted for six months, where workers refused to work until their demands were met. To the surprise of the managers, they were able to keep the plants running at near capacity only with the help of the computer systems. They did not need any workers. The computers did the work for them. The unions were defeated and eventually had to cave in to keep at least some jobs (Noble 1984, p. 65).

Overall, computers significantly reduced the number of workers needed to run processing industries. Largely due to computers, the productivity of the continuous processing industry increased by 250% between 1947 and 1966 (Noble 1984, p. 63). Nowadays the chemical and petroleum industries are by far the manufacturing industries with the largest revenue per employee, with 10 times more revenue per employee than the next industries (U.S. Census Bureau 2007).

15.2 COMPUTER-CONTROLLED MACHINE TOOLS

Processing industries computerized rapidly due to the ease of handling the product. Discrete industries faced more difficulties making and handling complex part geometries. It would have been difficult to finance all the research needed to make computerized part manufacturing work, were it not for a big push by the U.S. military after World War II.

The problem of making complex geometries is to move the tool in the path of the geometry. Thomas Blanchard solved this for his Blanchard lathe using a mechanical link (see Section 9.4). Already in 1921, this mechanical link was replaced with electrical sensors and actuators. In the 1930s, a hydraulic link was used at the Cincinnati Milling Machine Company. By the end of World War II, tracer control machines sensing a shape and controlling a tool electrically were used in production (Noble 1984, p. 82).

The first to use computers in manufacturing was John Parsons, now considered the father of numerical control. After World War II, Parsons was making helicopter rotor blades for the U.S. military. One of these blades failed, and the pilot was killed. To improve the strength of the blades, Parsons proposed to use stamped metal parts rather than wooden parts inside of the rotor. Together with his employee Frank Stulen, they used an IBM computer in 1946 to calculate the stress on the rotor blade, resulting in a shape defined by 17 points. In a next step, they had the computer interpolate an additional 200 points between the 17 points. As for machining the shape with 200 points, Parsons resorted to a rather low-tech solution. One employee read the *x–y* coordinates from the computer, while the other moved the tool to these coordinates to drill a hole for the stamping tool (Olexa 2001). The shape was then smoothed with a file.

In effect, this was the very first prototype of a computer-controlled machine, soon to be known as numerical control (NC) machines. The computer calculated the locations, and the—in this case human—actuators simply moved the machine to these locations. Using human actuators, however, is a very time-consuming, boring, and slow process. This led Parsons naturally to the idea of automating the tool control, too. The technology of the 1950s, however, was still inadequate to solve the problem. While computers were able to calculate positions with high accuracy, the electric motors operating the tools were not accurate enough. The force of a motor is nonlinear, making it difficult to move a motor with precision. More troublesome, a tool cutting metal is pushed in the other direction by the part. Overall, the precision of a motor-driven tool was rather lousy, and Parsons did not have the funding to develop a solution—yet.

In 1949, the U.S. Air Force asked Parsons for help in making jet parts and provided funding for the further development of numerical control. He then approached one of the top knowledge institutions in the United States for help, the Massachusetts Institute of Technology (MIT). Little did he know that very soon, MIT would aggressively push him out of his own invention in order to claim it as their own. But at the beginning, everything was fine. Parsons, the air force, and MIT agreed to develop two numerically controlled *Card-a-matic Milling Machines*, to be given to Parsons after completion.

MIT completed the NC machines in 1952, and that's when the trouble for Parsons started. MIT did not deliver them to Parsons as agreed. Rather, they wanted to present it themselves to the Air Force. Parsons was not even invited to this presentation and had to approach the air force to be invited to the presentation of his own product. At this presentation, he felt clearly that he was *about as welcome around here as a bride's mother on a honeymoon* (Olexa 2001). During the presentation, he was told that this technology included nothing worth patenting. He strongly disagreed and patented his inventions, receiving his patent just in time before MIT itself filed its patent application for these inventions, naturally claiming that these were invented by MIT (Olexa 2001). In any case, MIT, especially Professors Gordon S. Brown and Jay Wright Forrester, was now taking all the credit and had forced Parsons out of his own project. Parsons could not even get advertising material on the ongoing MIT developments (Noble 1984, p. 106ff). He even was fired from his own company after financial difficulties.*

MIT presented the first two-axis NC milling machine for industry in 1953 (Noble 1984, p. 134). The first commercial production of NC machines was by Bendix Corp in 1954 and 1955, under license of Parsons. Funded by the Air Force, MIT developed a five-axis numerical controlled machine in 1955, more axes than a human could possibly control at the same time. Now, the Air Force was finally convinced that these products were ready for industry use and could replace the tracer control machines used so far. The resulting order of 105 NC machines for the Air Force kick-started an industry of NC machine toolmakers. Soon, four different companies in America were producing NC tools for the air force (Noble 1984, p. 201ff). The first machines were installed in 1958, including combinations of multiple machines. Some even had an automatic tool change, where the machine changed tools itself as programmed, hence being the first of many machining centers.

As it turned out, machining in a laboratory at MIT and machining on a real shop floor are quite different. Rather than a clean workplace, shop floors are dirty, hot, and often vibrating from the action of the nearby

---

* Different from other inventors throughout history pushed aside by more ambitious copycats, Parsons did receive royalties for his patents—even though the big money was made somewhere else. His contribution was eventually recognized in 1985 with the National Medal of Technology, and he was completely rehabilitated in 1993 when he was added to the National Inventors Hall of Fame for inventing numerical control.

machines.* The workspace usually included lots of metal chips, chemicals, oil, and water. All of these were very incompatible with the sensitive electronics of the 1960s. Additionally, machines at MIT were in the care of highly trained PhDs, whereas shop-floor maintenance was more acquainted with sledge hammers than electronic devices. Finally, there was a lack of programmers to program these machines, resulting in training by trial and error. And, oh yes, there were errors. Lots of things went wrong. Machines broke frequently due to the environment; lack of maintenance slowed down repairs; and when machines worked, they often produced scrap if the tool moved a different path than planned. In this case, you were lucky if the machines only trashed the part, but they also often broke the cutting tool, and in the worst case, the machines cut into themselves, creating a $100,000 piece of steel junk. Overall, in 1959, these machines ran only 20% of the time, standing 80% of the time due to defects and other problems. Naturally, the companies that were the first to buy NC machines bore the brunt of the problems of the new technology, and many of them went bankrupt (Noble 1984, p. 220ff).

Nevertheless, over time, these problems were sorted out, machines became more robust, maintenance learned how to handle electronics, and programmers collected experience. Running an NC machine was nevertheless cumbersome. A programmer set up the computer for the shape of the part. The computer then calculated the path of the tool and stored the data either on punched paper tape or magnetic tape. This tape was then loaded into the machine, where the machine finally produced the parts. The next logical step was to connect the computer to the machine directly. Rather than carrying punched or magnetic tape around, a cable connected the computer to the machine, and the program was uploaded from the computer to the machine. This improvement happened during the end of the 1960s, resulting in direct numeric control (DNC) machines. Finally, the computer was no longer a separate box somewhere else but was integrated as part of the machine, resulting in computer numeric control (CNC) machines, albeit nowadays, the terms NC and CNC are used interchangeably for all types of numerical control.

However, it is doubtful if NC machines really were worth their money or just an expensive gadget. In fact, at least until the 1980s, NC machines

* I once worked in the offices above a workshop with multiple 1000-ton machine presses cranking out car parts. It is not pleasant if your coffee cup constantly has waves inside. On the plus side, you notice any problem downstairs right away if the pattern of the vibrations changes.

were probably less cost-effective than traditional manufacturing (Noble 1984, p. 341). While they reduced direct labor cost, they increased indirect labor cost with maintenance and programming. Additionally, NC machines were not only more expensive than hand-operated machines; they were also much more quickly outdated. Before 1950, the average machine in the United States was more than 10 years old, with 15% being more than 20 years old. In comparison, a 20-year-old NC machine nowadays would have computers running the disk operating system (DOS), or if you are lucky, Windows 3.1, with 64 kilobytes of memory. It would be difficult to find someone to program it nowadays and even more difficult to find spare parts. NC machines need replacement due to obsolescence much more often than traditional machines.

Yet, NC machines spread throughout industry, despite cost disadvantages compared to the traditional approach using skilled labor. The reason was—like so often in history—distrust by the management toward labor. Even though production using NC machines was more expensive, it took control away from labor and gave it to management. The capital investment was to replace labor (Piore 1968). NC machines were advertised falsely as requiring no employees. As the assembly line and interchangeable parts replaced skilled fitters with unskilled labor, NC machines replaced skilled (and therefore headstrong) machinists with unskilled operators for loading and unloading parts. To most managers, the benefit of controlling labor outweighed the risk of NC machines.

However, as before in history, controlling labor remained elusive. While there was more control for management, the operators at the machines—while being unskilled—were not dumb. For example, some machines allowed the operator to set the speed of the NC machine, and soon, all machines were set to 80% speed. Management in turn then programmed a higher speed to begin with, disconnected the knob, or just changed the scale, so a speed of 100% was displayed as 80%. The operators found out soon enough and slowed down machines in a different way. Stops due to broken circuit boards or metal chips in electric components increased, and the overall speed was reduced again (Noble 1984, p. 262f). Even nowadays, machining needs unskilled labor, and while their official bargaining position is weaker, at the end, it is the people at the machines who control production.

During the 1950s, there was even a technical attempt at a compromise. Rather than computers controlling the machines, the actions of a skilled operator were recorded and then played back on different machines without operators. Such record-playback machines (RC machines) were proposed and built by different inventors. The concept, however, was flawed. Making a good recording for a part was difficult, and operators were unable to control five-axis machines. However, the real failure was in the minds of people. Unions opposed the RC machine since they saw it as part of the trend toward NC, and managers opposed it since they did not want to give an inch to labor (Noble 1984, p. 93f). It remained an interesting but naive experiment.

The popularity of NC machines increased, although initially slower than expected. In 1963, only 0.01% of all machines were NC machines, and in 1973, still only 1% (Noble 1984, p. 213). Since then, the use of NC machines has grown significantly as reliability improved and prices decreased. In 1987, over 13% of all machines in the United States were NC machines (U.S. Congress Office of Technology Assessment 1990, p. 154). Nowadays, most machine tools sold are CNC machines (Heinrich 2001).

While NC technology was developed in the United States with funding from the U.S. Air Force, most NC machines are now produced in Japan and Germany. NC machines in the United States were geared toward the U.S. military, having all the bells and whistles the government wanted and could afford. For industry, however, these machines were usually overengineered and hence overly complicated and less reliable. This also caused a higher price than needed. Germany, and later, Japan, focused more on civilian industry, producing simpler, cheaper, more robust, and easier-to-use machines (Noble 1984, p. 222). By 1977, 30% of all machines produced were made in Germany. In the 1980s, Japan overtook Germany with even cheaper and easier-to-use machines. A Japanese NC machine could be bought for less than half the price of a comparable U.S. machine (Heinrich 2001).

Overall, the NC machine changed manufacturing profoundly. Almost in parallel, a slightly different invention using computers changed the assembly process—industrial robots. As with NC machines, the technology was born in America, but the technology leaders are now in Japan and Germany.

15.3 THE HELPING HAND—INDUSTRIAL ROBOTS

The word *robot* in the Czech language means *serf labor.* Czech writer Karel Čapek used the term for a science fiction story in 1920, with *robots* being artificially created biological servants. Writer Isaac Asimov picked up the term around 1940 and used it for mechanical assistants. Since then, reality has caught up with fiction, and numerous humanoid robots have been built. However, the technology of humanoid robots is still far from mature.

Industrial robots, on the other hand, are mostly not humanoid but basic manipulators of tools or products. They are nowadays almost omnipresent in advanced industries. With the progress of computer technology in the 1950s, it was an obvious idea to use computers for controlling manipulators. The first to act upon this idea was George Devol from Kentucky, applying for a patent for a *programmed article transfer* in 1954 and receiving his patent in 1961. He devised a computer-controlled hydraulic arm to transfer parts from one location to another.

Having an idea is the first step, but getting from the idea to a marketable product is quite another task. Most of all, Devol needed money for development, and he contacted numerous firms. When he finally was about to receive funding from one firm, this firm was bought by another company that saw no profit in robotics, and the funding did not materialize. However, Devol's main contact at that firm, chief engineer Joseph Engelberger, continued to believe in the idea. Together, they secured funding for their startup, Unimation, established in 1956.

The robot they developed looked very different from modern industrial robots. It did not have joints like a modern robot or look loosely like a human arm. Rather, it had a fixed arm, which could rotate and move up and down as well as slide in and out. In effect, it had more similarities with a crane than with modern robots. The programming was also not done by a computer, but the movement instructions were stored on a magnetic drum. Yet, the robot, Unimate, did what it was supposed to do—move parts from A to B (Figure 15.1).

Solving hundreds of problems, they secured their first sale in 1960, delivering the first Unimate industrial robot to a General Motors factory in Ternstedt, New Jersey. There, Unimate removed hot castings, quenched them, and passed them on to the next machine. The robot worked well, doing its job for the equivalent of 50 man-years before retiring to a museum (Engelberger 2000). Since the robot was clearly advantageous for dangerous and demanding transfers, other companies started to order, and sales increased. Nevertheless, the company made its first profit only in 1975, 19 years after its founding (Feder 1982). Strapped for cash for almost 20 years, Unimation sold licenses for its Unimate robot in

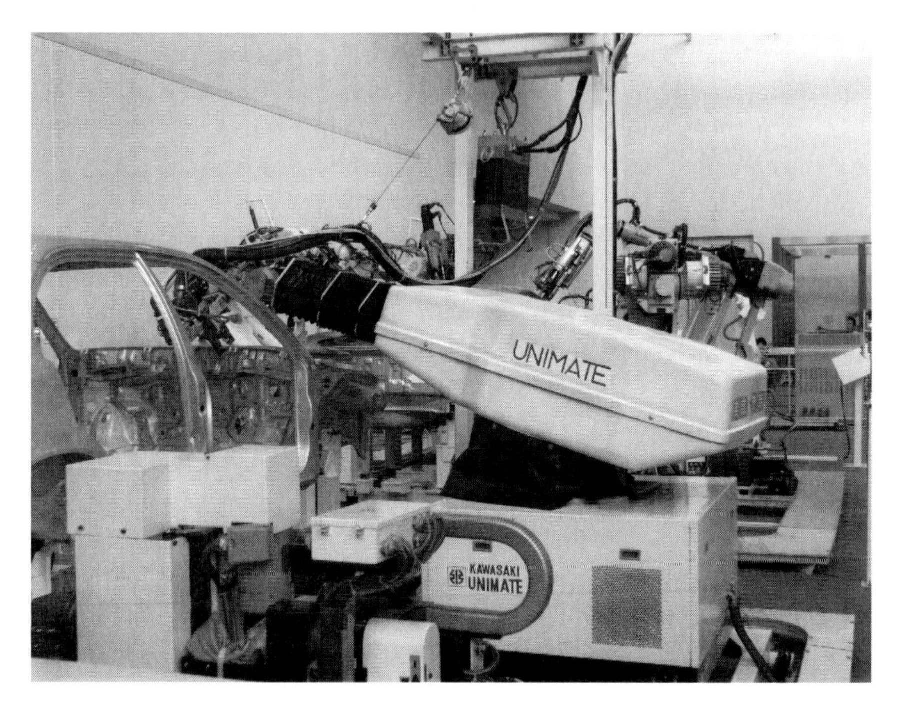

FIGURE 15.1
Unimate robot produced under license by Kawasaki, on display at the Toyota Commemorative Museum of Industry and Technology in Nagoya. (Photo by author, dated August 2012.)

1968 to companies in Japan and England, while continuing to manufacture robots in the United States.

Other companies and universities joined the concept and developed their own robots. The next important step in robotics was made—of all places—in a hospital. The Rancho Los Amigos Hospital in California developed an artificial limb to assist disabled people in 1963. This *Rancho Arm* looked much more like a human arm than Unimate, and it was the first robot controlled directly by a computer. Not to be outdone by a hospital, MIT also developed a robotic arm in 1968, able to lift the weight of a person. The robot had the—in my view—most unfortunate name Tentacle Arm. This was followed in 1969 by the Stanford Arm at Stanford University, the first electronically controlled arm used for assembly.

In 1973, German toolmaker KUKA introduced Famulus, the first industrial robot with six axes (Figure 15.2). This articulated robot is nowadays the most commonly used kinematic for industrial robots. The latest major step happened in 1981, when researchers at Carnegie Mellon developed

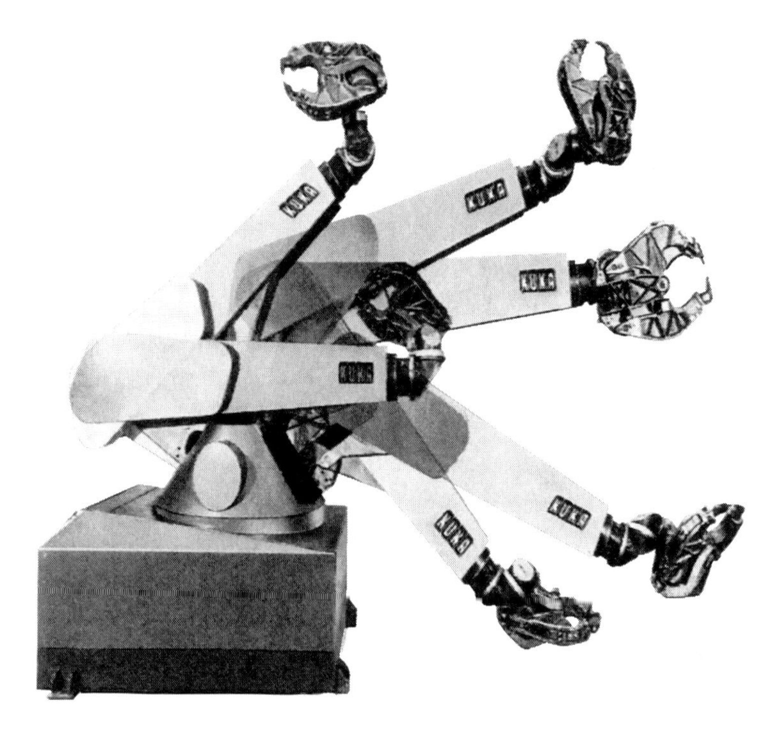

FIGURE 15.2
The first articulated robot with six axes, the KUKA Famulus from 1973. (Copyright KUKA Roboter GmbH. With permission.)

the Direct Drive Arm. In this robot, they moved the motors directly into the joints, rather than having many drive shafts going through the arm. This six-axis industrial robot is nowadays almost omnipresent in modern industry.

However, the articulated industrial robot is not the only kinematic used in industry. Another commonly used industrial robot is the SCARA* robot, developed in Japan in 1981. Having only four axes, the robot is ideally suited to pick parts up and place them elsewhere. Finally, the Delta robot with three arms forming a usually hanging tripod was developed at the École Polytechnique Fédérale de Lausanne in Switzerland in the 1980s. Also used for pick-and-place operations, the actuators are all in the base of the robot, controlling one of the three struts. Hence, the arms themselves are very lightweight, and a Delta robot is therefore usually significantly faster and cheaper than a SCARA robot.

* SCARA stands for Selective Compliance Assembly Robot Arm.

In the early years, industrial robots were a rarity. They were usually used only where the work was too dangerous or demanding for human workers. However, the use of industrial robots took off during the 1980s, and sales started to increase. This was in no small part thanks to GM under CEO Roger Smith and his vision of robotic factories. Unfortunately, his vision had a hard crash landing with reality, nearly bankrupting his company. GM, like many companies before, had its fair share of labor trouble. At the same time, the automotive industry in the United States was under pressure from the Japanese with their new Lean production system (see Chapter 16). Roger Smith wanted to solve two problems: getting rid of workers who had the audacity to have their own opinion while at the same time improving productivity to compete with Japanese carmakers. He wanted to achieve this through the widespread use of robotics in his plants. He wanted to turn his factories into lights-out factories. The only people to be left in his factory were those supervising robots and computers.

Smith started to invest heavily in robotics. As the CEO of GM, he also had the financial means to really splash out. Throughout the first half of the 1980s, Smith invested the eye-popping sum of $45 billion into robotics to make his dream come true. The United States at that time was in a trade war with Japan, and his vision of robotics was well received. The press and public opinion saw him as one of the saviors of the U.S. economy. In 1984, he was voted to be the Automotive Industries Man of the Year and Advertising Age's Ad Man of the Year, and awarded as the Best CEO in America (Finkelstein 2004).

Yet troubles started to roll in. For one thing, the unions were unsurprisingly very much against a worker-free factory. Labor relations took a dive, strikes increased, and morale suffered (Finkelstein 2003). But, even worse, as labor relations went down, costs went up. The more money GM spent on robotics, the more costs rose rather than sank. There were not enough skilled engineers and technicians to install, program, and maintain all these robots. Those that were available had much higher wages than the workers they were supposed to replace through robotics. Problems with robots multiplied like wildfire. Most robots did not work reliably, and utilization rates in factories dropped. Some plants had a utilization of only 50%, and productivity declined. Output in some plants dropped to half of its former nonautomated levels.

Even if the robots were working, problems continued. Painting robots painted themselves rather than the car, welding robots welded the doors shut, and defects and scrap rose to astonishing proportions. Some robots were scrapped shortly after installation. As one GM senior executive put it, with robotics, *all you do is automate confusion* (Finkelstein 2003).

Nevertheless, Roger Smith stayed true to his vision. Even after spending $6 billion in 1983, $9 billion in 1984, and $10 billion in 1985 without improvements, he continued to invest (Finkelstein 2003). Of course, there were managers critical of his vision, but only a few dared to raise their voices. Throughout his tenure, Roger Smith cultivated his circle of yes-men. Managers with inconvenient opinions were fired or sent to remote locations, ending their future career at GM (Finkelstein 2004). Unsurprisingly, Roger Smith was told only what he wanted to hear.

Despite lots of upbeat reports from GM managers, reality begged to differ, and the downfall continued. Market share dropped from an excellent 48% to 36% and continued to fall, along with the profits. GM made 11.7 cars per employee per year, while at the same time, Ford made 16.1 and Toyota made a whopping 57.7 cars* per employee per year (Finkelstein 2004).

In 1990, GM finally pulled the plug and terminated the tenure of Roger Smith. The investment of $45 billion into robotics was a loss. For this money, GM could have bought both Toyota and Nissan, which would have doubled their market share. Instead, GM lost big time, eventually filing for Chapter 11 bankruptcy in 2009. After the disaster, the press called CEO Roger Smith one of the worst American CEOs of all time. Fickle is the public opinion, and Roger Smith took only a few years to go from being the best CEO of America to the worst CEO of America.

Nevertheless, while the $45 billion did not help GM, it helped the robotics industry. GM's failed vision was also a big learning-by-doing program for the robotics industry. From the 1980s onward, industrial robots became more common, although most firms wisely avoided a full-out robotization.

Another attempt at increased robotization was, for example, *Halle 54* (Hall number 54) at Volkswagen in Wolfsburg. Starting in 1983, this assembly hall was designed for maximal automation, and was probably the most automated line in the world at its time (Jürgens, Malsch, and Dohse 1994, p. 71). However, this approach also lacked flexibility and robustness and overall was, at best, a mixed success (Kropik 2009, p. 26).

* However, Toyota has a much higher percentage of outsourced parts; hence, the number is somewhat distorted.

In any case, the robotics industry is now mainly concentrated in Germany and Japan. Currently, there about 1.4 million industrial robots in use worldwide. Sales increase almost 15% per year (IFR Statistical Department 2015), and robotics will continue to grow throughout the industry.

Even the idea of the lights-out factory is not far off. There are already a few factories where robots assemble products with no human interaction. Probably the most prominent example is the Japanese robot manufacturer FANUC with its plants at the FANUC headquarters at the base of Mount Fuji. At this plant, robots assemble other robots around the clock, for up to 30 days without human interaction. Workers only come in to haul the finished goods away. Other than that, the lights are turned off, as are the heating and air-conditioning (Null and Caulfield 2003).

The effort in establishing a lights-out factory is still immense, and it is telling that a company manufacturing robots was the first to achieve this. However, as computers get smarter, more and more high-tech factories will run autonomously, and the use of industrial robots will increase even more.

15.4 COMPUTERIZED PRODUCTION PLANNING

Scientific management aims to understand manufacturing and management by the numbers, and this soon extended to material supply. Already in 1913, the first paper was published on economic order quantities to determine the ideal amount of stock (Ford 1913). The body of research on this topic grew quickly. The goal was to reduce inventories and improve availability of material. Inventory is dead money. The material has usually already been paid for but not yet sold to the customer, and hence, it is just money sitting around. Reducing inventory frees up cash for other investments. One problem was the complexity of modern products. A car, for example, can easily have 10,000 different components, and within a plant, hundreds of thousands of different components may have to be tracked. Computers are ideally suited to store and process large data sets, and it is to no surprise that computers were used for inventory control already in the 1950s.

In 1964, Joseph Orlicky developed a computer system that not only kept track of inventory but also helped with planning production and orders for new materials. This was soon followed by similar systems by competitors. The computer-based approach became known as material requirements planning, usually abbreviated as MRP. The system took the expected orders; calculated through the bill of materials, available inventories, and production times; and ordered raw materials accordingly. The system suggested a production schedule to have the goods ready just when the customer needs them while reducing costly inventories at the same time. Since a computer can calculate all these numbers with great accuracy, the software appealed to management. These computer systems run with the precision and accuracy of a Swiss clock.

Unfortunately, the real shop floor is anything but a Swiss clock. In many cases, switching to MRP caused widespread chaos. MRP calculated when to produce what in order to satisfy the expected customer demand. However, this precise calculation is usually thrown off balance due to unforeseen events like late shipments, wrong shipments, defective parts, discrepancies between real inventory and MRP data, errors in the bill of materials, missing parts, defective machines, absent workers, order changes, and many more. In sum, reality differed significantly from what MRP expected, and chaos ensued.

While it is unlikely that all of these problems pop up in the same product, with hundreds or thousands of different products made each day, these problems are the daily bane of every shop-floor manager. To be fair, part of the problem is also due to the increasing complexity of modern products. For example, a shoemaker before the industrial revolution had comparatively few materials on hand: a few types of leather, yarn, nails, and some wood. If one was missing, he simply adjusted a different material. This is no longer possible with complex modern products using interchangeable parts, and MRP tries but often fails to handle this complexity.

Due to the many problems with MRP, software vendors added new features. During the 1980s, they developed a new, improved system able to also calculate machine capacity. Additionally, the system tracks financial data. This system is known as manufacturing resource planning. Since this also abbreviates to MRP, it is usually called MRP II to avoid confusion with MRP I. Unfortunately, the situation on the shop floor has not improved much, and many companies nowadays using MRP or related systems still do not calculate capacity constraints.

After MRP I and MRP II, a new system was introduced during the early 1990s, business resource planning, abbreviated BRP, although it is not quite clear what the difference was. In any case, the situation on the shop floor did not improve, and BRP was a short-lived *burp*, pun intended.

The next step during the late 90s integrated all functions of an enterprise into one, including sales, product design, and human resources to name just a few. Hence, it is named enterprise resource planning, abbreviated as ERP. While the product eventually delivered most of its promises in other areas, the confusion on the shop floor remained unchanged.

This was followed by yet another system during the early 2000s. This system is now web based and called ERP II. Unfortunately, the web-based approach did not change things on the shop floor. The latest stage of the evolution is called enterprise resource management, ERM, and is even more integrated and better than ERP. Yet, the situation on the shop floor remained unchanged. And I expect that cloud computing, which is currently advertised for such applications, will not improve the shop floor either.

The inflow of new acronyms with often marginal differences led to confusion in the nomenclature, and the whole shebang is often simply called MRP. Alternatively, the system is also often named after the vendor. The German vendor Systemanalyse und Programmentwicklung (for System Analysis and Program Development), or SAP, is by far the largest software vendor for ERP systems. Sales of MRP systems started to increase significantly during the 1970s and have been growing ever since. Currently, the worldwide market for MRP systems is almost $50 billion per year (Martens and Hamerman 2011).

Yet overall, the product failed to deliver its promises. One of the goals of MRP was to reduce inventory, or to increase turnover. Turnover in the United States during the 1940s was around seven, meaning that about seven times per year on average, all inventory was sold and replaced with new material. In other words, about one-seventh of your annual sales were sitting around on the shelves tying up cash. MRP wanted to improve this number and increase the turnover. However, there is no sign of any increase in turnover despite widespread use of MRP. Turnover increased only when the just-in-time approach from Japan started to be applied in the 1990s, but this was independent of MRP (Hopp and Spearman 2001, p. 173).

On the other hand, production planning also faced a continuously increasing complexity. For example, the Ford Model T had less than 100 different part numbers. However, in a modern car, you would barely be able to build a steering wheel with only 100 part numbers, and for the entire car, there are easily 30,000 part numbers* or more. In my view,

* This includes the part numbers that go into subassemblies.

while MRP did not deliver the promised salvation in manufacturing planning, it prevented it from getting worse.

Implementation is also very costly. A useful MRP system can cost easily $80,000—per user. This does not yet include the personnel needed to train users, maintain the software, and program custom tools. Different surveys showed that less than 10% of the companies installing MRP got a return of investment within 10 years. In fact, the average return of investment for MRP systems is negative $1.5 million, meaning that for every MRP system installed, the company loses, on average, $1.5 million. Ninety percent of MRP users are unhappy with the performance of their software. Some companies even pulled the plug after investing up to $250 million into MRP (Hopp and Spearman 2001, p. 173ff).

Examples of large MRP failures are numerous. Hershey missed the crucial Halloween sale in 1999 due to problems with a new $112 million MRP system. Whirlpool lost Home Depot as a $650 million customer after a botched MRP implementation in 1999. Nike invested $400 million into a new MRP system in 2000, only to lose $100 million in sales after problems, with their share price dropping by 20%. Hewlett Packard had similar problems in 2004, and despite strong contingency planning to cover unforeseen problems, reality was much worse. Hewlett Packard could not deliver and lost about $160 million due to the software change. Cadbury Schweppes had the opposite problem: MRP inflated their UK inventory in 2006, and they had to sell these perishable goods at a discount, with a total loss of £12 million. These are only a few examples, with more popping up regularly. All these aforementioned losses do not include additional substantial legal costs that followed most of these problems.

So why do companies still use MRP? There are a number of different reasons. For one, the idea of a computer taking care of all your problems is still very enticing to managers, with the vendors naturally advertising the successful implementations of their software. While MRP is not solving all problems, it can also be argued that MRP prevented them from becoming worse with the increasing complexity of modern products. Secondly, MRP works reasonably well for less frequently changing data, for example, in human resources or cost accounting. Additionally, if the customer uses, for example, SAP, the supplier has a direct benefit in order management if they use the same compatible system. In addition, many modern managers do not know anything else but an ERP system for production management. Finally and probably most significantly, implementation of MRP is usually a one-way road. Once MRP is part of the business system of the company, it is incredibly hard to go back. There are very few examples of production lines where MRP was replaced by paper-based kanbans (see Chapter 16). Once you use MRP, you are stuck with it.

Overall, the MRP systems used today for production are still based on the methods by Joseph Orlicky in 1964, assuming a perfect world. It is a folly to believe that merely dumping this complexity onto a computer will solve these problems; it will just make them harder to fix. To address the problems with control of a shop floor, it is necessary to make the system robust against problems. There is such a system; however, it was created not in the United States but in Japan: the Toyota Production System.

BIBLIOGRAPHY

Engelberger, J., 2000. Sounds Like a Robot to Me. *NZZ Folio.*

Feder, B.J., 1982. He Brought the Robot to Life. *NY Times.*

Finkelstein, S., 2003. GM and the Great Automation Solution. *Business Strategy Review* 14, 18–24.

Finkelstein, S., 2004. Case Study: GM and the Great Automation Solution, in: *Why Smart Executives Fail: And What You Can Learn from Their Mistakes.* Portfolio Trade, New York.

Ford, H.W., 1913. How Many Parts to Make at Once. Factory, *The Magazine of Management* 10, 135–136.

Heinrich, A., 2001. The Recent History of the Machine Tool Industry and the Effects of Technological Change. University of Munich, Institute for Innovation Research and Technology Management.

Hopp, W., Spearman, M.L., 2001. *Factory Physics*, 2nd ed. McGraw Hill Higher Education, New York.

IFR Statistical Department, 2015. Industrial Robot Statistics—World Robotics 2015 Industrial Robots. IFR International Federation of Robotics. Available at http://www.ifr.org/industrial-robots/statistics/.

Jürgens, U., Malsch, T., Dohse, K., 1994. *Breaking from Taylorism: Changing Forms of Work in the Automobile Industry*, Auflage: New. Cambridge University Press, Cambridge, UK; New York.

Kropik, M., 2009. *Produktionsleitsysteme in der Automobilfertigung.* Springer Science & Business Media, Dordrecht, Netherlands.

Martens, C., Hamerman, P.D., 2011. *The State of ERP in 2011: Customers Have More Options in Spite of Market Consolidation.* Forrester, Cambridge, MA.

Noble, D.F., 1984. *Forces of Production: A Social History of Industrial Automation.* Transaction Publ., New York.

Null, C., Caulfield, B., 2003. Fade to Black the 1980s Vision of "Lights-Out" Manufacturing, Where Robots Do All the Work, Is a Dream No More. *CNN Money.*

Olexa, R., 2001. The Father of the Second Industrial Revolution. *Manufacturing Engineering* 127, Issue 2.

Piore, M.J., 1968. The Impact of the Labor Market Upon the Design and Selection of Productive Techniques Within the Manufacturing Plant. *The Quarterly Journal of Economics* 82, 602. doi:10.2307/1879602.

U.S. Census Bureau, 2007. *2007 Economic Census*. United States Census Bureau, Suitland, MD.

U.S. Congress Office of Technology Assessment, 1990. Making Things Better: Competing in Manufacturing (No. OTA-ITE-443). Congress of the United States—Office of Technology Assessment, Washington, DC.

16

The Toyota Production System and Lean Manufacturing

To produce only what is needed, when it is needed and in the amount needed.

Taiichi Ohno (1912–1990)
Father of the Toyota Production System

The next step in the evolution of manufacturing comes from Japan: the Toyota Production System. Japan had its first contact with the west in 1543 through Portuguese traders. They also introduced firearms, which soon became a decisive weapon in Japanese military conflicts. Japanese gunsmiths quickly surpassed western makers in gun technology and quality (Perrin 1995). However, the Portuguese brought not only goods and technology; they also introduced Christianity to Japan. The new religion was popular not only among the commoners but also among aristocracy.

Eventually, the ruling daimyo saw the Christian activities as a threat to his power and enacted a number of countermeasures. Christianity was banned in 1587, and after the Shimabara Rebellion of 1637/1638, almost 40,000 rebels were executed. Christianity went underground. To also prevent the introduction of other religions, Japan entered a state of isolation. No foreigner was allowed to enter Japan, and no Japanese was allowed to leave on penalty of death. Trade was severely limited to four ports, one for Korea, one to trade with the Ainu people of Hokkaido, one for the Ryūkyū Kingdom (modern-day Okinawa), and one in Nagasaki for the Dutch—the Portuguese being no longer welcome. Only samurai were allowed to carry weapons, and the manufacture and use of firearms fell out of fashion

(Diamond 1997, p. 257). Japan was almost completely isolated, with no further development of its medieval technology.

For the next 200 years, numerous western ships tried to trade with Japan. The Japanese rejected their requests, sometimes politely, sometimes by force. Europe, however, continued to invent and advance. In 1853, Commodore Perry and his four U.S. Navy ships sailed into Tokyo harbor, threatening to open fire with their latest Paixhans gun, the first gun capable of firing explosive shells. In the face of this military superiority—and after a year of consideration—the Japanese agreed to all of Perry's demands, ending a 200-year period of isolation.

People have a tendency to see their own nation as better than any other. Hence, for the Japanese, it was quite a shock to see how the rest of the world advanced technologically during their self-chosen isolation. The forced opening of Japan led to a period of turmoil. Unemployment rose, and frictions between Japanese and foreigners increased, resulting in killings of foreigners and subsequent naval bombardments of Japanese cities. Japan was also hit financially, as the Japanese currency was based on gold and silver coins, with gold being about five times more valuable than silver. Internationally, however, gold was 15 times more valuable than silver, and lots of foreigners made a killing by exchanging silver for gold in Japan. Within a short time, over 60 tons of gold left Japan roughly at half price (Metzler 2006, p. 16f). Combined with a couple of devastating earthquakes, this turmoil toppled the political system. The shogunate ruling over puppet emperors for centuries ended in 1868, and the political and military power returned to the newly crowned 15-year-old Emperor Meiji.

What follows is one of the most remarkable catch-ups in history. In less than 40 years, the Meiji restoration turned Japan from a medieval society to a modern industrialized society. Political reforms took power away from local lords, and samurai were no longer allowed to carry weapons. The government undertook immense efforts to advance technology and sent its scholars abroad to study French administration, the Prussian army and police, the British Navy, and U.S. banking (Cameron and Neal 2002, p. 265ff). Scores of experts were invited to come to Japan and teach engineering, medicine, agriculture, law, economics, military organization, science, and many other topics. At one point, the government spent about one-third of its budget on foreign experts.

These efforts had a dramatic effect and fast-tracked Japan through the industrial revolution. Output of iron and coal multiplied. Steam engines were introduced to Japan. The merchant fleet soon included hundreds of steam ships, and the country built thousands of kilometers of railway tracks. Compulsory education started, and Tokyo University was founded. Public health and life expectancy increased. Within 100 years, Japan accelerated from a medieval society to one of the top three economies in the world. Its military was able to present a serious challenge to the rest of the world, being defeated only by the United States during World War II.

16.1 THE FOUNDING OF TOYOTA

It was in this environment of technological and industrial growth that a young carpenter's son, Sakichi Toyoda (1867–1930), grew up and eventually became the founder of Toyota Industries.* Sakichi Toyoda (Figure 16.1) is known in Japan as the king of inventors, racking up a total of 85 patents, although the latter ones had significant input from his children and relatives. He was most active in the field of looms. His first invention was an improvement of a wooden handloom in 1890, where the shuttle was operated with only one hand, doubling efficiency.

Sakichi Toyoda achieved a major breakthrough in 1896 with a loom that stopped automatically when a thread broke. Up to then, this was a major quality problem in weaving. Workers had to continuously monitor looms for broken threads. Failure to catch a broken thread early would lead to a weaving defect that would mar the fabric. Sakichi Toyoda's system stopped the loom automatically whenever a thread broke.

This system was the start of one of the fundamentals of the Toyota Production System, where a machine or process stops whenever an abnormality is detected. The goal was to make the machine idiotproof (in Japanese called *baka yoke*). However, making something idiot proof subconsciously implies that the operator is an idiot, and nowadays, the much more polite term *mistake-proof* or *poka yoke* is preferred. At Toyota, the overarching approach is called *Jidouka*. The word itself means automation, but Toyoda changed one Japanese character slightly from *motion* to *work*. This did not change the pronunciation but added the component character for *man*. In the west, this *automation with a human touch* is known as *autonomation*.

---

* Disclaimer: I worked at Toyota Central Research & Development Laboratories in Nagoya, Japan, for five years. Toyota was my first employer, and I remember these times very fondly.

FIGURE 16.1
Sakichi Toyoda, king of inventors.

Another famous method developed by Sakichi Toyoda was the *five whys*. Whenever a problem surfaced, he tried to ask *why* five times to get to the bottom of the problem. Hence, rather than just fixing the symptoms, he tried to understand the root cause of the problem in order to solve it. This method is also now part of the Toyota Production System.

Unlike many other inventors, Sakichi Toyoda was able to cash in on his developments, founding a large number of companies. He opened his first company, Toyoda Shoten, in 1894, followed by a company for steam-powered looms in 1898, Toyoda Loom Works in 1907, a spinning mill in 1914, Toyoda Spinning and Weaving in 1918, and Toyoda Automatic Loom in 1926, to name just a few of the establishments. Keeping his companies aligned, however, turned out to be difficult, and there were numerous legal battles over patent rights and sometimes outright refusals to deliver goods between his companies (Wada 2004).

Probably his greatest achievement was the Model G loom in 1925. This loom was able to run completely automatically, requiring no supervision. Operators only had to restock shuttles with yarn occasionally for the automatic shuttle changer. At its time, the Model G loom was the world's most advanced loom, significantly improving both quality and productivity. One unskilled worker was able to supervise 30 to 50 looms. To satisfy the market demand for this loom, Toyoda built its first assembly line in 1927, where the looms moved automatically from station to station.

In 1929, Toyoda licensed his Model G loom to the Platt Brothers in the United Kingdom for the sizeable sum of £100,000. Initially, the Platt Brothers praised the loom as a miracle loom. However, they could not manufacture it reliably and sold only 200 altogether (Mass and Robertson 1996). The Platt Brothers accused Toyoda of sending outdated blueprints, and indeed, the speed of design changes at Toyoda was faster than their documentation. The bigger problem, however, was very sloppy manufacturing practices at the Platt Brothers. For example, the shop floor was not even, machines were poorly set up, and final adjustments were done rather sloppily.

Regardless of the reason for the problems at the Platt Brothers, the deal provided Toyoda with a sizeable chunk of cash. Legend has it that in 1930, on his deathbed, Sakichi begged his son Kiichiro to use the money to found the Toyota Motor Company. While this is a beautiful story to tell, it is unfortunately not true. Every cent of the money from the Platt Brothers was spent as a bonus for the Toyoda employees on the 100th day of Sakichi's death as compensation for earlier wage cuts (Wada 2004). Instead, there were a number of different reasons why Toyoda moved to the automotive business.

His eldest son, Kiichiro Toyoda, was a university-trained mechanical engineer with international experience from the Platt Brothers. He would have been well suited to take over his father's business. However, relatives of his son-in-law Risaburo Kodama lent a sizeable amount of money to Toyoda. Due to this obligation, Sakichi adopted Risaburo after he married Sakichi's daughter Aiko.* By Japanese law, this turned Risaburo Kodama into Risaburo Toyoda. Hence, he was the firstborn son and the first in line to inherit the automatic loom family business, which he did (Sato 2008, p. 43). Therefore, the newly established automotive business may have been a possibility for Sakichi to provide his own flesh and blood Kiichiro with his own enterprise.

Both Sakichi and Kiichiro visited the United States. They were very impressed by the large numbers of cars they saw. Kiichiro also went to

* Such adult adoptions are common in Japan and are a normal part of the Japanese culture.

England twice to visit the Platt Brothers. In 1922, the Platt Brothers were doing very well, but when Kiichiro returned in 1929, their company was marred by financial problems and unemployment (Wada 2004). This left a lasting impression on Kiichiro. He saw much more potential in the automotive business than in looms. He started researching automotive technology as an alternative only one month after he returned from England despite the reservations of his adopted brother-in-law Risaburo (Cusumano 1985, p. 58). The first prototype of a Toyota engine was produced in 1933, and the first car prototype, in 1935 (Mass and Robertson 1996).

The financing of the automotive branch was also done the traditional way, with one-fourth coming from investors and three-fourths of the money loaned by banks. In 1937, Toyota Motor Company* was split from Toyoda Automatic Loom and established as its own independent company. As the president of his own automotive company, Kiichiro Toyoda set out to build nothing less than the best car in the world. Its first model, the Model A, had a Chrysler body, a Ford frame, a Ford rear axle, a Chevrolet front axle, and a Chevrolet engine. Like China nowadays, it included a lot of copied technology. Some parts were also sourced from producers in Japan that copied the foreign components (Cusumano 1985, p. 63ff). In any case, Toyota claims that they modified the parts to avoid patent restrictions—but you would get the same answer from China nowadays, too. Nevertheless, when entering a new business field, it is definitely a viable strategy to learn as much as you can about your competitors' technology. Toyota did learn fast. While neither the technology (except for the hybrid gasoline–electric Prius model in 1993) nor the design of Toyota was particularly outstanding, Toyota's production system excelled far beyond anything ever seen before. The Toyota Production System is the world's benchmark in manufacturing organization and copied all over the world under the label Lean manufacturing.

Besides the concept of autonomation developed by Sakichi Toyoda, the second pillar of the Toyota Production System according to Toyota is just-in-time (JIT). The idea originated from Kiichiro Toyoda. It is said that he got the idea when he missed a train in England. The train left just on time, but Kiichiro was slightly late. From this, he developed the concept that

* The family name is Toyoda, although the name of the automotive company is Toyota. Based on a competition in 1936, they changed the second-to-the-last letter from a *d* to a *t*. This is easier to pronounce internationally. Additionally, in Japanese katakana characters, *Toyota* is written with eight strokes, which is considered a lucky number.

materials should arrive just when they are needed, not earlier or later. He named this *just-on-time delivery*—with a slight grammatical hiccup, *just in time*, and introduced it on the shop floor as early as 1936 (Toyota Motor Corporation 1999, p. 97).

Kiichiro Toyoda even wrote a detailed manual in 1937 about his ideas. Besides his vision of JIT, he also went into the details of material flow. According to Eiji Toyoda, the manual was 10 centimeters thick, which would approach 1000 pages (Shimokawa et al. 2009, p. 226). However, writing 1000 pages is a daunting task, let alone writing a comprehensive manual of 1000 pages. Honestly, I think this may be exaggerated. In any case, there were probably few people—if any at all—who actually read the full 1000 pages. In any case, Kiichiro Toyoda set up the second pillar of the Toyota Production System with his JIT approach. Yet, while he had the idea, he lacked the tools to implement it. Manufacturing was still planned beforehand, and all the little problems in reality prohibited the factory from adherence to its schedule. Kiichiro Toyoda's concept remained merely a vision.

In any case, for the next few years during World War II, the problem of having materials on time was heavily surpassed by the problem of having any materials at all. Like other nations, Japan changed to a planned economy during the war to handle the constant shortages of materials. The largest effort at Toyota went into obtaining materials—legally or otherwise. Disposition of parts turned into barter to get the needed materials. Combined with a workforce consisting of geishas, nuns, and criminals, productivity took a nosedive. It did not help that due to a legislative error, the criminals received the largest food rations (Reingold 1999, p. 30). Shortages were so severe that Toyota was forced to produce military trucks with only one headlight and only two brakes at the rear axle (Toyoda 1987, p. 74). Luckily for Toyota, the war ended just three days before a major bombing run was scheduled for its main factory, and the Toyota factories had only minor damage at the end of the war.

However, the problems were not yet over for Toyota. Inflation reduced liquidity; customers paid late if at all; materials accumulated during the hoarding in World War II sold only slowly; and in 1950, the company faced bankruptcy. The only solution was to radically cut the labor force.

However, after the war, the U.S. occupational government promoted unions in Japan. These unions could have complicated the cutting of the labor force at Toyota. However, this problem resolved itself rather quickly. As it turns out, the communist party of Japan soon dominated these newly established unions. Remember, this was around 1950, and McCarthy* was just about to start the communist witch hunt that would bear his name. A communist-dominated union was definitely on the wrong foot with the U.S. occupational government in Japan, which soon took corrective action (Cusumano 1985, p. 138ff). Afterward, the unions in Japan were and still are only a shadow of any western union. Each company has its own union, often dominated by middle management. Hence, Japan has over 60,000 separate unions, and the three primary umbrella organizations are all but powerless. Therefore, during the financial crisis of Toyota in 1950, the company was able to keep strikes to a minimum, 1760 employees left voluntarily, and company president Kiichiro Toyoda resigned. Shortly thereafter in 1950, the Korean War started, and the increasing orders from the U.S. military made Toyota profitable again.

This renewed prosperity finally allowed Toyota to build a completely new way of manufacturing that would take the world by storm. The man most responsible for creating this Toyota Production System was Taiichi Ohno.

16.2 TAIICHI OHNO AND THE TOYOTA PRODUCTION SYSTEM

Taiichi Ohno (1912–1990) (Figure 16.2) was born in Japanese-occupied Manchuria. His father named him after his work making fire-resistant bricks (Ohno and Mito 1988, p. 135). After moving back to Japan and graduating from Nagoya Technical High School, he joined Toyoda Spinning & Weaving in 1932. His early work involved creating standard work descriptions. One of Toyoda's main competitors, Nichibo, was able to provide better quality at lower cost, and Ohno was sent to learn their methods. Whereas Toyoda back then used large lot sizes with different buildings for different processes, Nichibo preferred to work in small lot sizes and integrated manufacturing lines. Hence, Nichibo may have been one of the inspirations for Ohno in the creation of the Toyota Production System (Fujimoto 2001, p. 60). Soon thereafter in 1943, Ohno moved to the Toyota Motor Company.

* Joseph McCarthy (1908–1957) was a Republican U.S. senator from Wisconsin. He is now mostly remembered for being the force behind the repression and intimidation of anything remotely connected to communism between 1950 and 1956.

FIGURE 16.2
Taiichi Ohno, the main driver behind the Toyota Production System, in his element. (Copyright Toyota Motor Company. With permission.)

Within three years, Ohno was put in charge of the machining shop. This was, back then, a traditional workshop with a group of machines operated by skilled craftsmen. By 1947, he had his machines arranged in sequence of operation with individual workers operating multiple machines. He also reduced lot sizes (Ohno 1988, p. 11). By 1953, one operator was in charge of 5 to 10 machines (Smalley 2006). However, this was only the beginning, and over the next years, Ohno and his assistants would completely revolutionize manufacturing organization.

The underlying principle of this revolution was based on Kiichiro Toyoda's JIT, but Ohno developed the tools to put it into practice. Up to that point, most manufacturing was planned beforehand. Managers or owners tried to foresee customer demand and then planned how much goods of which type to produce. This production program was then pushed through manufacturing. The big, overwhelming flaw of this approach was that nobody is able to predict what exactly the customer will buy. Any prediction of customer demand is, by nature, incorrect, sometimes more, sometimes less. Hence, any planned manufacturing will have a constant mix of too much inventory of some products while being out of stock of others. At the same time, the shop floor will be overwhelmed with regular production orders, mixed with countless rush orders to supply customers with products that are out of stock. It is a recipe for chaos; it is a recipe for inefficiency; it is a recipe for lost sales.

Ohno reversed this logic and aimed to maintain a stock of inventory, reproducing only what the customer pulled out of the stocks. Hence, as it is called nowadays, he moved from a push system to a pull system. One key part of a pull system is a supermarket. The principle is based on retail supermarkets developed in the United States during the first half of the twentieth century. The customer walks through the shelves with the goods and takes whatever he/she needs. The supermarket merely aims to replenish the goods sold to maintain constant availability. In manufacturing, a supermarket is, similarly, an organized inventory with an upper limit of goods produced. Ohno first heard about supermarkets from a classmate in high school, who brought pictures back from a visit to the United States (Sato 2008, p. 67). In 1948, he implemented his first material supermarkets at Toyota. By the time he saw his first retail supermarket in the United States in 1956 (Reingold 1999, p. 51), most of his machining workshop was organized with supermarkets (Smalley 2006). It is important to note that he was not the only one experimenting with supermarkets in manufacturing. For example, Lockheed Corporation also used supermarkets for jet parts in 1954 (Sato 2008, p. 68).

One unsolved problem was how to get information from the supermarket back to manufacturing. Initially, from 1953 onward, they simply wrote the number and type of product taken out of the supermarket on a scrap of paper and gave it to the first process for production. Over time, these scraps evolved and turned into permanent cards with color coding and detailed information (Cusumano 1985, p. 289). These cards also returned to the supermarkets with the produced goods. Effectively, these cards went around in a circle. When the customer took parts from the supermarket, the cards went back to production, passed through production together with the products, and eventually ended up in the supermarket again with reproduced goods, ready for the next cycle. By 1956, these cards were a regular part of operations in Ohno's machine shop (Smalley 2006). Only by 1964 were those cards given a proper name, which represents the idea of the Toyota Production System to the west like no other method of Toyota. These cards were called *kanban*.

By 1954, Ohno was able to extend his methods also to his suppliers (Ohno 1988, p. 32). Already since 1939, Toyota was active in supplier development through Toyota subcontractor discussion groups (Cusumano 1985,

p. 252). Ohno leveraged this to also implement pull systems using kanban and supermarkets for his suppliers. Through this, he reduced their inventory from two months' worth of supply to a daily delivery (Cusumano 1985, p. 279). This helped Toyota enormously when the 1973 oil crisis hit the industry. Due to its Lean system, Toyota survived the crisis in much better shape than any competitor (Ohno and Mito 1988, p. 84).

With kanban and supermarkets, the fundamentals of the pull system were complete. For mass-produced goods, a pull system excels over a classical push system. Stocks can be lower, reaction times faster, delivery performance better, and the system overall much more flexible than any push system. Hence, these pull systems are far superior. Most mass-producers around the world try to implement pull systems in their factories—albeit sometimes with mixed success, as we will see in Section 16.4. However, the change was not an easy one. Throughout all these changes, Ohno encountered stiff resistance from shop-floor operators for taking away their freedom (Cusumano 1985, p. 306). He encountered even more resistance from colleagues and other managers who believed in the traditional way. Their pride was also hurt by this young upstart outperforming them (Ohno and Mito 1988, p. 74). Until the early 1960s, the system was known as the Ohno system, or less flattering, the *abominable Ohno system* (Ohno and Mito 1988, p. 97). However, he did have the support of his boss, Eiji Toyoda, relative of Sakichi Toyoda and future president of Toyota Motor.

Nevertheless, it was a very difficult time for Ohno. Now, Japanese have a reputation for being very polite, avoidance of conflicts, and an overall quiet and demure manner. An average Japanese would not have been able to pull this off even with the support of his boss. Luckily for Toyota, Ohno was anything but an average Japanese. Whenever he encountered resistance to his plans, he yelled, kicked people, handed out impossible tasks, and generally was very forceful. His coworkers were scared of him and avoided him as much as possible, and many of them had trouble sleeping at night because of him (Reingold 1999, p. 41f). Without this attitude, he probably would not have been able to implement his vision.* However, his system performed much better than that of his colleagues, and Ohno rose through the ranks at Toyota. Thus, his area of responsibility increased, and he could apply his system on an increasingly larger scale.

* Please keep in mind that while there is probably an abundance of bosses around the world who yell, scream, and mentally abuse their workers, Ohno also knew what he was doing—which cannot be said of every boss.

Yet the evolution of the Toyota Production System did not stop after supermarkets and kanban. Another important aspect was actually imported from the United States. During the war, statistical process control tools were developed in the United States but all but abandoned after 1945 (see Section 14.2). Two experts in quality control, William Edwards Deming and Joseph Juran, moved to Japan and started to teach these methods. While few in the United States cared about these anymore, their classes were in high demand with Japanese managers on all levels. By 1947, these methods were also introduced at Toyota (Toyota Motor Corporation 1999, p. 100), helping Toyota and other Japanese companies to improve their product quality—which, back then, was still not up to U.S. standards. Toyota improved these methods further, leading to Total Quality Control (TQC), or Total Quality Management (TQM) as it is known in the United States. Deming is still revered in Japan, where he received government honors. The most significant industry quality award of Japan is named after him.

Another cornerstone of the Toyota Production System is the idea of continuous improvement, called *kaizen* in Japanese. Eiji Toyoda brought a Ford employee suggestion booklet from his 1950 visit to Ford. This manual described how Ford encouraged its employees to provide improvement ideas. Eiji then introduced a very similar system at Toyota. While Ford itself abandoned the idea soon thereafter, it flourished at Toyota (Shimokawa et al. 2009, p. 239). While similar systems in the west result in about one suggestion per employee per year, Toyota gets up to 50 ideas per employee per year. Part of the continuous improvement is the relentless elimination of waste, but this is also not new. Already, Henry Ford sent workers on two-week assignments to find and eliminate waste (Brinkley 2003, p. 151).

Yet another idea developed in the United States during World War II but all but abandoned soon thereafter was Training within Industry, a structured approach to train and manage workers (see Section 14.2). The methods, however, made it to Japan and were introduced to Toyota during the 1950s (Fujimoto 2001, p. 70). Especially, the job instruction training on how to teach workers new skills was almost completely based on the U.S. approach. Job relations pertaining to worker–supervisor interaction were taught until 2000. Job methods teaching workers how to improve their work was replaced by the 1950s with a similar course by Shigeo Shingo. Nevertheless, when Toyota opened its first full plant in the United States, named NUMMI (see Chapter 1), the Japanese managers pulled out an old U.S. Training within Industry manual to aid the training effort for the new U.S. workers (Huntzinger 2007).

Also introduced in 1950 was the *line stop*. The principle of autonomation developed by Sakichi Toyoda stopped a system whenever there was an abnormality. Ohno applied this idea to assembly lines (Cusumano 1985, p. 280). Whenever a worker had a problem he could not fix during his allowed cycle time, he stopped the line, and supervisors rushed to him to help him.* This idea of stopping the line is often considered an original Toyota invention. However, Henry Ford already used the same method as early as 1930 (Norwood and Sheeler 1931, p. 10). While this initially led to many stops, it also showed where the process was not yet stable enough. By improving these problems, the lines soon were operating much more efficient than an assembly line that was not allowed to stop. This line stop was soon enhanced with an *Andon* light system, where a green light means everything is in order, yellow indicates small problems, and red means the line has to stop.

Another idea came from Germany. Already between the wars, the Junkers aircraft plants arranged their work content into parts of equal duration. This allowed them to maintain a more constant flow throughout their assembly process. After the work was done, each aircraft moved one slot forward at the same time. This system was called *Taktverfahren* (*cyclic method*). During the war, multiple exchanges of personnel and technology between the German and Japanese axis powers brought this method to the Mitsubishi aircraft plants in Japan. From there, it got to Toyota, and hence, in Lean manufacturing, the average time between parts is named after the German word *Takt*, meaning pulse, stroke, timing, beat, or cycle (Baudin 2012).

Probably the most famous tool of the Toyota Production System besides kanban is the quick changeover method. Initially, Toyota had difficulties in its sheet metal stamping shop. Ford had one or more machines for every part, and they could continue to crank out the same part for weeks on end. Toyota, on the other hand, did not have many stamping machines. Even if they did, these machines would have been mostly idle since Toyota, during the 1940s, produced only few cars. Hence, they used the same machine to make multiple parts. This, however, required them to change tools to make different parts. A stamping tool is a multiton block of steel that has to be placed with high precision to make good parts. Failure in alignment

* If a worker stopped the line in the West, supervisors would also rush to him, though not to help, but rather, to yell at him and ask why the heck he was stopping operations. Hence, workers in the West with multiple years of exposure to Western management would do anything but stop the line.

would produce scrap parts, or at worst, destroy the tool altogether. With a changeover time of two to eight hours, this required lot sizes equal to half a month of inventory (Reingold 1999, p. 45), building up excessive amounts of stock.

Toyota started a program to bring the changeover time down from multiple hours to less than 10 minutes. Hence, while at Toyota, this is called *Quick Die Change* (QDC), in the West, this method was coined *Single-Minute Exchange of Die* (SMED) by Shigeo Shingo. Toyota did as much preparation as possible while the machine was still running and then improved the actual changeover time through the use of fixtures and standards. While it took decades, by 1970, they were able to change the tool within only three minutes (Smalley 2010). This allowed much smaller lot sizes, less inventory, and better flexibility. The method itself is not new, as similar approaches were part of Training within Industry (Isao 2006), used by machine tool maker Danley Corporation (Smalley 2010), or mentioned as *external setup* by Gilbreth (Ignizio 2009, 22%). Toyota, however, excelled at the implementation of this method. In the west, this is sometimes seen as one cornerstone of the Toyota Production System, albeit at Toyota, it is only one tool of many.

While the evolution of the Toyota Production System never stops, the system was well-rounded by 1980. Yet the name Toyota Production System (TPS) was used only from 1970 onward (Shimokawa et al. 2009, p. 1). An internal Toyota Production System manual was written only in 1973 (Smalley 2006).

During the 1970s, the Toyota Production System grew famous, but many people believed it to be a Japanese cultural phenomenon that could not be replicated outside of Japan. Here Toyota proved them wrong, too. In 1984, Toyota finally decided to open a plant in the United States. While Toyota did have smaller enterprises in different countries, including truck bed manufacturing in the United States, it was the last major Japanese carmaker to open a plant in the United States (Shimokawa et al. 2009, p. 230). Even then, rather than going for a full plant, it approached U.S. companies for a joint venture. Since their first choice, Ford, declined their request, Toyota joined forces with General Motors (GM). Together, they reopened the GM plant in Freemont, California, naming it New United Motor Manufacturing, Inc., better known as NUMMI. Before its closure, Freemont was the worst plant at GM, with low quality, high cost, high absenteeism, and significant employee relationship issues. GM eventually closed the plant in 1982. Hence, it was a big bet to reopen this plant, especially since they rehired 80% of the employees of the previous plant. Yet, Toyota pulled it off, and NUMMI soon was the best-performing plant in all of GM (Womack 1990, p. 83). Quality improved drastically, cost was reduced, and employee morale skyrocketed. In overall performance, it was only a sliver below Toyota's own plants in Japan. (See Chapter 1 for more details.) This also proved that the Toyota Production System does not depend on the Japanese culture but also works in other countries.

The Toyota Production System gave Toyota a decisive advantage over its competitors. By implementing a pull system using supermarkets and kanban, combined with small lot sizes, they were able to significantly reduce inventories. A normal plant in the west often has two weeks' worth of raw materials, whereas Toyota has only about two hours' worth of material. Hence, they have less capital tied up in materials, also requiring less space. Quality problems are noticed much earlier and hence are solved more quickly. The average Toyota worker creates two to four times the value compared to a U.S. worker, not because the former works harder but, rather, more efficiently (Cusumano 1985, p. 196). This makes Toyota a very profitable enterprise while still offering cars at a lower price and better quality compared to similar U.S. or European models.

Taiichi Ohno, the key driver of the Toyota Production System, ended his career at Toyota Motor in 1978 as executive vice president, before becoming chairman at Toyoda Gosei, which produces plastic and rubber parts. This is surprising because for his achievements, he should have become chairman at Toyota Motor, and they should have erected a statue in his honor in Toyota city. Rather, he was shoved off to be chairman at a minor company within the Toyota group making plastic parts. Rumor has it that top management at Toyota was unhappy that he talked about the Toyota Production System, whereas the top executives would have preferred this business model to be less widely known. Another possible reason is that his forceful personality yelling at, kicking, and terrorizing coworkers may have made him too many enemies in management. Yet a third possible reason is that legend building at Toyota focuses on the members of the Toyoda family, especially Sakichi, Kiichiro, and Eiji Toyoda. Taiichi Ohno usually gets only a passing mentioning among other engineers in the official Toyota publications. In any case, while he did not develop this system on his own, he surely was the driving force behind the biggest revolution in manufacturing since Henry Ford.

16.3 THE WEST WAKES UP

The radical new approach of the Toyota Production System soon influenced other Japanese carmakers. Yet, due to the language barrier, the method was nearly unknown in the West. After all, business in the United States was doing well, and companies were making good money. Hence, there was no need to look for other ways. This changed drastically when the Organization of Arab Petroleum Exporting Countries (OAPEC) put an oil embargo in place, starting the 1973 oil crisis.

The price of oil quadrupled within a few months. As oil is a critical raw material for the transport and production of almost all goods, prices increased, and demand went down. This hit hard for Western manufacturers with the traditional system using large quantities of raw materials, work in progress, and finished goods. Stocks of unsold goods increased drastically, tying up cash while not earning any profit. Especially, carmakers had difficulties selling their standard gas-guzzlers to the public while gas prices quadrupled.

The Japanese automotive industry, on the other hand, fared better. The Toyota Production System required significantly less inventories. Hence, it was much easier to slow production without accumulating mountains of unsold goods (Clarke 2002). Additionally, they had small fuel-efficient cars in their portfolio, which were more popular than the American gas-guzzlers.

What did the Japanese industry have that the mighty U.S. economy did not? This question soon occupied thousands of researchers and started the era of looking east. There was initially little information available. The first English-language publication appeared in 1977, describing the kanban system (Holweg 2007). Ohno himself wrote a Japanese book on the Toyota Production System,* albeit the publication was delayed to 1978 due to restrictions by Toyota (Holweg 2007). By the early 1980s, the English-language research on the Toyota Production System and its tools multiplied (Waring 1994, p. 160).

Probably the most significant research was a large study by the Massachusetts Institute of Technology, comparing the productivity of

* However, if you read the books carefully, you will find out that Ohno talks very little about the details of the Toyota Production System but much more about other companies, like 7-Eleven.

automotive plants of different makers in different countries, starting in 1979. The results were staggering. Almost by any measure, Japanese automotive plants significantly outperformed Western plants. Vehicles per employee, cost per vehicle, floor space per vehicle produced, reach of inventory, rework, and quality all clearly showed that Japanese automobile manufacturers, especially Toyota, operated much more efficiently than Western makers. The study published a report, *The Future of the Automobile*, in 1984, but much more famous is the resulting best-selling book *The Machine That Changed the World*, published in 1991.

Almost immediately, denial set in. GM was very much upset that the study showed its plants as underperformers. European makers did not like the fact that they spent more effort on quality control than on building the car in the first place. A top executive at Ford concluded in 1977 that they already knew everything that Toyota did. Western managers believed that the difference was due to low labor cost in Japan, or unfair trade barriers, or cultural particularities of Japan, or simply dumb luck (Holweg 2007).

Only after Toyota started its joint venture with GM in 1980 did the message sink in. The joint venture NUMMI, formerly one of the worst plants of GM, performed on a par with Japanese plants, even though they used the same workers. NUMMI had no low-labor cost, no trade protection, and lots of American employees. No, NUMMI used the advantages of the Toyota Production System in America, with American laws, employees, and suppliers. And it worked!

The industries worldwide started a wave of Lean production. The Opel production system started in 1990, followed by the Porsche production system, the Ford production system, and the Chrysler operating system in 1994; Skoda in 1996; Audi in 1997; GM in 1999; Daimler and Bosch around 2000; and Volkswagen in 2001 (Clarke 2002). Overall, there is a consensus in industry that the Toyota Production System—also known as Lean manufacturing—is the right way to improve performance.* Unfortunately, many tried to implement within six months what Toyota needed 50 years to achieve, and most are still struggling with the concept today, 20 years later.

* The term *Lean production* was coined by John Krafcik in 1988, as an international expression for the Toyota Production System. After all, GM, Ford, and others could not really admit that they were simply using nothing more than the ideas of a Japanese competitor.

16.4 LEAN MANUFACTURING—THE NEW RELIGION?

During World War II, the United States used a number of small islands in Melanesia as air bases. This impressed the locals very much, as this technology was far superior to their Stone Age tools. Steel birds came and opened their bellies, and people and cargo came out. After the war, the Americans packed up and left, stopping the supply of cargo planes. To have the steel birds return, the locals decided to do exactly what the Americans did. They built mock airplanes and control towers, with a person sitting inside in front of a straw radio wearing wooden headphones. Locals paraded on the former military grounds, carrying wooden stick-like rifles; lit up signal fires on the runway; and waved landing flags. They even built new airstrips. Unfortunately, despite them doing exactly what they observed from the Americans, no planes arrived. While these *cargo cults* may look amusing to us, unfortunately, there are many similarities between the cargo cults and Lean manufacturing.

While most practitioners agree that Lean manufacturing is the best approach to improve production, most attempts fail. Enormous efforts have been put into Lean transformation projects that show only very few results. Comparative studies find that between 50% and 95% of all Lean projects fail or have only minimal results (Ignizio 2009; Richter 2011, 13%). The reasons for this waste are manifold, but I believe there are different core reasons for this mess:

- Lack of understanding
- Lack of employee involvement
- Lack of focus, resources, and time
- Lack of process confirmation

A lack of understanding shows most often in the use of buzzwords. The manifold culture of the Toyota Production System is reduced to a few catchy phrases. Probably the first phrase to roll over the shop floors was *kanban*, part of the information flow in a Lean pull system. Since Toyota does kanban, Western management decided that they also need kanban. They furthermore believed that they understood kanban. However, this is akin to somebody who reads a book on how to ride a bike without ever using one. Both Lean and biking are learned by doing! Hence, many uses of kanban are similar to the cargo cult. Pieces of paper on the shop floor are called kanban, and it is expected that these magically make the plant better.

A lack of employee involvement means that management made the decision without discussing it with the shop-floor personnel. All the wealth of knowledge available with the workers on the shop floor is rarely used, often leading to a small oversight that negates the whole project.

Lack of focus, resources, and time means that the manager making the decision is of course too busy to do it himself/herself. Most managers underneath do not have the time either. Hence, it is often delegated to an intern or, if the company has more money, to an external consultant. While there are lots of smart interns and consultants, they often do not understand the product and system of the company in enough detail. Additionally, they are usually under significant time pressure to show results. This leads to a half-baked implementation that does not work and quickly falls apart.

A lack of process confirmation happens if management is satisfied with a good presentation rather than an actual working system. Don't get me wrong; of course management believes that the system is working. However, many Western managers often have an abysmal lack of knowledge about what is really happening on their shop floor. It may have worked in theory, or even may have worked once in reality, but it takes time and effort to firmly establish the system. Stop the effort after the first presentation, and everything will fall apart again. If you buy a machine, say a coffee maker, it will make you very similar coffee each time you use it. An organization, however, is in constant flux, more akin to the cleanliness of your dwelling. Cleaning up once and keeping it clean are two completely different problems. One is easy; the other one requires constant effort and attention.*

Overall, this pressure by management to implement Lean without understanding, employee involvement, focus and resources, and process confirmation leads nowhere. And, just to be clear, you would need all of them. Having stellar employee involvement but neglecting anything else will also fail. This pressure to implement Lean results in many plants where only a few buzzwords are introduced together with some make-believe activities. Management is happy, although the mess is the same as before. Only, the workers' opinion about their bosses went down another notch.

* Yes, my apartment was clean once, too.

For example, let's look at Just in time (JIT), where material should arrive only when needed. I distinctively remember a guided tour at a large European carmaker, where the tour went through warehouses with piles of material up to the roof, supposedly all arriving just in time. Pressure to implement JIT also provides lots of income for third-party warehouses across the street from the plant. Piles of material are stored there, paid for and owned by the manufacturer but somehow not appearing in their books.* In addition, delivery is, of course, just in time by a forklift or truck getting the stuff from the warehouse across the street. It looks like JIT, but it does not come with any of the advantages.

Yet another trend is to reduce inventory. After all, Toyota is known for its low inventories. Hence, management decides to reduce inventory. What they overlook is that inventory is necessary for a working production system. As the production system becomes more stable, it is possible to reduce inventories. However, reducing inventories without improving the production system beforehand will only lead to missing materials. With one or two materials missing, you cannot produce the goods. The other materials needed, however, have arrived! Overall, you may end up with more materials than before.

Probably among the latest buzzwords in Lean manufacturing is *leveling*. While there are different ways to smoothen production, leveling often refers to a repeating production pattern. The advantage is that the supplier has to deliver the same quantities every week, allowing him/her to operate with less stock. However, leveling puts very high demands on the stability of the production system. If you decide to fix a two-week production plan, the plant should be able to stick to it. Unfortunately, most plants in the West are unable to do this. There are a constant lack of different materials, technical problems, rush orders, stockouts, plant closures, strikes, and other events that force a change of the plan. In fact, despite having seen many large-scale leveling projects with significant investments of time and money in different plants and companies, I have never seen a working beneficial leveling outside of Toyota. Even at Toyota, it is used more as a measure of production stability rather than a lever to create this stability. In the West, many plants, even entire corporations, repeatedly attempt leveling because top management decided to do so. Only in small, private

* Usually, it is not yet really paid for, but there is an obligation to purchase these materials, so sooner or later, they will have to be paid for. Naturally, this service of storing goods and up-fronting the money is also paid. Hence not only is the material still there; it is probably more expensive than before.

circles do the Lean experts of the corporation admit that leveling does not work anywhere within their companies.

Another method that is currently much promoted is *kata*. The Japanese term originally stands for a choreographed movement in martial arts to teach students behavioral reflexes. With respect to Lean, kata is a series of questions that should be asked in every improvement project: (1) understand the challenges; (2) understand current condition; (3) define the short-term target; and (4) move toward the short-term target. This approach is generally reasonable and based on previous methods like Training within Industry. Yet, like many other Lean approaches, it is still way too often believed that simply using the word makes production better.

I could give you many more examples where a buzzword is pushed through the organization simply because Toyota (supposedly) does it. The result is a wasteful dog-and-pony show (Clarke 2002) of employees doing rituals without understanding (Richter 2011). I have seen many plants where simply mentioning Lean manufacturing will cause the shop-floor operators to reject the project based on past experience with nonsense Lean projects. While the Toyota Production System and Lean manufacturing are very powerful, in many cases, they are applied very sloppily, increasing waste rather than reducing it.

16.5 GURUS AND SNAKE OIL SALESMEN

Lean manufacturing has been a hot topic in manufacturing since around 1980. However, the Western industry did not and often still does not have expertise in Lean. This resulted in a large demand for people skilled in the Toyota Production System and Lean manufacturing. The laws of supply and demand worked, and the demand for Lean expertise was filled. Unfortunately, while there are consultants skilled in Lean, there are also a lot of people who promise more than what they can deliver.

A common way to appear knowledgeable is through the liberal use of buzzwords and foreign-language terms. Why use English when you can use Japanese words. Use *heijunka* rather than *leveling*, *kaizen* rather than *improvement*, *muda* rather than *waste*, *gemba* rather than *shop floor*, and *kanban* rather than...well actually, that word is now understood everywhere in industry, so the effect has worn off. Some native Japanese consultants take this one level up by talking completely in Japanese and working through a translator, even though they are able to speak English. In both cases, it sounds very impressive, but in reality, it is very confusing.

Another strategy is to make unreasonable demands. Rather than reducing inventory, the goal is to reduce the inventory to zero. The goal for the lead time is zero, for waste is zero, for setup times is zero, for breakdowns is zero, and for lot size is...well, okay, let's be generous...one. Naturally, it is impossible to reach these goals. Toyota aims to minimize inventories with respect to the production system, but even Toyota has slightly increased its inventories since 2005 to improve production.* Ohno explicitly stated that reducing inventory to zero is nonsense (Ohno and Mito 1988, p. 25). Yet this quasi-religious zeal still pushes Lean to extremes.

One way to fill the demand for Lean experts was to target Toyota employees and subcontractors. The often valid assumption is that having direct experience in the Toyota Production System helps in implementing similar systems elsewhere. Hence, headhunters searched for current and former employees of Toyota. Some companies collected ex-Toyota workers like stamps in an album.† Nevertheless, simply employing an ex-Toyota worker is no guarantee of success. Not everybody at Toyota knows the system in detail. Additionally, most companies cannot provide the environment to implement the system, and the established dominant corporate culture suffocates the small Lean movement.

Probably the first and foremost Lean guru was Shigeo Shingo. He worked in Japan as a consultant and conducted occasional trainings for Toyota related to the Training within Industry methods. During the late 1970s he became interested in the Toyota Production System, constantly trying to arrange numerous meetings with a more and more reluctant Taiichi Ohno. In 1980, he published a book about the Japanese production system without the consent of Toyota, containing some confidential information. Toyota, in response, abruptly ended a 25-year business relationship with Shingo‡ (Isao 2006).

Shingo moved to the United States and started consulting for U.S. companies about the Toyota Production System. Now, people like to brag. If you are a consultant, this can be very profitable. If you are the first Lean consultant in the United States and have worked at Toyota, you can make a killing. Hence, Shingo, with the help of his promoter Norman Bodek,

* Nevertheless, they are still much Leaner than most Western companies are.

† Having been a Toyota stamp myself in different companies' albums, I can tell you that this is not the worst thing that can happen to you.

‡ Officially, it was of course a *mutual decision*.

started to turn himself into a brand. His Japanese book was translated to English. However, the title suddenly changed from *Toyota Production System* to *Shingo System* (Shingo 1988). Shingo claims that a small presentation in 1945—which is no longer available—was the foundation for the Toyota Production System. He claimed that he invented the quick changeover method—when all he did was participate in a single workshop for one of the last presses to be improved, when the method was already fully developed (Smalley 2006). However, he did give the method its western name, SMED. Bodek got him an honorary doctorate degree from Utah State University (Bodek 2004). In his own books, he added himself repeatedly in the timeline of manufacturing. He supposedly developed the theory of flow layout; introducing scientific thinking; developing quick changeover; introducing preautomation; and inventing poka yoke, the non-stock production system, and many more (Shingo 1988, p. 433ff). In reality, he did not really contribute anything new to the Toyota Production System.

However, he did help popularize Lean manufacturing in the West. For this, he is still revered as a guru, especially in the United States. He is also sometimes and incorrectly described as the inventor of the Toyota Production System. He supposedly even taught Taiichi Ohno Lean. The leading award in the United States for Lean manufacturing is named after him. His books have staggering prices. Small books with 300 pages are bestsellers for almost $100, while a 100-page pamphlet on SMED goes for $31.95, even though my Japanese contacts at Toyota consider these books very poorly written. The Japanese original books, in any case, are much cheaper and also often out of print, as he is all but unknown and insignificant in Japan.

16.6 WHAT ELSE HAPPENED BESIDES LEAN?

Naturally, there were also different methods and approaches tried in the west other than Lean. Without going into too much detail, we will look at some of the more well-known ones: Volvoism, Goldratt's theory of constraints, *Factory Physics*, and Six Sigma.

Volvoism was a manufacturing approach developed at Volvo in Sweden. Assembly lines were often criticized for dehumanizing work. Volvo experimented with group work, first in Kalmar in 1984 and then in the Uddevalla plant in 1990. Rather than the car coming down the assembly line, a group of about 10 workers assembled the vehicle on one spot (Womack 1990, p. 101). In effect, this was a reversal to car manufacturing before Henry Ford's assembly line.

While the satisfaction of the workers was somewhat higher than in assembly lines, productivity was not. The uneven demand for materials and parts clashed with the standard material transport (Clarke 2002). Quality did not improve. Even the stress levels of workers remained about the same. Workers commented that the work was not that much different from an assembly line (Lantz 1995).

Eventually, Volvo decided that the experiment did not work out and, in 1987, changed Kalmar back to a conventional assembly line. Uddevalla was closed in 1993, after only three years of operation (Clarke 2002). Mercedes thought about copying Volvoism in its Rastatt plant but eventually decided against it. While the experiment did not work out, it is an example of the tendency in Europe, especially Germany, to see the workplace more as a social institution, whereas in the United States, the hire-and-fire mentality is much more widespread.

Yet another method propagating throughout the industry was the Theory of Constraints developed by Eliyahu M. Goldratt (1947–2011). Like Shingo, Goldratt was a consultant, and his work was less focused on academic rigor but more on making money. Goldratt studied the management of constraints and how to schedule production subject to these constraints.* He created a software package that claimed to develop the optimal scheduling for bottlenecks. After this claim of optimality was rejected in a legal trial, he created another software package with the most unfortunate name, DISASTER (Trietsch 2005). However, his methods were not really new. Most of his approach was known before in the Soviet Union and the Western world as the *Critical Path Method* and *Program Evaluation and Review Technique*. Goldratt simplified these methods, while neglecting to give credit to others.

Yet his approach was very popular. For one thing, it was not Japanese. Goldratt promoted his methods from 1978 onward, right during the start of the Lean wave. For some Westerners, it was hard to swallow that Japan, utterly defeated with two atomic bombs in World War II, could

* In this, he was active in the same field as me. However, my research is focused on bottleneck detection (Roser, Nakano, and Tanaka 2001, 2002; Roser, Lorentzen, and Deuse 2015), while Goldratt just assumed you would find the bottleneck somehow.

outsmart the industrial giant United States in manufacturing. A competing approach by an Israeli was much more welcome to the hurt U.S. pride.

Secondly, Goldratt wrote easy-to-read books. He definitely had the skill to express complex topics in easy-to-understand words. His most famous book, *The Goal*, was not a business book—of all things, it was a love story! He explained his methods through a fictional plant manager, Alex Rogo, who used the theory of constraints to save not only his plant but also his marriage and his Boy Scout excursion. In my view, the love story is rather cheesy and, by itself, not remarkable. However, for a business book it is very easy reading. It is probably the ease of reading rather than the intellectual content that makes this and some of Goldratt's other books business bestsellers. The experts in industry are split on Goldratt. For some he was a genius and leading academic; for others, he was an impostor that could sell.

As Goldratt avoided mathematical methods, he was understandably rather unpopular with researchers who focused especially on these mathematical methods. While operations research was stuck in highly advanced and highly impractical math, others aimed to develop mathematical methods with focus on practical application. Probably the most well-known is the book *Factory Physics* (Hopp and Spearman 2001). The authors, Hopp and Spearman, are pretty much the opposite of Goldratt. While Goldratt dislikes math, it is the key to factory physics. While Goldratt is very easy reading, *Factory Physics* is something you have to chew your way through by force. Naturally, Hopp and Spearman were rather skeptical of Goldratt's methods. Once, Spearman even joined a seminar by Goldratt under a false name. He then dropped the ruse and took Goldratt's approach apart on the spot in front of the other seminar participants. I would have loved to see that!

In any case, *Factory Physics* and other books like *Optimizing Factory Performance* (Ignizio 2009) aim to provide a framework for manufacturing, providing equations where possible while still maintaining a tradeoff for usefulness. As such, it is a continuation of scientific management, albeit there is still much more to explore.

Another of the still-popular methods developed outside of Japan is Six Sigma. The method is based on statistical process control and was developed at Motorola in 1986. In 1995, GM picked up the method. The assumption is that every manufacturing process has variations, which can be described statistically. The smaller the variations are, the more products will be within the tolerance limit. This variation is measured by calculating the standard deviation, usually mathematically expressed using the Greek letter sigma, σ. Motorola decided that their goal for tolerance limit would be Six Sigma. This would yield 99.99999980% good parts, or only 0.009 defects per billion parts. To give you an example, Six Sigma would mean that out of all the 314 million people in the United States, only one would be sick.

While it is easy to demand this level of quality, it is much harder to achieve. The Six Sigma method itself has eased up a bit on this demand. Six Sigma somehow stands now for 4.5 sigma, easing the precision to only 99.99966% or 3.4 defective parts per million. For this book, this would allow me a whopping two typos.* However, Six Sigma is a very high and costly requirement. To make matters worse, this requirement is set regardless of the cost or benefit of the requirement. It completely ignores the cost of failures. Take, for example, a pen. Of course, we would like for our pens to work. Yet the failure of an inexpensive disposable pen is much cheaper than the cost of achieving 99.99966% good pens. Sure, it is possible, but a normal consumer would not buy the pen, because it would simply be too expensive.

Hence, it is unsurprising that Six Sigma fails to achieve its goals. In fact, most Six Sigma projects fail, and many corporations are pulling out of Six Sigma (Ignizio 2009, 18%). In response, Six Sigma adapted the much more successful Lean manufacturing, naming the reborn method *Lean Six Sigma*. Like many Lean implementations in the West, however, Six Sigma has many shortcomings and misunderstandings compared to the Toyota Production System. Maybe even more than other non–Six Sigma Lean implementations, it uses a much-formalized approach that, to me, often values some standard tools more than the problems they are trying to solve.

While Six Sigma would have normally gone the way of many other buzzwords, it had one stroke of genius. With Lean or Toyota, no matter how much experience you have, there is no Lean degree. With Six Sigma, you can get such a degree, and it has a catchy name, too. You can be a *Six Sigma Green Belt*, *Six Sigma Black Belt*, or *Six Sigma Master Black Belt*. These courses are selling like hotcakes in industry. For only $99, you can get a Six Sigma Green Belt online, and for not much more, even a Master Black

* With all due respect for the valuable work of my copyeditors, I believe that this book will not conform to Six Sigma standards. After all, it will be nearly impossible to find and remove all the mistakes I have added in the first place.

Belt. I seriously doubt that this will give the participants much knowledge on Lean manufacturing, but it certainly does impress some employers.* I don't quite get the point, but if it helps you in your career, go for it. Just do not expect me to be impressed.

Overall, there is a broad consensus in industry that Lean manufacturing is the right way to go to improve manufacturing. Yet many companies have problems walking down that path, often demanding too much in too little time. Lean manufacturing is not easy, but done successfully, it can truly make a difference. It worked very well for Toyota, it works well for Honda, it turned Trumpf around, and it saved Porsche from bankruptcy. Lean manufacturing also works for companies outside the automotive industry. Variations of Lean manufacturing are Lean banking, Lean healthcare, Lean administration, Lean services, Lean military, and others. Lean manufacturing is the most advanced method for manufacturing organization that we have available. Yet, in many companies, Lean is merely a selection of tools and buzzwords, without the underlying mindset and attention to detail necessary to make it truly work.

BIBLIOGRAPHY

Baudin, M., 2012. Takt time—Transfer from Germany to Japan in World War II. *Michel Baudin's Blog.*

Bodek, N., 2004. *Kaikaku: The Power and Magic of Lean: A Study in Knowledge Transfer.* 98 pp. PCS Press, Vancouver, WA.

Brinkley, D., 2003. *Wheels for the World: Henry Ford, His Company, and a Century of Progress*, 1st Edition. Viking Adult, New York.

Cameron, R., Neal, L., 2002. *A Concise Economic History of the World: From Paleolithic Times to the Present*, 4th ed. Oxford University Press, New York.

Clarke, C., 2002. Forms and Functions of Standardisation in Production Systems of the Automotive Industry: The Case of Mercedes-Benz (Doctoral). Freie Universität Berlin, Berlin.

Cusumano, M.A., 1985. *Japanese Automobile Industry: Technology & Management at Nissan & Toyota.* Harvard University Press, Cambridge, MA.

Diamond, J., 1997. *Guns, Germs, and Steel. The Fates of Human Societies*, 1st ed. W.W. Norton and Company, New York.

* I believe that with five years at Toyota, three years of Lean projects at McKinsey, five years of Lean expert training at Bosch, and years of teaching Lean through my university, I have significant experience in Lean manufacturing. While this duly impresses others in Lean discussions, however, it is far too often followed by the question *But, do you have a Six Sigma black belt?* Well, I don't, but if people keep on asking, me I'll eventually do a $300 online black belt course just to shut them up.

Fujimoto, T., 2001. *The Evolution of Manufacturing Systems at Toyota*, 1st ed. Productivity Press, Oxford, UK.

Holweg, M., 2007. The Genealogy of Lean Production. *Journal of Operations Management* 25, 420–437. doi:10.1016/j.jom.2006.04.001.

Hopp, W., Spearman, M.L., 2001. *Factory Physics*, 2nd ed. McGraw Hill Higher Education, New York.

Huntzinger, J., 2007. The Roots of Lean: Training Within Industry: The Origin of Japanese Management and Kaizen, in: Proceedings of the TWI Summit. Presented at the TWI Summit.

Ignizio, J.P., 2009. *Optimizing Factory Performance: Cost-Effective Ways to Achieve Significant and Sustainable Improvement* (Kindle Edition), 1st ed. McGraw-Hill Professional, New York.

Isao, K., 2006. Shigeo Shingo's Influence on TPS—An Interview with Mr. Isao Kato.

Lantz, A., 1995. Gruppenarbeit in der schwedischen Industrie: Ein Forschungsüberblick aus sozialpsychologischer Perspektive. Sozialforschungsstelle Dortmund—Publikationen; Arbeit—Zeitschrift für Arbeitsforschung, Arbeitsgestaltung und Arbeitspolitik; 4, 2, S. 142–169.

Mass, W., Robertson, A., 1996. From Textiles to Automobiles: Mechanical and Organizational Innovation in the Toyoda Enterprises, 1895–1933. *Business and Economic History* 25, 1–37.

Metzler, M., 2006. *Lever of Empire: The International Gold Standard and the Crisis of Liberalism in Prewar Japan*. University of California Press, Berkeley, CA.

Norwood, E.P., Sheeler, C., 1931. *Ford Men and Methods*, 1st ed. Doubleday, Doran, Garden City, New York.

Ohno, T., 1988. *Toyota Production System: Beyond Large-scale Production*, 1st ed. Productivity Press, Cambridge, MA.

Ohno, T., Mito, S., 1988. *Just-In-Time for Today and Tomorrow*. Productivity Press, Cambridge, MA.

Perrin, N., 1995. *Giving Up the Gun: Japan's Reversion to the Sword, 1543–1879*, Reissue. David R Godine, Boston.

Reingold, E.M., 1999. *Toyota—A Corporate History*. Penguin, London.

Richter, L., 2011. Cargo Cult Lean. *Human Resources Management & Ergonomics* 5, 84–93.

Roser, C., Nakano, M., Tanaka, M., 2001. A Practical Bottleneck Detection Method, in: Peters, B.A., Smith, J.S., Medeiros, D.J., Rohrer, M.W. (eds.), Proceedings of the Winter Simulation Conference. Presented at the Winter Simulation Conference, Institute of Electrical and Electronics Engineers, Arlington, Virginia, pp. 949–953.

Roser, C., Nakano, M., Tanaka, M., 2002. Shifting Bottleneck Detection, in: Yucesan, E., Chen, C.-H., Snowdon, J.L., Charnes, J.M. (eds.), Winter Simulation Conference. San Diego, CA, pp. 1079–1086.

Roser, C., Lorentzen, K., Deuse, J., 2015. Reliable Shop Floor Bottleneck Detection for Flow Lines through Process and Inventory Observations: The Bottleneck Walk. *Logistics Research* 8, Issue 7.

Sato, M., 2008. *The Toyota Leaders: An Executive Guide*. Vertical, New York.

Shimokawa, K., Fujimoto, T., Miller, B., Shook, J., 2009. *The Birth of Lean*. Lean Enterprise Institute, Inc., Cambridge, MA.

Shingo, S., 1988. *Non-Stock Production: The Shingo System of Continuous Improvement*. Productivity Press, Cambridge, MA.

Smalley, A., 2006. A Brief Investigation into the Origins of the Toyota Production System. Art of Lean. Available at http://marekonlean.files.wordpress.com/2010/09/origins-and-facts-regarding-tps.pdf.

Smalley, A., 2010. A Brief History of Set-Up Reduction: How the Work of Many People Improved Modern Manufacturing. Art of Lean. Available at http://artoflean.com/files/A_Brief_History_of_Set-Up_Reduction.pdf.

Toyoda, E., 1987. *Toyota: Fifty Years in Motion*, 1st ed. Kodansha America, New York.

Toyota Motor Corporation (ed.), 1999. *Great Dreams, Days of Passion: A Photographic History of Toyota's Early Days*. Toyota Motor Corporation, Toyota City, Aichi-Prefecture, Japan.

Trietsch, D., 2005. Why a Critical Path by Any Other Name Would Smell Less Sweet? Towards a Holistic Approach to Pert/Cpm. *Project Management Journal* 36, 27–36.

Wada, K., 2004. Kiichiro Toyoda and the Birth of the Japanese Automobile Industry: Reconsideration of Toyoda–Platt Agreement (CIRJE F-Series No. CIRJE-F-288). CIRJE, Faculty of Economics, University of Tokyo.

Waring, S.P., 1994. *Taylorism Transformed: Scientific Management Theory since 1945*. The University of North Carolina Press, Chapel Hill, NC.

Womack, J.P., 1990. *The Machine That Changed the World: Based on the Massachusetts Institute of Technology 5-Million-Dollar 5-Year Study on the Future of the Automobile*, Later Printing. Scribner, New York.

Section IV

The Cutting Edge

17

Where Are We Now?

The worker is not the problem. The problem is at the top! Management!

W. Edwards Deming (1900–1993)
Quality guru

Much of what we call management consists of making it difficult for people to work.

Peter Drucker (1909–2005)
Management guru

As you can see from the preceding chapters, manufacturing has come a long way since *Homo habilis* manufactured the first stone tools 2.6 million years ago. For the most part of this period, progress was rather slow. It took many millennia from the first manufacturing technique of cutting stone to the second of joining components. Similarly, it took generations from the first full-time specialists during the Neolithic revolution to the first use of water and wind power, the first standards of the Harappan culture, and the rising prestige of manufacturing with the guilds of the medieval age.

Technology changed only slowly, for two reasons. For one, most effort was needed simply not to starve to death. Secondly, there were few activities to generate new technical knowledge. Training was mostly based on an apprentice system, where the youngster learned the same tricks of the trade from the master. New ideas were often discouraged, from Roman emperors destroying the knowledge of unbreakable glass to medieval guilds breaking advanced tools.

Only during the last 200 years did technology and manufacturing really accelerate to become faster, better, cheaper. The industrial revolution turned many people's lives upside down. While some came out better than others, in the long run, everybody benefitted. Nowadays, major changes often come within one generation or less. The speed of the introduction of water- and steam-powered spinning and weaving left traditional craftsmen no time to adapt. Electricity and electric motors became widespread within a few decades. Interchangeable parts, bicycles, automobiles, and the assembly line revolutionized society. A single person could have watched the first flight of the Wright brothers in 1903 and witnessed mankind landing on the moon in 1969. In less than 70 years, we went from the first powered flight to landing on the moon.

The technological advances continue, with networked computers being at the current forefront of change. Imagine going back in history only 30 years and explaining to someone that you have a device in your pocket that can instantly access almost all information mankind has ever generated and lets you communicate from nearly anywhere with almost anybody on the planet (your smartphone). He/she would be most thoroughly amazed.* Transplanting someone from only a few decades ago into the current age would render this person nearly unable to live properly due to these changes.†

These effects are due to new inventions in science but also the ability to manufacture these goods faster, better, cheaper. Society generates wealth not because of what we do but because of how efficient we do it (Pearson 1992, p. 207). Improvements in manufacturing increase efficiency and make products affordable. For example, it is estimated that in 2000 BCE, 1 kilo of iron was worth the modern-day equivalent of about $10,000. In this case, the material value of your car alone would be in excess of $10 million. Luckily, due to improved manufacturing, the price went down significantly. Only 1000 years later, the price of a kilogram of iron was reduced to $500. During the early Middle Ages, it was $70. Nowadays, you can get a kilo of iron for a mere 50 cents, and your car probably contains less than $500 worth of iron (Marsh 2012, p. 15).

* He/she would also be shocked that you use it mostly to look at cat pictures and inform the rest of the world that you just ordered coffee.

† It is not only a theoretical example but also a very practical problem for newly released long-term prisoners. Simple things like the scanner at the supermarket and automatic doors make reintegration into society even more difficult for people who spent decades behind bars (Ross and Richards 2002).

This book details the history of manufacturing. However, history on its own is merely an academic exercise of things past. The main benefit of understanding history is not only satisfying our curiosity but, more importantly, learning from it. Understanding past successes and mistakes helps us to have a more successful future. Hence, this chapter summarizes the lessons learned from history and tries to understand the current trends in manufacturing, before the next chapter gives an outlook into the future.

17.1 SIGNIFICANCE OF MANUFACTURING

Through the industrial revolution, manufacturing gave us the biggest jump in productivity ever. Technologies like interchangeable parts and assembly lines are crucial for production and, hence, for most of the wealth that we enjoy nowadays. Global prosperity rests on the basis of manufacturing improvements.

Yet, while manufacturing is critical for growth, its importance declines in advanced economies. In the first world, the service industry is much larger than manufacturing. Economies move from agriculture to manufacturing to service. This is both for the number of people working in manufacturing and for the contribution to gross domestic product (GDP).

It is not a surprise that manufacturing employment declines in advanced nations. Rising labor costs lead to labor-saving methods. Especially in advanced nations, mechanization and automation take over the tasks of human workers. For example, even major manufacturing nations like Germany or South Korea have reduced manufacturing-related employment by about 10% within the last 10 years (Manyika et al. 2012, p. 8). In all developed nations, manufacturing employment dropped from 71.5 million people in 1980 to 51.5 million by 2010 (Marsh 2012, p. 237). The trend shows no sign of stopping. Yet, since low-skilled jobs are easiest to cut, average wages are increasing.

What is most interesting, however, is the development of manufacturing as part of the GDP. In agricultural societies, there is both little manufacturing and little wealth. For those countries, manufacturing is only a small share of the GDP. As nations industrialize, manufacturing becomes important. As the GDP grows, so does its share of manufacturing. At the height of industrialization, manufacturing may contribute up to 35% to the economy. However, as the economy continues to grow, the share of manufacturing falls back to 15%. The service industry takes over as the motor of growth (Manyika et al. 2012, p. 8). Hence, manufacturing is important to industrialize but becomes less prominent afterward (Kazmer 2014).

Yet, less prominent does not mean less significant. For one, manufacturing generates enormous demand for services and helps the service industry to prosper. The lion's share of trade and transportation services is generated by manufactured goods. The same applies to research and development, which is also completely dominated by manufacturing (Manyika et al. 2012). In the United States, for every dollar generated in manufacturing, another $1.34 is generated through additional economic activity. This makes manufacturing the biggest creator of growth synergies of all economic sectors in the United States (The Manufacturing Institute 2012, p. 3).

At the same time, prices of manufactured goods are increasing much less than other products and services. For example, between 2000 and 2010, inflation increased prices in the United States by 22.5%. During the same time, the cost of durable goods did not increase at all. Instead, it fell by 18.2% (Marsh 2012, p. 233).

Manufacturing has an additional benefit of stabilizing economic growth. In times of recession, one way to cut cost is to lay off people. In manufacturing, however, people are often only a small part of the overall cost. The bigger part of the expenses is raw materials and investments in machines and buildings. While materials in the supply chain can be reduced, machines and buildings, on the other hand, have already been paid for. There is little cost saving by not running a machine. Hence, even in a recession, it often makes economic sense to continue to produce on a smaller scale rather than to stop operations completely. Countries with a large share of manufacturing, for example, Germany, often survive recessions much better than primarily service-based economies.

As a rough estimate, even for developed economies, manufacturing should contribute about 20% of the GDP for a healthy economy. The United States has only around 12%. While this is still better than France with 11%, Great Britain with 10%, and Australia with 7%, it is much less than Germany with 22%, or Austria and Japan with 19% (World Bank 2015). The U.S. government is trying to strengthen the manufacturing sector, although so far, the trend of the manufacturing share of the economy decreasing in the United States continues unbroken.

17.2 LESSONS LEARNED FROM HISTORY

Throughout history, there were a number of developments in manufacturing that worked extremely well, for example, the division of labor and interchangeable parts. Over time, these provided a gargantuan increase in productivity. These are nowadays widely copied and used almost universally throughout industry. Other developments, on the other hand, were less successful. They may or may not have worked very well for some companies, but most companies copying these technologies failed to achieve the desired benefit. In the next few paragraphs, we will explore what worked, what didn't, as well as the reasons for these discrepancies in success.

One of the first technics used to improve manufacturing productivity was the division of labor. The tasks were spread among different people, making each individual more proficient and hence more efficient. This happened at the latest during the Bronze Age but possibly much earlier. The next major advancement was mechanization, starting with ancient Egyptian lathes, followed by medieval technology (see Section 5.3). Naturally, during the industrial revolution these machines were responsible for additional major performance improvements (see Chapter 7 onward). Of course, mechanization was closely coupled with utilizing the power of animals, wind, and water (see Section 4.1). Later even larger sources of power became available through steam power and electricity (see Chapter 8 and Section 11.1).

This, in turn, enabled the use of interchangeable parts (see Chapter 9). From then on, parts were no longer custom-made and laboriously fitted, but were generic and interchangeable. This standardization of parts also improved overall quality. Eventually, interchangeable parts allowed the use of assembly lines (see Chapter 13) and robotics (see Chapter 15). The majority of manufacturing in the world uses most or all of these methods to improve efficiency and productivity.

On the other hand, numerous well-described and published manufacturing methods failed to achieve their lofty goals. Taylorism, despite all the effort added, provided lots of understanding but little actual improvement for most companies (see Chapter 12). Material requirements planning (MRP) systems used to organize the material flow and production schedules have a very doubtful record of accomplishment (see Section 15.4).

The largest benefit from the theory of constraints or Six Sigma (see Section 16.6) probably goes to consultants selling their services in these fields, without much benefit to actual production. The Toyota Production System and its derivative Lean manufacturing worked well for some companies, including Toyota, but many other companies fail (see Chapter 16). This is despite a broad agreement in industry that Lean manufacturing is the way forward to improve production. The majority of enterprises put much effort in Lean methods but often with little results to show for it.

Why is it that some methods worked extremely well for industry, whereas others mostly fail to be copied successfully? In my view, there is one crucial difference. The improvements that worked are mostly mechanical in nature. Mechanization, power generation, assembly lines, interchangeable parts, and robotics are all hardware changes. The only possible exception is the division of labor. Yet even this is often fixed through the design of a particular workspace or machine. Once set in place, human workers have few opportunities to deviate from the system.

On the other hand, methods that were difficult to copy successfully are all organizational in nature. Taylorism, MRP, the Toyota Production System, Lean manufacturing, Six Sigma, and the Theory of Constraints are all changes more often geared toward people rather than hardware. Operating a machine will produce similar parts every time with little difference regardless of the operator using the machine. The machine does not care if it works or idles; it simply repeats the task it was designed to do so. Men, on the other hand, do indeed have their own minds. They often do things differently depending on their whims. They often disagree with instructions or were not told the instructions in detail. They forget guidelines, make errors, and deviate from what they should do if their supervisor is not looking. We can bend steel to our will, but we are ill-suited to work with people.

Having workers do what management wants them to do is a constant uphill struggle and major source of conflict in industry. Naturally, workers see it differently and often have to struggle with managers who lack both an understanding of the possibilities of the system and social skills. Of course, this does not prevent them from having an abundance of unfulfillable demands. Hence, industry often has major difficulties in dealing with people—both upward and downward in hierarchy.

Throughout history, numerous technical advances were made not to increase productivity but to get rid of people. Louis-Nicolas Robert invented the continuous paper machine with the primary purpose to get rid of workers (see Section 7.6). Honoré Blanc developed interchangeable parts to control strong independent gun makers (see Section 9.1). Taylor considered an intelligent gorilla a better worker than the people he had (see Section 12.2). Ford complained that with a pair of hands, he always got a human being, too (see Section 13.5). Chemical plant managers were thrilled to discover that through automation, they did not need workers and hence could ignore their *unreasonable* demands (see Section 15.1). One of the reasons for the development of computerized machine tools was the depowering of workers (see Section 15.2). General Motors (GM) CEO Roger Smith invested heavily in robotics with the goal to get rid of workers, envisioning a lights-out factory (see Section 15.3). Manufacturing history is a lot of history of workplace conflict, with managers trying to solve the conflict by getting rid of the workers.

Nevertheless, we still need workers. Even advanced lights-out factories and semiautomated chemical plants need maintenance and installation. Yet, most companies still treat workers as machines and lack suitable conflict management methods. The most frequent employee interaction is still the manager informing the worker what he/she is supposed to do. Now, this may work with machines, but it gives, at best, mixed results when dealing with people. For a manufacturing company to be successful, this means improving its work relations. For example, the Toyota Production System works so well because Toyota has developed a culture where shop-floor objections are taken seriously and the input of the worker is truly welcome.

Overall, how to deal with people both upward and downward in hierarchy will distinguish the excellent companies from the merely mediocre. This is closely related to how companies make decisions and who they include in the decision-making process. As this is mostly still very unstructured, the decision-making process may be one of the large potentials for improvement.

17.3 THE BIG POTENTIAL: DECISION MAKING

In the past, manufacturing has made a number of significant steps forward. From division of labor to interchangeable parts, assembly lines, and Lean manufacturing, each of these steps made manufacturing more efficient. After all, manufacturing is all about efficiency and avoiding waste. The question is what is still out there? What other big potentials may be found in manufacturing to become faster, better, cheaper?

In my personal experience, one huge, if not enormous, source of waste and inefficiency is corporate decision making. Overall, decision making in industry is lousy at best (Baale and Bergholz 2005). Anybody with experience working in a larger corporation probably knows of numerous decisions that were outright insane. Management sometimes even makes decisions that they know are detrimental for the corporation, yet they make them anyway because they were forced to do so by higher management or risk being labeled as a troublemaker. After all, disagreeing with management too frequently is not good for the career. Most people value their career more than the success of the company.

Examples of bad business decisions in manufacturing are numerous. One company with declining sales was looking for a plant to close. They identified one plant in a small town that was long since the worst of their enterprise, losing money every year. This plant should have been closed long ago and would have been the prime candidate for closure—except for the small fact that the wife of the CEO grew up in the small town where the plant was located. Nobody even dared to suggest that plant to the CEO for closure, and another plant had to take the bullet. At the time of writing, Volkswagen (VW) has a scandal with intentionally manipulated emission tests of their diesel engines, dubbed as *Dieselgate* by the media. Saving a few bucks by knowingly cheating on government emission tests will now cost the company dozens of billions of dollars. When GM CEO Roger Smith went headfirst into robotics (see Section 15.3), there were a few managers at GM who urged caution. However, it quickly became clear that this was a career-ending move at GM. Unsurprisingly, the upper management turned into yes-men.

Another example is herd behavior. Some other company does it successfully; hence, it is copied without understanding. One prime example is the Toyota Productions System. Due to its success, it is copied widely but, in many cases, without true understanding. Often, the shop floor knows what works and what does not, but management has its *vision*, which is then implemented regardless of usability. An enormous amount of waste has been generated in manufacturing in the name of Lean. Similar examples can be found in offshoring, where companies merely follow the herd and move manufacturing to Asia, often with disastrous results.

Yet another example often discussed in media is mergers and acquisitions. Top CEOs try to shine through mergers and acquisitions, aiming to increase total revenue (and possibly their paycheck). Yet, 50% to 75% of all mergers fail (Finkelstein 2004), often destroying billions in the process. Of those that do not fail spectacularly, most do not generate any value. True successful mergers are few and far between. Spectacular failures are much more frequent. Quaker Oats bought Snapple in 1994 for $1.7 billion and sold it only three years later for a trifling $300 million, a loss of $1.4 billion. In 1998, Daimler merged with Chrysler, which the Daimler CEO Schrempp called a *marriage in heaven*. Yet this marriage was troubled from the beginning, ending in divorce in 2009. It is estimated that the whole deal cost Daimler about €40 billion. Yet these are only some of the largest failures in manufacturing mergers.

Management is difficult. The higher up a manager is, the more complex the decisions that are required, while at the same time, the quality of information goes down. Yet, decision making is often much worse than it could be. Enormous resources, effort, and money are wasted through bad decision making. As W. Edwards Deming (see Section 16.2) put it, *The problem is at the top; management is the problem* (Deming 2000). Consulting guru Peter Drucker believed that only 3% to 5% of all U.S. companies are well managed (Waring 1994, p. 103). Sydney Finkelstein has identified key habits of management failures, including perceived infallibility, elimination of opposition and development of yes-men, overestimation of abilities, underestimation of problems, and inflexibility to adapt (Finkelstein 2004).

Overall, it is clear that a fish rots from the head down, and management is its own worst enemy. The only reason that many companies with such a management survive is that usually, the competition is no better. You do not need to be good; you only need to be better than the competition.

The question is how to improve? There is a vast amount of management literature, with numerous approaches to decision making. In theory, good decision making is nicely described. The problem is the practical implementation. Have you ever tried to tell the top executive of a large corporation what he is supposed to do? It is...kind of...difficult. Top management does not like opposition from below. Western management grows the sort of managers who make bad decisions. Current managers prefer to promote subordinates who are like them. They also prefer to promote subordinates who have a good record of accomplishment, or more precisely, who *appear* to have a good record of accomplishment. This is often known as the Peter Principle, where managers rise to the level of their incompetence. In effect, promotion often goes to managers who can hide their mistakes, successfully blame others, appear to follow orders, and usually agree with their bosses. Taiichi Ohno and his radical new ideas would have never made it in a traditional established Western company.

With the managers we have now, this inefficient decision making will probably continue. Since information usually goes through multiple levels of hierarchy before it reaches the top, it has been filtered multiple times to weed out any negative connotations and to focus on the positive aspects. At the top, the data no longer say much about reality, muddling the decision-making process even more.*

There is an urgent need for better data. There is an urgent need for better understanding of the shop floor. Better agglomeration of large quantities of information, nowadays called big data, is one possibility. Improved analysis of large data sets related to manufacturing and sales can improve decision making. A much more relevant approach emphasized by Toyota is frequent visits by executives on the shop floor and talking directly to the operators. Managers occasionally need to take samples of information directly from operations to validate the information they receive through hierarchy. This is easier said than done, since most middle managers try very hard to hide the reality and often stage a dog-and-pony show for top executives.† In some plants I know, the only time when the plant is renovated, walls are painted, and the bathrooms are fixed is for an upcoming CEO visit. Shop-floor personnel then spend half a day polishing and cleaning machines so everything looks good. Naturally, in the unlikely case that the CEO asks a question to the operators, they have to refer him to the managers in charge, lest they hear the inconvenient truth. Yet, without useful data, managers cannot make good decisions. Overall, decision making is a huge potential, yet one that is difficult to tap. As long as we have the managers we have, we will not get much change.

---

* For example, the nickname of the former Daimler headquarters was *bullshit castle*, due to the large amount of…ahem…incorrect information going around.

† For a sarcastic list on how to fudge your bosses' plant visit, read my blog post *How to Misguide Your Visitor—Or What Not to Pay Attention to During a Plant Visit!* on http://www.allaboutlean.com/misguide-plant-visit/.

17.4 NEED FOR SPEED

One fundamental change in manufacturing within the last 20 years is the importance of speed. For most manufacturing enterprises, the speed with which new products are introduced into the market is increasing. Product life cycles are getting shorter. While only a few decades ago, products were available for multiple years, production of similar products now stops after a few months. By the time the supply chain is filled up with raw materials, semifinished goods, and final products, it has to be emptied again for the next version.

In the long tern, the significance of actual production may decline, and logistics may become more important for success. It matters less how cheaply it is manufactured. It will be more important how fast it will get to the customer. However, probably most important is the efficiency with which the transition from one product to the next is managed. Since there are usually multiple parts going into one product, it is critical to manage the flow of the different components without too many leftover components that are no longer needed or a lack of new components for the new version.

In the past, many manufacturers often pursued a strategy of low-cost labor, moving manufacturing abroad to Asia and other countries. Yet, the main markets are usually not in Asia but rich first-world countries in America and Europe. Shipping goods from China to the United States is very cheap, but it takes months for the vessel to reach the West Coast and another few days for goods to be distributed throughout the nation. Hence, all new products coming from China will be delayed by months. It is even worse if different stops are involved. A component produced in Europe, shipped to China for assembly, and continuing to the United States for sales extends the supply chain by half a year. This is half a year's delay to adapt to customer demand or to implement model changes. Furthermore, it is half a year's worth of stock that does nothing while taking a tour of the world.

Naturally, airfreight is much faster, reducing the delivery time to a week, including customs clearance. On the other hand, the price for airfreight is significantly more expensive than transport by ship. Yet for long supply chains, small fluctuations in either supply or demand often create shortages of parts in manufacturing or sales. Hence, manufacturers are often forced to use airfreight in order to receive urgently needed parts on time.

I have seen cast-iron parts worth $100,000 being delivered by air with the transport cost being twice the value of the parts. Another company had a daily helicopter flight for months from East Europe to Germany to deliver some O-rings that where in short supply. However, the alternative of stopping an automotive assembly line would be even worse. In fact, some cynics say that airlines only survive financially due to such frequent supply-chain hiccups of manufacturers requiring expensive airfreight.

As a countertrend, many companies already prefer to manufacture close to their customers to increase speed and flexibility. For the United States, this is often in Mexico; for Europe, it is often Eastern Europe. Some companies even accept higher labor cost in exchange for speed. Zara fashion, for example, is made in Portugal and Spain. Hence, Zara has low-cost copies of the latest haute couture already in its European stores while the competition is still loading in Shanghai.

Managing your supply chain to cope with model changes, upturns, and downturns is a key competitive factor in modern manufacturing. The value of shorter, flexible supply chains is slowly becoming visible to management, even though traditional cost accounting is still blind to this value.*

Another relevant factor is to know where your parts are and when they will arrive. Unfortunately, there are frequently differences between what you think you have and what you really have. Parts may have been scrapped, computer errors subtracted two parts where one should have been subtracted, shipments were sent to the wrong warehouse, etc. Logistics managers are always dreaming of a system that reliably tells them what they really have.

One possible solution in the works is the Internet of Things. For tracking parts this is usually implemented using small radio-frequency identity (RFID, Figure 17.1) chips that allow each part to talk to the computer directly. Rather than relying on a computer database, the system would electronically count the parts it has on its shelves. While this would not solve all stock differences, it has the potential to reduce these problems. Prices for RFID chips are now low enough that manufacturers are starting to use these in production. Nevertheless, as with all new technologies,

* Traditional cost accounting often assumes a static situation. There are no ups and downs, but only a constant and steady flow of parts. It rarely includes the cost of urgent airfreight, or old parts that need to be scrapped, but rather lives in a near-perfect stable world. While that would be nice, reality begs to differ.

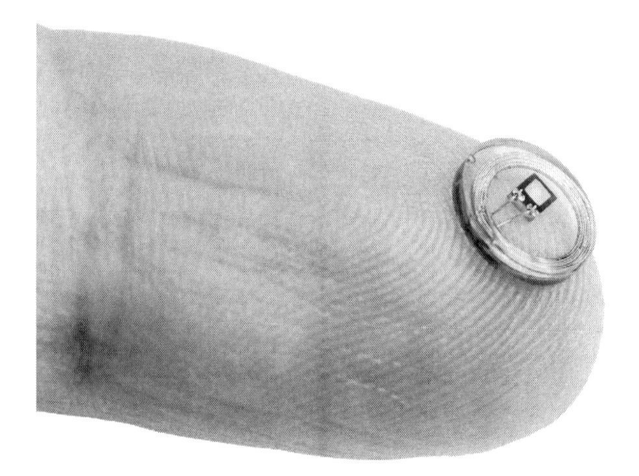

FIGURE 17.1
RFID chip.

implementing RFID will probably be not as easy and simple as the technology vendor advertises. I have seen RFID systems where the computer system took so long to process the RFID information that the workers switched back to typing the numbers by hand, negating any potential benefit.

Overall, speed is one of the keys for competitiveness in many manufacturing businesses. The trend of the last few decades is likely to continue. In the future, it will be much less company against company and more supply chain against supply chain (Richter 2011).

17.5 NEED FOR FLEXIBILITY

Another continuing trend is mass customization, where a product is manufactured using conventional mass-production techniques but every product is customized for an individual customer. Before the industrial revolution, pretty much everything was customized. No matter if it was shoes, clothing, furniture, or food, almost everything was custom-made by craftsmen for individual orders and to customers' specifications. This changed dramatically during the industrial revolution. The prime example is Ford (see Section 13.5), where millions of identical Ford Model Ts cruised the roads. In fact, unless you are very wealthy, most of the things you own are still mass-produced. For almost everything you own, there are thousands of people who have the very same product. This applies to your computer or mobile phone, kitchen gear, furniture, and even the clothes you wear. Probably hundreds of people are wearing the very same shirt as you right now.

Of course, people are not standardized. Hence, standardized products do not always fit your life. Your furniture—unless it is custom-made—probably does not fit perfectly in your room, leaving some gaps at the walls. Electric equipment may lack features you would have liked but instead has features you do not care about. Even your clothing is probably a best fit from the available products but not a perfect fit.*

Hence, there is a desire for customized products. We do not want mass-marketed stuff, but we want our own personalized product. Naturally, we still would like to have this at or near mass-produced prices. The challenge is for producers to achieve this mass customization while keeping the prices still affordable for most customers. Industry is already evolving to fulfill these needs, and—naturally—there are already buzzwords in use talking about flexible manufacturing or agile manufacturing (Sanchez and Nagi 2001).

One example is Essilor, which makes lenses for glasses. Every individual lens is custom-made to match the eye of a particular customer. Yet thousands of lenses are produced each day. Another example would be cars. While the basic model is still mass-produced, the multitude of options make nearly every car a unique vehicle.

For example, dividing the total revenue of Toyota by the total number of unique products sold would leave only around $200,000 per vehicle combination, or only a handful of identical vehicles on average. An average BMW car exists only twice in the world if all options are factored in. The Chemical company BASF has, on average, sales of $500,000 per product. Essilor sells about $33 worth of every unique lens, meaning that nearly every product is unique. Even for clothing, Inditex—the mother company of Zara—sells, on average, only $300,000 worth of clothing for one item (Marsh 2012, p. 58).

The problem is how to achieve this flexibility while keeping the prices low. For some products, of course, customers are less price sensitive. Custom-mixed breakfast cereals, for example, sell well even though they are triple the price of standard cereals. This is different for more expensive

* In my case, for example, my right foot is half a size larger than the left one; hence, one shoe always fits slightly less than the other.

products. The prime examples are cars. Custom-made cars are available, for example, a Rolls Royce or a Bentley. However, these luxury vehicles are out of reach for most people. Hence, for more complex products, different approaches toward flexibility are used. The goal is to keep as much as possible the same and change only the interface with the customer.

The automotive industry realizes this with two slightly different approaches, the platform system and the modular system. While these sound almost similar, there are major differences. A platform strategy aims to keep the underlying part of a car identical as much as possible. VW pursues this platform strategy. For example, the basis for the VW Golf is the same as for the VW Touran, VW Caddy Life, VW Jetta, VW Tiguan, Škoda Octavia, Seat Leon, Seat Toledo, Seat Altea, Audi A3, Audi TT, and Audi Q3. This does not even count the numerous versions available for each model. The advantage is that the underlying technology has to be developed only once. Merely the interface to the customer, for example, the dashboard, seats, and body style, has to be changed to give each model its own flair, including the perceived level of luxury.

Toyota, on the other hand, pursues a different strategy through its concept of modularity. While Toyota also has platforms, there are much fewer models per platform. Toyota has one of the clearest model strategies in the industry, where each market segment is serviced with one and only one model. This contrasts starkly with the multitude of different brands competing against each other within the same company as for Volkswagen or GM. Rather than reusing the entire platform, Toyota's modular strategy reuses individual components. For every new model, Toyota aims to develop no more than 20% new parts. Hence, more than 80% of the parts going into a new model are taken from previous models. For example, the air-conditioning design is identical in many different Toyota models, regardless of their platform. Similarly, the profile of the rubber seal around the doors is identical for almost all car models at Toyota, with the entire seal differing only in length depending on the door. Mercedes, on the other hand, sees value in having an individually designed rubber seal for its different models but of course has to pay the price for the complexity.

Toyota also invested heavily in flexibility. For example, new assembly lines, by default, have to be able to accommodate multiple different models. Changeover frequencies are increasing and lot sizes decreasing (Toyota Motor Corporation 1988, p. 327). As a result, Toyota greatly increased its flexibility. Until the mid-1980s, Toyota wanted to produce at least 10,000 vehicles per model and month to achieve profitability. From the 1990s onward, this was reduced to only 2000 vehicles per model and month (Shimokawa 1994, p. 107), and it will be even lower in the future.

This modular concept removes complexity from the value chain. Hence, fewer parts have to be designed, managed, and held in stock. Overall, there also will be fewer stockouts stopping the production line. Through flexibility, capacity utilization is held high even for a changing product mix. Overall, manufacturers aim to provide perceived individuality, with low-cost mass-produced technology underneath of the hood. In any case, flexibility is one of the current key manufacturing trends that will determine the economic success of companies.

17.6 NEED FOR LABOR RELATIONS

Finally, the last significant focus in manufacturing is labor relations. The labor market in most western nations has two very different sides. Unskilled labor is usually readily available, with many workers unemployed. Skilled labor, on the other hand, is in short supply, and many positions are open despite good salaries. It is estimated that there will be a worldwide shortage of 40 million skilled workers by 2020 (Manyika et al. 2012). U.S. companies often need more than six months to find a suitable candidate for a position requiring skills (Economist 2013).

Here, companies have to compete on more than just the price. For many skilled workers nowadays, job satisfaction and work–life balance are more important than the salary. Companies aiming to hire skilled employees therefore have to provide a suitable working environment. Sabbaticals, home offices, and flexible working times are all examples of this trend.

Even for medium-skilled jobs, companies have to provide more than just a salary. Take, for example, Toyota. Until a few years back, assembly lines at Toyota usually had two shifts of eight hours each, separated by a four-hour break. The four hours were used as a capacity buffer. If for some reason, the output did not meet the daily quota, the line worked overtime to catch up and to keep the output stable. Hence, this 8/4/8/4 model gave Toyota maximum flexibility.

However, this also lea to grueling late-night and early-morning shifts, disturbing not only the biorhythm of the workers but also their families and social lives. Because of such working conditions, already in the 1990s, Toyota had problems finding new workers. Even worse, workers whom

Toyota saw as lifetime employees started to leave for other companies in droves. For example, in 1991, within one year, three-fourths of the people hired left the company (Clarke 2002).

Hence, Toyota changed the shift pattern. They changed the 8/4/8/4 shift pattern to two adjacent eight-hour shifts starting around 6:00 a.m. and ending shortly after midnight. This includes four breaks, three 10-minute breaks plus a 45-minute lunch break. This new shift pattern allowed much more reasonable working hours. Toyota sacrificed flexibility for work satisfaction to keep its workers motivated and its lines staffed.

Another example where Toyota changed for social reasons is the assembly-line layout. Before, the line was over one kilometer long, having hundreds of workers. Any problem at one spot led to the entire line being stopped using the Andon cord (see Section 16.2). While Toyota encouraged its workers to do this, they still felt embarrassed for inconveniencing hundreds of other people. Hence, they sometimes avoided pulling the cord, leading to larger problems later on.

To solve this problem, Toyota cut the assembly line into five segments, decoupling these segments with a buffer of a handful of cars (Clarke 2002). Hence, pulling the Andon cord inconvenienced much fewer people. The worker was also more likely to have a social relationship with the inconvenienced colleagues, which created more understanding and eased the feeling of guilt. For similar reasons, Toyota also started to introduce quality checks at the end of each segment rather than at the end of the entire line (Becker 2006, p. 29). Any quality problem coming up would be within a much smaller group with social connections, again easing the feeling of guilt for the workers.

Overall, decision making, speed, flexibility, and skilled workers are key elements for manufacturing. This is not to say that other aspects have become irrelevant. Of course, quality or cost will still be important. Yet, having good decision making, a satisfied workforce, a flexible production system, and a fast response time will become more and more significant for economic success.

BIBLIOGRAPHY

Baale, O., Bergholz, W., 2005. *Das deutsche Führungsproblem: Kompendium der Arbeitsfreude in Staat und Wirtschaft.* Deutscher Taschenbuch Verlag, Munich.

Becker, H., 2006. Phänomen Toyota. Springer, Berlin; Heidelberg.

Clarke, C., 2002. Forms and Functions of Standardisation in Production Systems of the Automotive Industry: The Case of Mercedes-Benz (Doctoral). Freie Universität Berlin, Berlin.

Deming, W.E., 2000. *The New Economics: For Industry, Government, Education*. MIT Press, Cambridge, MA.

Economist, 2013. What to Do Now: Shape Up—For Offshored Jobs to Return, Rich Countries Must Prove That They Have What It Takes. *The Economist*.

Finkelstein, S., 2004. *Why Smart Executives Fail: And What You Can Learn from Their Mistakes*. Portfolio, New York.

Kazmer, D., 2014. Manufacturing Outsourcing, Onshoring, and Global Equilibrium. *Business Horizons* 57, 463–472.

Manyika, J., Sinclair, J., Dobbs, R., Strube, G., Rassey, L., Remes, J., Roxburgh, C., George, K., O'Halloran, D., Ramaswamy, S., 2012. *Manufacturing the Future: The Next Era of Global Growth and Innovation*. McKinsey Global Institute, McKinsey Operations Practice, Washington, DC.

Marsh, P., 2012. *The New Industrial Revolution*. Yale University Press, New Haven, CT.

Pearson, S.A., 1992. Using Product Archeology to Identify the Dimensions of Design Decision Making (MS). Massachusetts Institute of Technology.

Richter, L., 2011. Cargo Cult Lean. *Human Resources Management & Ergonomics* 5, 84–93.

Ross, J.I., Richards, S.C., 2002. *Behind Bars: Surviving Prison*. Alpha, Indianapolis, IN.

Sanchez, L.M., Nagi, R., 2001. A Review of Agile Manufacturing Systems. *International Journal of Product Research* 39, 3561–3600.

Shimokawa, K., 1994. *The Japanese Automobile Industry: A Business History*. Continuum International Publishing Group—Athlone, London; Atlantic Highlands, NJ.

The Manufacturing Institute, 2012. *Facts about Manufacturing*. The Manufacturing Institute, Washington, DC.

Toyota Motor Corporation, 1988. *Toyota: A History of the First Fifty Years*. Toyota Motor Corporation, Tokyo, Japan.

Waring, S.P., 1994. *Taylorism Transformed: Scientific Management Theory since 1945*. The University of North Carolina Press, Chapel Hill, NC.

World Bank, 2015. *Manufacturing, Value Added (% of GDP, 2014 Data or Latest)*. The World Bank, Washington, DC.

18

Things to Come

Any sufficiently advanced technology is indistinguishable from magic.

Arthur C. Clarke (1917–2008)
In Profiles of the Future: An Inquiry into the Limits of the Possible

Having learned the lessons from the past in the preceding chapter, there is one remaining question left open. Where will the journey bring us? What are the developments in manufacturing that will come upon us? This chapter looks into the future and tries to identify things to come. Please be aware that predicting the future is difficult, and by the time you are reading this, these predictions may already be outdated by reality. The following sections describe scenarios of how the future could develop. These options are possible or even likely, although it is difficult to say when they will happen. However, in my view, it definitely would be worthwhile to be prepared for the following developments in manufacturing.

18.1 THREE-DIMENSIONAL PRINTING

One technology that has the potential to revolutionize manufacturing is three-dimensional (3-D) printing (Figure 18.1). The first to use this technique was Charles Hull, around 1984. He selectively cured a polymer using ultraviolet light. Since then, many different 3-D printing approaches have been developed.

There are two main approaches common in industry. The first approach adds layers of material, for example, thermoplastics, to build up the part. The second approach solidifies a powder or liquid, for example, melting

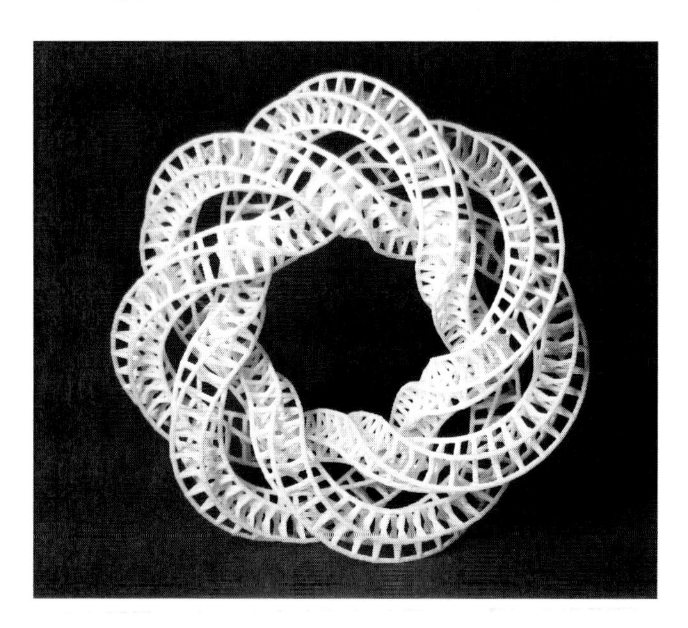

FIGURE 18.1
Example of a 3-D-printed part. (Image by fdecomite and licensed under the Creative Commons Attribution-ShareAlike 2.0 Generic license.)

metal powder using a laser, also to build up layers for the part. Other less frequent techniques also use, for example, regular printing paper cut into shape and glued on top of other pieces of paper. Three-dimensional printers have been developed that print in color or even use different materials in different areas. A part can be printed stiff in one area and flexible in another area, with a controllable gradual change of stiffness. Endless printers can produce parts of nearly infinite length. Three-dimensional printing has even been used with biological cells, aiming to print missing organs from scratch.

Three-dimensional printing has different advantages. Most significantly, a part can be created directly from a digital model. The part can have a complex internal structure that would be impossible to make using traditional techniques. For example, cooling channels in a mold can have any shape, able to follow the contour of the mold, whereas traditional techniques are mostly limited to straight holes. A good trade-off between weight and stiffness is possible by printing an internal structure similar to honeycombs or tetrapods. Hence, 3-D printing is frequently used in the aerospace industry. Multiple parts can be printed in situ without the need for assembly. Entire clocks have been printed in one piece. All you need to do is shake out the remaining powder and wind it up. Downsides are still slow manufacturing speeds, rather rough surface quality, and a limited ability to mix materials.

While the technique was initially confined to laboratories, prices have come down significantly. Nowadays, 3-D printing can be competitive not only for unique parts but also for lot sizes up to a few hundred parts. Currently, around 28% of all expenditures for 3-D printing are for final parts, rather than research.* It is expected that this will rise to 80% by 2020 (Economist 2012a), although others believe that this will take until 2040 (Marsh 2012, p. 61).

However, once 3-D printing is mainstream, it will reduce the significance of the actual manufacturing and enhance the importance of design. This can be compared to the music industry. Up to the turn of the century, production of music gave control to record labels, initially through records and later through CDs. Since then, however, music has gone digital, and you do not need the manufacturing process in between. You can download (legally or otherwise) the music files directly to your music player. This took power away from the record industry and gave it to the artists and also to the consumer. The middleman is no longer necessary.

This also may happen to other products. With a 3-D printer, products can be printed anywhere, requiring no special expertise. Digital designs can be shared, improved through the collaboration of the crowd, and pirated as easily as a music file. Of course, there are also limitations. A car, for example, is most likely to be too complex to ever be printed in one go and would still require assembly of components. Nevertheless, 3-D printing has the ability to shake up manufacturing within the next 30 years.

Visionaries are taking this idea even one step further. Three-dimensional printing basically sticks together chunks of material, creating rather rough products overall. However, imagine you could arrange individual atoms and molecules to build up parts. The possibilities would be endless. Since all manufactured products are built from matter, they are fundamentally all more or less precise arrangements of molecules.

Imagine you had a small nanobot that could arrange molecules for you. The first thing it should do is build another nanobot. These two nanobots would then build another two, and so on, until an army of nanobots could shape any part that you can think of. All you would need is the design and some raw materials, and your bots could get started.

* However, if you visit a 3-D printing exhibition, you will notice that most displayed parts are fantasy figures from Dungeons and Dragons or other entertainment products rather than industry parts.

Others believe that the idea is fundamentally infeasible and flawed. For one thing, it would require a lot of atoms to be arranged. For example, the human body consists of about 10^{26} atoms, a one followed by 26 zeros. Building up the atoms would take a very long time. More critically, however, are the physical effects on an atomic level. Atoms cannot be stacked like bricks but may interact with each other like tiny magnets. Moreover, since the nanobot is made of atoms, too, the material will interact with the bot (Kaku 2011, p. 201). Overall, the exercise may be akin to stacking confetti while your fingers are covered with honey. With luck, you may be able to get two pieces of confetti aligned. Yet to build something meaningful, you need to stack countless pieces of confetti amounting to about the same weight as the entire earth. The idea is titillating, yet, at best, still far out in the future. Optimists expect nanobots by 2100 at the earliest, if at all (Kaku 2011, p. 199). Pessimists say it is impossible.

18.2 DOWNFALL OF THE AUTOMOTIVE INDUSTRY

The automobile is the most complex mass-produced product and probably the most expensive item most people will buy besides a house. Around 8 million people in the United States work in automotive manufacturing or automobile-related services, a whopping 4.5% of all private employees. For Europe, this is even higher, with 12 million people representing 8% of all private jobs (Economist 2012c). An average car has around 30,000 parts including subcomponents. Over 80 million cars are produced annually for a very competitive market.

This combination makes car manufacturing the most challenging manufacturing operation in the world. The organization has to deliver parts reliably both in quantity and in quality. Any hiccup in the delivery of parts, and the manufacturing process will stop. In addition, every line stoppage costs from $2000 to $120,000 per hour for idling workers and machines. Unexpected line stoppages will quickly erase the profits of an automotive company. A superior production system is crucial for automobile companies to compete, and it is no surprise that many advancements in manufacturing in recent history were developed by automobile companies.

However, there are major changes foreseeable for the future of the automotive industry. There are a number of smaller changes that will challenge automobile companies in the next few years. These are mostly related to new materials and technologies necessary to produce modern automobiles. Different new power train technologies are currently coming on the market. The most significant one is the gasoline–electric hybrid pioneered by Toyota, but electric-only vehicles are also slowly appearing. Other techniques, for example, fuel cells, are under development. As for other materials, there is also a trend to replace metal with aluminum or high-end plastic material. Most challenging here are the ultrastrong and ultralight carbon fiber composite materials.

These are all new technologies that automobile companies have to master if they want to stay competitive. With traditional techniques, for example, sheet metal or cast iron, carmakers have over 100 years of experience. There is much less experience with electric vehicles, aluminum, or carbon fiber composites. The race is already on for automobile companies to learn these technologies, apply them to their products, and become faster, better, cheaper. The success of automobile companies will depend heavily on their ability to learn faster than the competition. There will be some companies that bite the dust, but overall, the automobile industry will master this transformation.

Yet, all of these changes are the small fry compared to the disruptive technology that could erase no less than half of the worldwide demand for automobiles. And, in most likelihood, the automobile industry will introduce this devastating change themselves. Currently, the average automobile, for most of its life, is anything but mobile. The average car does nothing more than occupy a parking lot for over 95% of the time. At best, it is storage for shopping bags or suitcases. Overall, it is a very expensive item both in purchase and in operation for very little actual usage.

Sharing a vehicle with other people will increase this utilization. Car sharing in European cities has a utilization of up to 22% (Cornet et al. 2012), making much better use of the vehicle. Economically speaking, sharing a car is the sensible thing to do. Unfortunately, there are quite a number of inconveniences with car sharing. Most critically, you have to first walk to the parking lot where the vehicle is located. After use, you have to walk the same way back. Even in big cities, this is often an inconvenient 5- or 10-minute walk, not to mention having to carry the stuff you want to bring along. Hence, only around 2.5% of all drivers in European cities actually use car sharing, and then often only as a second option next to their own vehicle.

However, this will change drastically with the self-driving car. Rather than going to the parking lot, you would order the car via your mobile phone to pick you up at the curb in five minutes. While on the road, you do not need to drive but can read a book, watch a video, or simply take a nap. The ride will also be much safer than with a human driver. Human drivers often get distracted by answering a call, texting, or simply changing the radio station. Overall, human errors are the cause of 90% of all accidents (Economist 2012b). Computers will not be distracted. While computers are not perfect either, they are much less likely to be involved in accidents. The ride will also be cheaper. Computers can optimize their speed for fuel efficiency. On highways, they can trail in the slipstream of preceding cars at a very close distance, while having a fast-enough reaction to stop if needed. Upon arrival, there is no need to look for a parking lot; you merely get out, and send the vehicle on its way. Overall, you will have the convenience of a taxi for a price below having your own car. In this case, the question is, do you still need your own car?

Most likely not. The convenience and economic benefits of self-driving shared cars will probably convince large parts of the population to ditch their own vehicles. This will drive utilization of vehicles up. One shared car can replace up to 32 individual vehicles. Already, 500,000 fewer cars are sold in America due to car sharing, a number expected to rise to over 1 million by 2020 (AlixPartners 2014). In urban environments, 90% of all cars could be replaced by shared cars while maintaining the same level of mobility (Fagnant and Kockelman 2014).

This will be devastating for automobile producers. Where before, they sold three cars, now they sell only one. Of course, not everybody will abandon their private cars. Furthermore, shared cars will rack up kilometers faster and hence reach the end of their usable life earlier, although computers can drive much more gently than humans and hence expand the durability of a car. As for distance driven, current trends are toward driving less (Economist 2012c), but with the convenience of car sharing, distances may increase again (Cornet et al. 2012) due to convenience and mobility for people without a driving license.

Nevertheless, in all likelihood, with self-driving shared cars, half of all vehicles in advanced economies are no longer needed. Therefore, half of all car factories are no longer needed. Neither are its workers. Rather than producing over 80 million cars per year worldwide, demand will drop to 40 million or less, devastating the industry in a below-bottom-line fight for survival. Additionally, car-sharing companies buying hundreds of thousands of cars every year will be able to strike a much better bargain with carmakers than any individual customer will. Hence, not only will the quantity of cars sold go down; the profit margin for each car will also erode. Rather than being the pinnacle of industry, automotive manufacturing will become a mere supplier for car-sharing companies.

Out of 8 million people in the United States related to the car industry, 4 million may no longer be needed. Taxi drivers and truck drivers will be as rare as carriage drivers nowadays. Probably hardest hit will be trains and public transport, which rely on economy of scale for profit. Who would ride a train if a self-driving car offers much more convenience and privacy for little additional cost? Overall, manufacturing will be reduced, and unemployment will rise. Social disruptions are likely.

Already, the mythos of the car as a fashion item is waning. For many young people, the mobile phone is already a much more trendy accessory than the dinosaur-burning necessity of a car. Unfortunately, this is little understood by automobile CEOs, who are mostly car geeks. They love cars and simply cannot fathom that many people do not need their own set of wheels. Like horse breeders 100 years ago considered a stinky noisy car no match for the beauty of an equine, car executives value the beauty of driving over being at the mercy of a mindless computer. Yet, while there are still horses around nowadays, they are very few, and merely for sports or hobby. The same will happen to cars. As horses are prohibited from most highways, so will be human-operated cars. Yet, automobile CEOs will ignore the signs. The inventory of unsold cars will eventually go up while companies fight for survival in a strongly declining market. It is possible that steel prices will crash because brand-new cars will be melted down again since there are no buyers left.

The first self-driving cars are already on public roads. Google vehicles have driven themselves hundreds of thousands of kilometers safely (Brynjolfsson and McAfee 2012, 17%) (Figure 18.2). Most carmakers and some universities experiment with self-driving cars. Assisted driving for stop-and-go traffic, staying in lanes, or preventing rear-end collisions is already a standard feature for high-end cars. The technology of self-driving cars will become mainstream within the next 10 to 20 years, with all the disruptions for the industry. This will affect not only carmakers but also logistics companies, airlines, train services, insurance companies, and many more. As a side effect, a car will start to look more like a camping van, where the passenger can either sleep or work during the ride.

Probably the biggest handicap for implementation is the legal risk. At one point, a robotic car will kill someone by mistake. While at the same time, hundreds of thousands of people were *not* killed due to self-driving cars, the legal system will be looking for justice and—more importantly—for

FIGURE 18.2
Google self-driving vehicle prototype. (Image by Google.)

someone to pay the lawyers' fees. Although the legal system will eventually come to cope with this problem, American class action lawsuits and public trials will make this a difficult process.

Significant use of self-driving cars will possibly happen first outside of the U.S. My guess would be Japan, with both an affinity for new technology and high labor cost. Possible among the first uses are not passenger cars but, rather, trucks. Truck drivers need to rest, delaying delivery and idling the truck for two-thirds of the time. Computers could not only eliminate the cost for the biological driver but also increase speed and utilization. After the technology has proven itself in transportation—including the not-to-be underestimated debugging process and its legal consequences—the public will eventually accept self-driving passenger cars. It will be very beneficial to the users, but devastating for producers of vehicles.

18.3 RISE OF ROBOTICS

As the automotive industry will be decreasing in importance due to increasing abilities of computers, another field will increase: robotics. Since the failed experiments of GM CEO Roger Smith (see Section 15.3), robotics has improved, and robots are nowadays mainstream in many manufacturing operations.

However, robots have also started to appear as consumer goods. Robotic helpers have already started to enter the household. While still rather primitive, robotic lawn mowers and vacuum cleaners are already working in countless private homes.* Yet robotics is still in its infancy. Humanoid robots are still mere showpieces, barely able to walk unassisted. Figure 18.3 shows a humanoid robot prototype during a robot competition. However, once these initial difficulties have been sorted out, a robotic maid may be a new trendy consumer good to ease our household chores. Rather than merely cleaning the floor, it could do the laundry (including ironing, folding, and putting it in the wardrobe), dust shelves, fix lights, clean up after your toddler, clean the dishes, even cook if you want. It could pick up deliveries by other robots, shop for household necessities, and pretty much take away all burdens of the household.

Sure, a human helper can do that already is and currently still cheaper than a robotic maid. The big worry in most cases is finding a reliable, confidential, and trusted helper to let into your home rather than merely anybody willing to do it. The feeling of being in full control will probably help sales of robotic helpers. At one point in the future, the most expensive mobile thing you buy may no longer be a car (since you do not buy cars anymore, as per Section 18.2) but, rather, a robotic maid. This may happen sooner than we think. Robotic maids may become mainstream by 2025 in technophile countries like Japan and South Korea.

In manufacturing, robots will also continue to spread. In advanced economies, repetitive manual work is done more and more by robots rather than humans. Even former low-cost countries like China are starting to invest heavily in robotics. One of the largest electronics manufacturers in China, Foxconn, planned to reduce manual labor through the investment of 1 million robots (Economist 2011), although technical problems prevented this from happening so far. Like so often before in history, labor cost seems to be less important here than control of labor. Particularly at Foxconn, numerous workers killed themselves due to the difficult working conditions, generating worldwide bad publicity for Foxconn. Robots, on the other hand, are unlikely to throw themselves off a roof. Even if they did, nobody would care. As for the workers, the official phrasing of Foxconn is that they will move the workers higher up in the value chain,

---

* Including mine. For the last 10 years, robotic vacuum cleaners have been cleaning our home on a daily basis. While they have some difficulties with corners and tight spots, on average, our home is much cleaner than if we would do it by hand—which, admittedly, would not be that often.

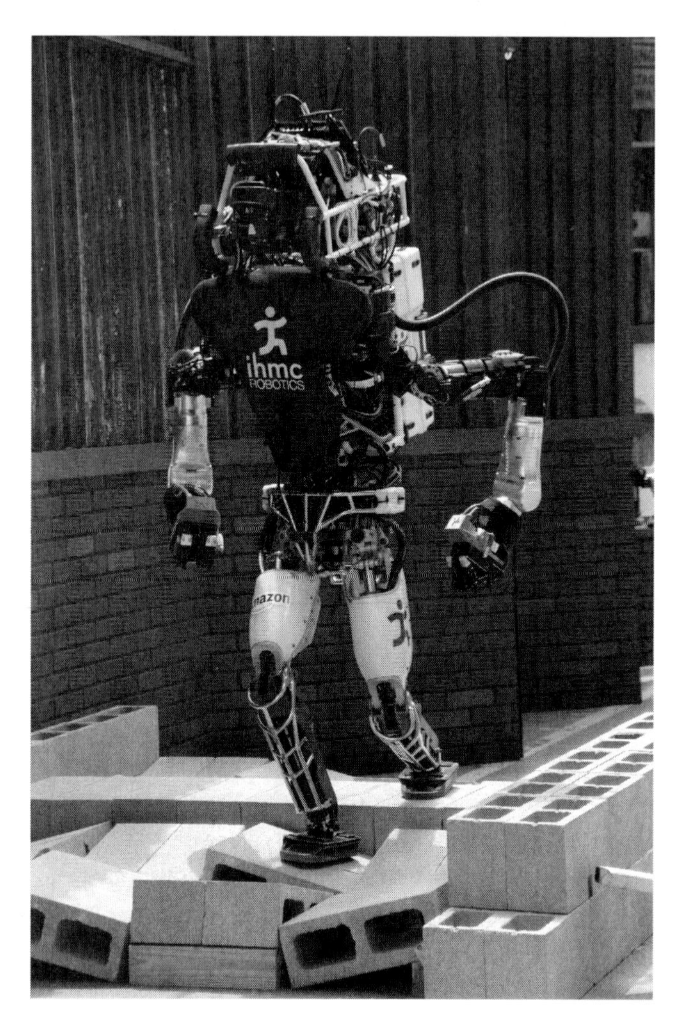

FIGURE 18.3
Autonomous humanoid robot by team Institute for Human and Machine Cognition (IHMC) at the 2015 Defense Advanced Research Projects Agency (DARPA) Robotics Challenge. Out of 23 teams, only 2 were able to complete the challenge. This robot made second place. (Image by DARPA.)

but it is probably more likely that they are not moved up but, rather, moved out of the value chain altogether.

This, of course, will be difficult for the common worker, who will lose his/her job and livelihood. Yet that is what has happened in the past already. For the last 30 years in the United States, wages for unskilled labor continue to decrease (Brynjolfsson and McAfee 2012, 47%), with workers trying to compete with the cost of robots. Unfortunately for the workers, the cost of robots is decreasing even faster. Robotics costs have fallen by half within the last 20 years in most advanced economies (Economist 2013b), with no stopping in sight.

Around 1800, the U.S. gross domestic product grew about 1% per year. In the twenty-first century, growth is about 2.5% per year. However, while the U.S. economy is increasing, this no longer translates into rising income. Between 1970 and 2000, the median income of households rose much more slowly than the gross domestic product. Since 2000, this crawl has come to a complete standstill and is currently reversing. The common middle-class worker no longer profits from industrial growth. Nearly all of the wealth generated goes to the top 20% income earners. The rich are getting richer, while the rest are left out (Brynjolfsson and McAfee 2012, 39%; Piketty 2014).

The biggest hindrance to more widespread use of robotics is their lack of fine motor control and hand–eye coordination. Simply putting a screw in a hole is difficult because the robot has difficulty identifying the screw, grabbing the screw out of a box, and then identifying the hole. Yet research is currently addressing these problems, and more skilled robots are designed every year. As the use of robots increases, so will the importance of the robotics manufacturers. By some estimates, the size of the robotics industry will exceed that of the automotive industry by 2050 (Kaku 2011). Robotics will then rule the world of manufacturing, both in production and in usage. However, depending on how the future plays out, robotics may rule much more than just manufacturing.

18.4 THE END OF WORK?

Robots have been squeezing out low-skilled labor for the last 30 years. This is due to improved mechanics but even more so due to improved computers. Within 15 years, processor speed has improved a thousand-fold. Yet, more importantly, the algorithms using these faster computers have improved even more. By one estimate, while processor speeds have improved by a factor of 1000, better algorithms are responsible for an improvement factor of 43,000, for a combined performance improvement by a factor of 43 million (Brynjolfsson and McAfee 2012, 24%). Overall computer performance doubles every seven months, with no signs of stopping.

A machine works best for mathematical problems or clearly defined decision problems. However, there are things that humans still can do better than computers. Humans are still better at pattern recognition and decision making. A computer can calculate numbers faster than anybody on earth but mostly cannot tell the difference between an apple and a table. A prime example of computers outsmarting humans is chess. In 1997, for the first time, a computer, Deep Blue, defeated the world's leading chess grandmaster, Garry Kasparov, in a standard tournament chess match. Since then, even commercial chess software plays at or above a grandmaster level. However, what few people know is that the very best chess player in the world is neither man nor machine but both. A team of computer-assisted chess players is still able to defeat any chess-playing machines (Brynjolfsson and McAfee 2012, 68%). It is the mix of computer strength and human intuition that provides far superior skills.

Nevertheless, computers are getting better at pattern recognition and understanding of everyday problems. Hence, it is just a matter of time until a computer will be able to design an improved version of itself. The point when robots and computers are able to build better robots and computers without human intervention is called technical singularity. This is a fundamental change in history. Rather than requiring a human, computers are able to improve themselves. Technical evolution would jump from carbon-based life forms to silicon-based...well...machines. The advance in knowledge and technology would increase—an intelligence explosion—and humans would be left behind. While scientists mostly agree that it will happen, they disagree widely about when this will happen. Estimates range from 20 to 100 years into the future (Kaku 2011, p. 65ff).

Low-skilled blue-collar jobs are already being squeezed by machines. Lately, however, even white-collar middle-class jobs are being replaced by computers. A computer already writes many basic newspaper articles on sports events (Economist 2013a). Even higher-class work is replaced by computers. For example, the legal industry in the United States requires reading of hundreds of court proceedings to find those that are relevant for the current case. Up to recently, small armies of lower-paid lawyers did this task. Yet, computers also replaced this work. A single pattern recognition software was able to do the work of 500 lawyers. This computer does the research both cheaper and faster. Additionally, comparing previous human work with computer results shows that human work was rather sloppy, with accuracy rates of only 60% (Brynjolfsson and McAfee 2012, 32%).

As the skills of computers increase, more and more tasks can be delegated to the machine, with better speed, productivity, and cost. Eventually, the age-old dream of many manufacturers will come true—they will no longer have to deal with other people. The people, on the other hand, are also the customers. The rich will get richer, while the soon-to-be unemployed middle class will lose its livelihood. Inequality will spread. There is a significant risk of disrupting the social fabric, with the newly poor rising against the rich upper class.

I can imagine three possible scenarios of how this would develop. In the first scenario we will become perpetual couch potatoes. All work, regardless if industrial, service, or domestic, will done by mechanical servants. The benefits will be shared among the masses to make sure nobody is lacking. Since there is no longer the need to do anything, more and more people will do exactly that: nothing! Within a few generations, humanity will be mostly a collection of couch potatoes with no motivation at all. Our lives will circle around eating, entertainment, and—possibly—sex.

In a second scenario, having increasingly smarter computers will hurt the human pride. No matter what you do, there is a computer that can do it better. It will feel like perpetual childhood, where the computer will praise you for your achievements yet can do things so much better than you. While nowadays, through training, you can eventually excel in some areas, in the future, computers will always be better. This will be very unsatisfying to humans. The pinnacle of evolution must not be outdone by a mere chip. Yet our brains are just no longer good enough to compete. The solution is to simply augment our abilities by technological means. Using computer–brain interfaces, we will have human intelligence coupled with the computation power of machines. We will become cyborgs: part man, part machine. Like the chess game example, these cyborgs may outperform both artificial and human intelligence. Over generations, the line between human and machine will blur. At one point, the human mind may even be completely uploaded into a computer, and we will no longer rely on the grey matter between our ears. By then at the latest, there will no longer be any difference between human and machine.

In a third scenario, artificial intelligence will replace not only human labor but humanity in its entirety! While in the past, soldiers could rebel and have rebelled against bad leadership, computers will follow orders to the letter. Industrial output will be unevenly distributed, and whoever can field the most battle-bots will be able to control the rest of society by force. In fact, most of society will not even be needed for the economy to function. Mankind may be destroyed in the fight of robot armies for power. It does not matter if these armies are controlled by a human or if computers themselves make the decisions. The last soldiers standing will not be humans. Evolution will have achieved the next step of silicon-based life forms, and carbon-based beings will go the way of the dinosaurs. While this sounds like science fiction, may I remind you that the biggest source of money in robotics research comes from the military, and they are not interested in snuggle-bots. While many people fear technology that they do not understand, the risk that artificial intelligence could lead to the extinction of humans also worries many experts and scientific leaders. This includes, for example, the former chairman of Microsoft, Bill Gates; theoretical physicist Stephen Hawking; and the CEO of Tesla Motors, Elon Musk (Cellan-Jones 2014; Rawlinson 2015).

In my opinion, a combination of the couch potato and the cyborg scenarios is most likely. Many humans love to compete but hate losing. Manufacturing will be done by robots and supervised by cyborgs. Productivity will grow and eventually will benefit most of society, even those couch potatoes who choose to do nothing. My only hope is that this transformation will not be violent like the Luddite uprising, nor lead to the extinction of mankind. In any case, I wish, my dear reader, that you will be surfing atop the waves of industrial, economic, and social change that surely are upon us. May you have a prosperous future!

BIBLIOGRAPHY

AlixPartners, 2014. *AlixPartners Car Sharing Outlook*. AlixPartners, New York.

Brynjolfsson, E., McAfee, A., 2012. *Race Against the Machine: How the Digital Revolution Is Accelerating Innovation, Driving Productivity, and Irreversibly Transforming Employment and the Economy*, Kindle edition. Digital Frontier Press, Lexington, MA.

Cellan-Jones, R., 2014. Stephen Hawking Warns Artificial Intelligence Could End Mankind. *BBC Technology News*.

Cornet, A., Mohr, D., Weig, F., Zerlin, B., Hein, A.-P., 2012. *Mobility of the Future—Opportunities for Automotive OEMs*. McKinsey & Company, Washington, DC.

Economist, 2011. Foxconn—Robots Don't Complain. Or Demand Higher Wages, or Kill Themselves. *The Economist*.

Economist, 2012a. Additive Manufacturing: Solid Print. *The Economist*.

Economist, 2012b. The Future of Driving: Seeing the Back of the Car. *The Economist*.

Economist, 2012c. Inside Story: Look, No Hands. *The Economist*.

Economist, 2013a. Reshoring Manufacturing: Coming Home—A Growing Number of American Companies Are Moving Their Manufacturing Back to the United States. *The Economist*.

Economist, 2013b. Innovation Pessimism: Has the Ideas Machine Broken Down? *The Economist*.

Fagnant, D.J., Kockelman, K., 2014. Development and Application of a Network-Based Shared Automated Vehicle Model in Austin, Texas, in: Proceedings of the Transportation Research Board Conference on Innovations in Travel Modeling. Presented at the Transportation Research Board Conference on Innovations in Travel Modeling, Baltimore, MD.

Kaku, M., 2011. *Physics of the Future: How Science Will Shape Human Destiny and Our Daily Lives by the Year 2100*. Penguin, New York.

Marsh, P., 2012. *The New Industrial Revolution*. Yale University Press, New Haven, CT.

Piketty, T., 2014. *Capital in the Twenty-First Century*. The Belknap Press, Cambridge, MA.

Rawlinson, K., 2015. Microsoft's Bill Gates Insists AI Is a Threat. *BBC News*.

Timelines

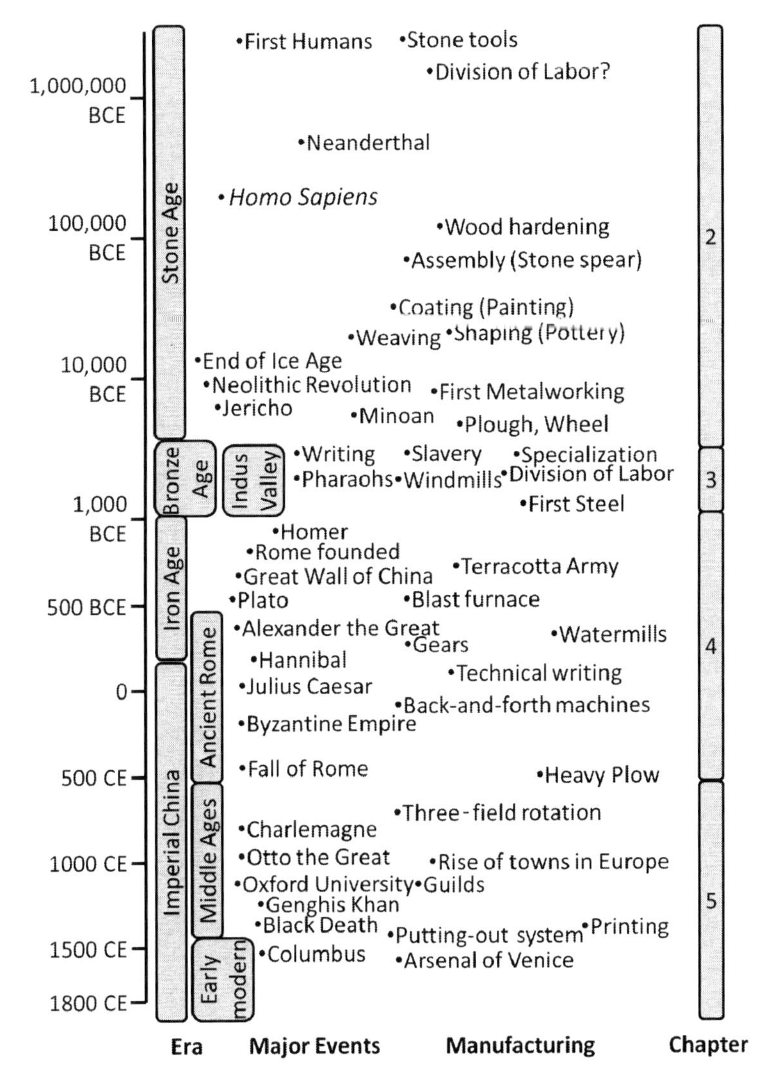

FIGURE A.1
Timeline of events corresponding to the chapters in Section I between ca. 2,600,000 BCE and 1800 CE. Please note that the scale before 1000 BCE is logarithmic; otherwise, the 2.5 million years of the Stone Age would be more than 99% of the timeline.

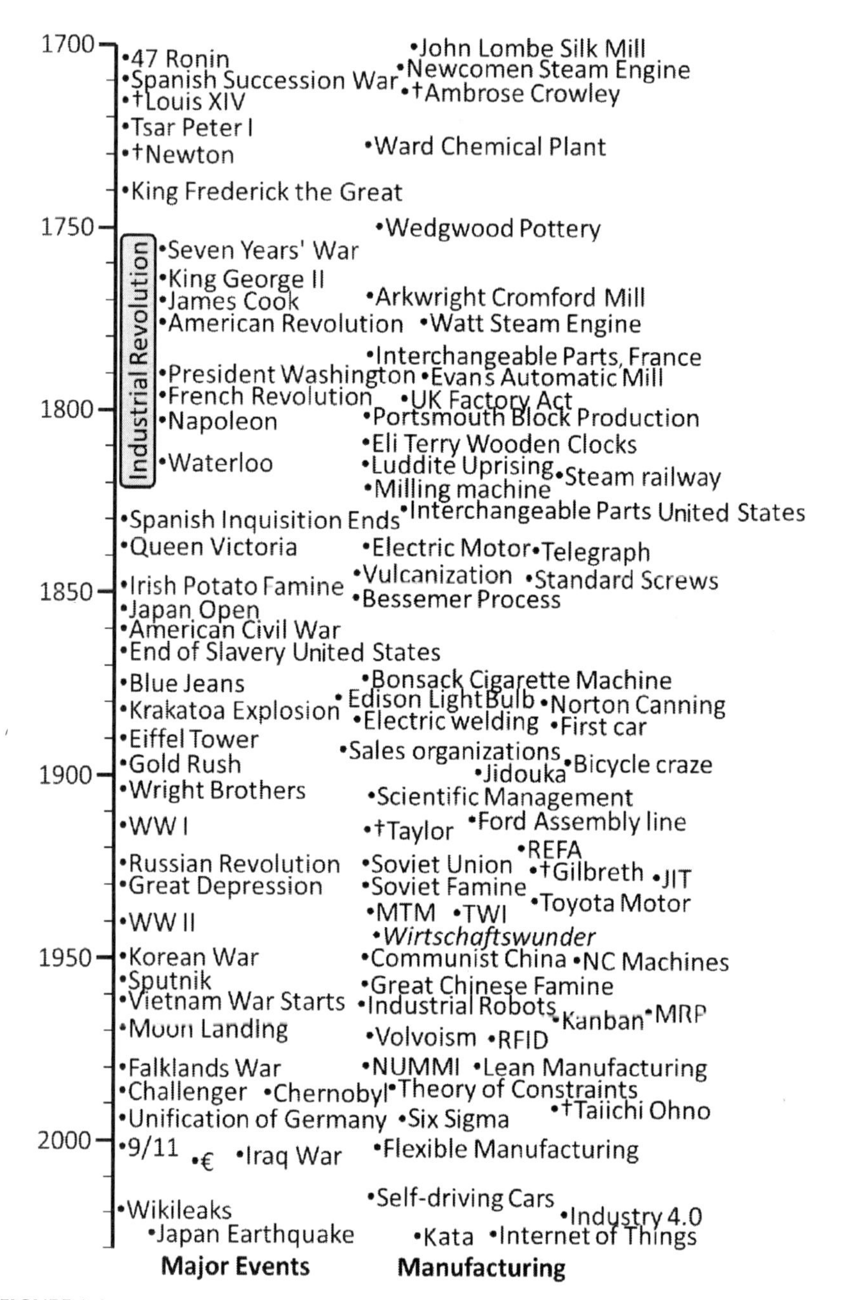

FIGURE A.2
Timeline of events corresponding to the chapters in Sections II and III between 1700 and today.

Index

Page numbers followed by f and t indicate figures and tables, respectively.

G

M

N

X

Y

Z

About the Author

Photo by Schönwälder.

Prof. Dr. Christoph Roser is an expert in Lean production and a professor of production management at the University of Applied Sciences in Karlsruhe, Germany. He studied automation engineering at the University of Applied Sciences in Ulm, Germany, and received his PhD in mechanical engineering from the University of Massachusetts, researching flexible design methodologies. Afterward, he worked for five years at the Toyota Central Research and Development Laboratories in Nagoya, Japan, studying the Toyota Production System and developing bottleneck detection and buffer allocation methods. Following Toyota he joined McKinsey & Company in Munich, Germany, specializing in Lean manufacturing and driving numerous projects in all segments of industry. Before becoming a professor, he worked for Robert Bosch GmbH, Germany, first as a Lean expert for research and training and then using his expertise as a production logistics manager in the Bosch Thermotechnik division. In 2013, he was appointed professor of production management at the University of Applied Sciences in Karlsruhe to continue his research and teaching on Lean manufacturing.

Throughout his career, Dr. Roser worked on Lean projects in almost 200 different plants, including automotive, machine construction, solar cells, chip manufacturing, the gas turbine industry, papermaking, logistics, power tools, heating, packaging, food processing, white goods, security technology, finance, and many more. He is an award-winning author of over 50 academic publications. Besides research, teaching, and consulting on Lean manufacturing, he is very interested in different approaches to manufacturing organization, both historical and current. His blog and website on Lean manufacturing and manufacturing history is at http://www.AllAboutLean.com.